Theory and Applications of NeutroAlgebras as Generalizations of Classical Algebras

Florentin Smarandache
University of New Mexico, USA

Madeline Al-Tahan
Lebanese International University, Lebanon

A volume in the Advances in Computer and Electrical Engineering (ACEE) Book Series

IGI Global
PUBLISHER of TIMELY KNOWLEDGE

Published in the United States of America by
IGI Global
Engineering Science Reference (an imprint of IGI Global)
701 E. Chocolate Avenue
Hershey PA, USA 17033
Tel: 717-533-8845
Fax: 717-533-8661
E-mail: cust@igi-global.com
Web site: http://www.igi-global.com

Library of Congress Cataloging-in-Publication Data

Names: Smarandache, Florentin, editor. | Al-Tahan, Madeleine Ali, 1986- editor.
Title: Theory and applications of neutroAlgebras as generalizations of classical algebras / Florentin Smarandache, and Madeleine Ali Al-Tahan, editor.
Description: Hershey PA : Engineering Science Reference, [2022] | Includes bibliographical references and index. | Summary: "Recognizing NeutroAlgebras and AntiAlgebras are a new field of research, this book uses the process of NeutroSophication of a classical algebraic structure to produce a NeutroAlgebra, while the process of AntiSophication of a classical algebraic structure produces an AntiAlgebra"-- Provided by publisher.
Identifiers: LCCN 2021044531 (print) | LCCN 2021044532 (ebook) | ISBN 9781668434956 (hardcover) | ISBN 9781668434963 (paperback) | ISBN 9781668434970 (ebook)
Subjects: LCSH: Partial algebras. | Neutrosophic logic. | Group theory.
Classification: LCC QA251 .T44 2022 (print) | LCC QA251 (ebook) | DDC 511.3/3--dc23/eng/20211006
LC record available at https://lccn.loc.gov/2021044531
LC ebook record available at https://lccn.loc.gov/2021044532

This book is published in the IGI Global book series Advances in Computer and Electrical Engineering (ACEE) (ISSN: 2327-039X; eISSN: 2327-0403)

British Cataloguing in Publication Data
A Cataloguing in Publication record for this book is available from the British Library.

For electronic access to this publication, please contact: eresources@igi-global.com.

Advances in Computer and Electrical Engineering (ACEE) Book Series

ISSN:2327-039X
EISSN:2327-0403

Editor-in-Chief: Srikanta Patnaik SOA University, India

MISSION

The fields of computer engineering and electrical engineering encompass a broad range of interdisciplinary topics allowing for expansive research developments across multiple fields. Research in these areas continues to develop and become increasingly important as computer and electrical systems have become an integral part of everyday life.

The **Advances in Computer and Electrical Engineering (ACEE) Book Series** aims to publish research on diverse topics pertaining to computer engineering and electrical engineering. **ACEE** encourages scholarly discourse on the latest applications, tools, and methodologies being implemented in the field for the design and development of computer and electrical systems.

COVERAGE

- Analog Electronics
- Qualitative Methods
- Microprocessor Design
- Chip Design
- Circuit Analysis
- Computer Hardware
- Sensor Technologies
- Optical Electronics
- Algorithms
- Computer Architecture

IGI Global is currently accepting manuscripts for publication within this series. To submit a proposal for a volume in this series, please contact our Acquisition Editors at Acquisitions@igi-global.com or visit: http://www.igi-global.com/publish/.

The Advances in Computer and Electrical Engineering (ACEE) Book Series (ISSN 2327-039X) is published by IGI Global, 701 E. Chocolate Avenue, Hershey, PA 17033-1240, USA, www.igi-global.com. This series is composed of titles available for purchase individually; each title is edited to be contextually exclusive from any other title within the series. For pricing and ordering information please visit http://www.igi-global.com/book-series/advances-computer-electrical-engineering/73675. Postmaster: Send all address changes to above address.

Titles in this Series

For a list of additional titles in this series, please visit:
www.igi-global.com/book-series/advances-computer-electrical-engineering/73675

Antenna Design for Narrowband IoT Design, Analysis, and Applications
Balachandra Pattanaik (Wollega University, Ethiopia) M. Saravanan (Vel Tech Rangarajan Dr. Sagunthala R&D Institute of Science and Technology, India) U. Saravanakumar (Muthayammal Engineering College, India) and Ganesh Babu T R (Muthayammal Engineering College, India)
Engineering Science Reference • © 2022 • 261pp • H/C (ISBN: 9781799893158) • US $225.00

Handbook of Research on Advances and Applications of Fuzzy Sets and Logic
Said Broumi (Laboratory of Information Processing, Faculty of Science Ben M'Sik, University Hassan II, Casablanca, Morocco & Regional Center for the Professions of Education and Training (CRMEF), Casablanca-Settat, Morocco)
Engineering Science Reference • © 2022 • 944pp • H/C (ISBN: 9781799879794) • US $395.00

Blockchain Technology and Computational Excellence for Society 5.0
Shahnawaz Khan (University College of Bahrain, Bahrain) Mohammad Haider Syed (Saudi Electronic University, Saudi Arabia) Rawad Hammad (University of East London, UK) and Aisha Fouad Bushager (University of Bahrain, Bahrain)
Engineering Science Reference • © 2022 • 309pp • H/C (ISBN: 9781799883821) • US $245.00

Handbook of Research on 5G Networks and Advancements in Computing, Electronics, and Electrical Engineering
Augustine O. Nwajana (University of Greenwich, UK) and Isibor Kennedy Ihianle (Nottingham Trent University, UK)
Engineering Science Reference • © 2021 • 522pp • H/C (ISBN: 9781799869924) • US $295.00

Emerging Nanotechnology Applications in Electrical Engineering
Ahmed Thabet Mohamed (Aswan University, Egypt)
Engineering Science Reference • © 2021 • 318pp • H/C (ISBN: 9781799885368) • US $225.00

IGI Global
PUBLISHER of TIMELY KNOWLEDGE

701 East Chocolate Avenue, Hershey, PA 17033, USA
Tel: 717-533-8845 x100 • Fax: 717-533-8661
E-Mail: cust@igi-global.com • www.igi-global.com

Editorial Advisory Board

Table of Contents

Detailed Table of Contents

In an attempt to generalize and create alternatives for classical algebraic structures, Smarandache in 2019 introduced the idea of treating binary operations on sets as partially well (totally outer) defined with axioms that are partially true (totally false) and even indeterminate for some elements of any given algebraic structure. The structures formed with these types of operations called Neutro(Anti) Operations are called Neutro(Anti) Structures and are named according to the axioms that are neutro(anti)-sophicated. In this chapter, the authors consider some properties of neutrosophic quadruple numbers with examples. In particular, they introduce and study the concept of NeutroQuadrupleRings and their substructures. Some basic definitions and a few important results are presented. It is shown that the intersection and union of any two NeutroQuadrupleSubrings of a particular class does not necessarily belong to the same class. Also, they give a necessary and sufficient condition for a neutrosophic quadruple ring (NQ(X), +, ·) to be a NeutroQuadrupleRing.

The process through which AntiAlgebraicStructures can be generated from any given classical structure is called AntiSophication. In this chapter, the authors introduce the concept of AntiHyperStructure and present some of its basic properties. It has been shown in our previous works that there are 19 types/classes of AntiHyperGroups. The chapter is devoted to the study of a particular class of these 19 classes, called a [2, 3] AntiHyperGroup, that is, a class in which associativity and reproductive axioms are totally false. Several examples and properties of [2, 3] AntiHyperGroup are presented. It is shown that the union of [2, 3] AntiSubhypergroups is a [2, 3] AntiHyperGroup and their intersection is not a [2, 3] AntiSubhypergroup. Also, it

AntiSophication is the process through which AntiAlgebraic structures can be generated from any given classical algebraic structures. The concept of AntiHyperStructures has been studied in previous work. The aim of this chapter is to study AntiSemihypergroups, AntiQuasihypergroups, AntiSubSemihypergroups, AntiHypergroups, and their properties. Some new definitions and theorems will be presented and supported by illustrative examples. In addition, the authors prove the existence of an infinite number of AntiSemihypergroups and AntiHypergroups.

NeutroAlgebras consist of algebras containing at least one NeutroAxiom. In the literature, there is a method for decision-making in a migrational situation that uses a NeutroAlgebra generated by the combining function in Prospector. This method is an important hit in showing both the utility and theoretical significance of NeutroAlgebras since it uses this abstract tool for solving a real-life problem. To the authors' knowledge, this is the only approach to NeutroAlgebras which is not exclusively theoretical. This chapter presents the rigorous mathematical proofs of the properties of that NeutroAlgebra because so far the theoretical approaches to the proposed method are minimal. The deeper mathematical knowledge of this specific NeutroAlgebra is a contribution to the general theory on this subject. The authors expect that this chapter will serve as dissemination of NeutroAlgebras among undergraduate and graduate students who may be interested in this issue.

This chapter improves and extends several definitions, properties, and applications of neutro-algebras. In classical algebraic structures, the laws of compositions on a given set are well-defined. In fuzzy algebraic structures, the laws of compositions on a given set are not well-defined exactly but with a membership value. This chapter examines those structures which have huge applications or strong potential for applications outside of mathematics, namely computer science or engineering. The study emphasizes that the nature of neutro-algebra and anti-neutro algebra differs from that of the classical algebra and fuzzy algebra. This chapter presents Venn diagram for neutro-function and establishes some relations between the partial algebra and neutro-algebra. The main contribution of the chapter is the introduction of the notion of neutro-group.

NeutroAlgebras and particularly NeutroGroups are novel algebraic structures where at least one of the classical axioms is satisfied by not all the elements of the set. In the theoretical literature consulted by the authors, most of the NeutroGroups examples are obtained from the well known Z_n groups endowed with either product or sum operations. This chapter aims to propose a method for the generation of NeutroGroups in a simple and automatic way from uninorms, with special interest in uninorms based on combining functions in both Mycin and Prospector expert systems. The use of theses uninorms suggests the possible application of NeutroGroups in fields such as decision making as can be found in the consulted literature, as well as for Expert Systems, among others. NeutroGroups of finite or infinite cardinality will be generated, where the axioms of NeutroAsociativity or NeutroInverse are fulfilled, and also indeterminacy is symbolically included within this theory. One family of non-commutative NeutroGroups is proposed.

There is a lot of ambiguous information in the real world that crisp values can't manage. The fuzzy set theory was proposed by Zadeh. It is an age-old and excellent tool for dealing with uncertain information. As a result, intuitionistic fuzzy set theory was suggested. However, these theories are incapable of dealing with all forms of uncertainty, such as indeterminate and inconsistent data in various decision-making situations. To address this shortfall, Smarandache proposed the neutrosophic set theory by introducing a degree of indeterminacy as an independent component. In this current decade, neutrosophic environments are mainly interested by different fields of researchers. Recently, Smarandache introduced theNeutroAlgebra and AntiAlgebras. NeutroAlgebras and AntiAlgebras represent a new research subject that is based on real-world examples. In this chapter, some properties of NeutroTopological space are introduced and studied with examples. Several definitions of NeutroInterior, NeutroClosure, and NeutroBoundary are defined, and the authors also studied its properties with examples.

The chapter aims to introduce the concept of NeutroHX-Group. Starting with HX-groups and using the newly defined concept of NeutroAlgebra, the authors define the NeutroHX-Group. Using the HX-groups associated with the dihedral group, they obtain the Chinese hypergroup associated with them. They present some examples to illustrate situations when the NeutroHX-Groups and AntiHX-Groups are obtained. Also, they determine a way to obtain commutative hyperstructure, considering the set of all HX-Groups with the dihedral group as support. The important result of this chapter is given by Theorem 1, where it is establishing a connection between NeutroHX-Groups and NeutroSemihypergroups.

The objective of this chapter is to introduce the concept of Neutro Ordered Rings by considering three Neutro Axioms (Neutro Abelian Group [additive], Neutro Semi group [multiplicative], and Neutro Distributive [multiplication over addition]) and total ordered relation or partially ordered relation. The concepts of Neutro Ordered Rings, Neutro Ordered Subrings, Neutro Ordered Ideals, Neutro Ordered Quotient Rings, and Neutro Ordered Ring Homomorphisms are presented. Moreover, several interesting results were reached, and a lot of illustrative examples were added.

A NeutroStructure is a structure that has at least one NeutroRelation or one NeutroAttribute, and neither AntiRelation nor AntiAttribute. In this chapter, the authors review and study the NeutroAlgebra structure and introduce several operations and properties of algebraic equations for image retrieval, such as NeutroContent-based image retrieval. It is a generalization of classical and fuzzy algebraic structures for image retrieval. An explanation of the new particular processes is provided by several results.

In this chapter, the authors define operations '+' and '×' on the substructure ideals of the ring. They find under what conditions the collection of ideals are NeutroAlgebra under + and ×. The study is carried out in the case of Z and Z_n, the ring of integers and ring of modulo integers, respectively, where n is a composite number. This study is extended to the case of polynomial rings R[x], Q[x], Z[x], and Z_n [x] where R and Q are real and rational, respectively. Some open problems based on this study are proposed.

The study of neutroalgebra has become one of the topics of great interest for researchers from different areas, where neutroGroup, neutro BCK-algebra, neutro Lie algebras, single valued neutro algebra, and others have been studied. This is why, in this work, the authors define the concept of quadripartitioned single-valued neutrosophic hypergroup as the initial concept of quadripartitioned single-valued neutrosophic hyper-algebra. They prove some properties on quadripartitioned single-valued neutrosophic hypergroup. The notion of NeutroQuadripartitionedSingleValuedHypergroup was defined, and some future research directions were proposed.

The focus of this chapter is on the study of NeutroSemigroups. NeutroSemigroups are alternatives to the classical semigroups where the closure law is allowed to be either partially well defined (with the degree of truth T) or partially indeterminate (with the degree of indeterminacy I) or partially outer defined (with the degree of falsehood F), and associative axiom is equally allowed to be either partially true (with the degree of truth T) or partially indeterminate (with the degree of indeterminacy I) or partially outer defined (with the degree of falsehood F). Elementary properties of NeutroSemigroups with interesting examples are presented.

Preface

The classical algebraic structures have all operations totally (100%) well-defined, and all axioms that are totally (100%) true (Smarandache, 2019; Smarandache, 2020a).

They do not leave room for partiality, or for operations that are only partially (< 100%) well-defined, and similarly for axioms that are only partially (< 100%) true (Smarandache, 2020b; Smarandache, 2020c).

But, in our real world, in the majority of situations, we deal with much partiality in all aspects of our lives (Agboola, 2020a; Agboola, 2020b; Agboola et al., 2020).

Therefore, most structures in today's world are NeutroStructures (structures dealing with partial laws (operations) and partial axioms (Hamidi & Smarandache, 2020; Rezaei & Smarandache, 2020; Smarandache et al., 2020).

The laws and regulations in our societies do not equally apply to all individuals, because an egalitarian society is a utopia (Agboola, 2020c; Agboola, 2020d; Agboola, 2020e; Smarandache, 2020d).

The laws apply in various degrees to the individuals, and also there are exceptions to the law (Ibrahim & Agboola, 2020; Mohammadzadeh & Rezaei, 2020; Smarandache, 2020e).

NeutroAlgebra vs. classical Algebra is like realism vs. idealism (Jiménez et al., 2021).

In this collective book we gathered many chapters that show the NeutroAlgebraic form of several classical Algebras (Al-Tahan, 2021b; Al-Tahan et al., 2021; Smarandache et al., 2021).

By the process of NeutroSophication one can transform a classical Algebra into a NeutroAlgebra (for example, a classical Group into a NeutroGroup, etc.), while using the AntiSophication a classical Algebra may be transformed into an AntiAlgebra (when an operation is 0% well-defined, or an axiom is 0% true, for example a classical Group into an AntiGroup (Kandasamy et al., 2003; Rezaei et al., 2021; Smarandache, 2021b).

ORGANIZATION OF THE BOOK

The book is organized into 17 chapters. A brief description of each of the chapters follows:

Chapter 1 is a review of reproduction of the evolution from the foundation of paradoxism (1980s), then of neutrosophy (1995) and its derivatives (neutrosophic set / logic / probability / statistics), then neutrosophic algebraic structures (2003) and ending with the NeutroAlgebra & AntiAlgebra (2019) as well as the NeutroGeometry & AntiGeometry (2021) of the previous work.

Chapter 2 introduces NeutroOrderHyperstructure by considering a partial order on NeutroSemihypergroups. Such an order when it satisfies the property of NeutroOrder, the whole system becomes a NeutroOrderSemihypergroup. More precisely, the authors define (for the first time) NeutroOrderSemihypergroups and illustrate their new definition by non-trivial examples. Also, they proved that there are infinitely many NeutroOrderSemihypergroups by showing that there is an NOHS of cardinality n for every $n > 2$. Moreover, they study some subsets of this new structure and investigate some of their properties. Finally, they leave some ideas for future related research.

Chapter 3 introduces and studies the concept of NeutroQuadrupleRings and their substructures. Some basic definitions and few important results are presented. It is shown that the intersection and union of any two NeutroQuadrupleSubrings of a particular class does not necessarily belong to the same class. Also, the authors give a necessary and sufficient condition for a neutrosophic quadruple ring (NQ(X), +, ·) to be a NeutroQuadrupleRing.

Chapter 4 discusses AntiHyperStructure and presents some of its basic properties. The authors study a particular class of these 19 classes of AntiHyperGroup, called a () AntiHyperGroup. Moreover, they provide some examples of the new concept and prove some related properties.

Chapter 5 generalizes classical Boolean algebra by introducing neutro Boolean algebra. In neutro Boolean algebra, the outcome of any given two elements under logical operations has three possibilities, such as appurtenance, non-appurtenance, or indeterminate. The authors provide some fundamental definitions and investigate some properties based on neutro Boolean algebra.

Chapter 6 addresses the issue of dealing with the heteroclinic data and its pattern finding is considered as one of major concern for the researchers. These types of data forms heteroclinic orbit which changes the space from equilibrium to non-equilibrium and vice versa. In this case several parallel lines can be drawn from the given point to measure the indeterminacy which used to be found in flight paths. It means these types of data do not follow the fifth law. To deal with them, the authors study NeutroGeometry to characterize them based on Euclidean, Non-Euclidean, and Indeterminate zones.

Chapter 7 discusses some properties of AntiSemihypergroups, AntiQuasihypergroups, AntiSubSemihypergroups, and AntiHypergroups. The authors present somenew definitions and theorems and support them by illustrative examples. In addition, they prove the existence of an infinite number of AntiSemihypergroups and AntiHypergroups.

Chapter 8 describes a method for decision-making in a migrational situation that uses a NeutroAlgebra generated by the combining function in Prospector. This method is an important hit in showing both the utility and theoretical significance of NeutroAlgebras since it uses this abstract tool for solving a real-life problem. It presents the rigorous mathematical proofs of the properties of that NeutroAlgebra because so far the theoretical approaches to the proposed method are minimal. The deeper mathematical knowledge of this specific NeutroAlgebra is a contribution to the general theory on this subject.

Chapter 9 improves and extends several definitions, properties, and applications of neutro algebras. In classical algebraic structures, the Laws of Compositions on a given set are well-defined. In fuzzy algebraic structures, the Laws of Compositions on a given set are not well-defined exactly but with a membership value. It examines those structures which have huge applications or strong potential for applications outside of mathematics, namely computer science or engineering. The study emphasizes that the nature of neutro algebra and anti-neutro algebra differs from that of the classical algebra and fuzzy algebra. Moreover, it presents Venn diagram for neutro function and establish some relations between the partial algebra and neutro algebra.

Chapter 10 proposes a method for the generation of NeutroGroups in a simple and automatic way from uninorms, with special interest in uninorms based on combining functions in both Mycin and Prospector expert systems. The use of these uninorms suggests the possible application of NeutroGroups in fields such as decision making as can be found in the consulted literature, as well as for Expert Systems, among others. NeutroGroups of finite or infinite cardinality will be generated, where the axioms of NeutroAsociativity or NeutroInverse are fulfilled, and indeterminacy is symbolically included within this theory. One family of non-commutative NeutroGroups is proposed.

Chapter 11 studies NeutroTopological space and provides some examples. Moreover, it defines some new concepts like NeutroInterior, NeutroClosure, and NeutroBoundary. Also, it studies their various properties.

Chapter 12 introduces the concept of NeutroHX-Group. Starting with an HX-group and using the newly defined concept of NeutroAlgebra, the authors define the NeutroHX-Group. Using the HX-groups associated with the dihedral group, they obtain the Chinese hypergroup associated with them. They present some examples to illustrate situations of obtaining NeutroHX-Groups and AntiHX-Groups. Furthermore, they establish a connection between NeutroHX-Groups and NeutroSemihypergroups.

Chapter 13 introduces the concept of Neutro Ordered Rings by considering three Neutro Axioms (Neutro Abelian Group (additive), Neutro Semi Group (multiplicative), Neutro Distributive (multiplication over addition)) and the partially ordered relation. The authors present interesting results and examples on Neutro Ordered Rings, Neutro Ordered Subrings, Neutro Ordered Ideals, Neutro Ordered Quotient Rings, and Neutro Ordered Ring Homomorphisms.

Chapter 14 introduces several operations and properties of algebraic equations for image retrieval, such as NeutroContent-Based Image Retrieval.

Chapter 15 defines operations '+' and '×' on the substructure viz. ideals of the ring. The authors find conditions for the collection of ideals to be NeutroAlgebra under + and ×. They extend the study case of polynomial rings by considering some special cases of them.

Chapter 16 defines the concept of quadripartitioned single-valued neutrosophic hypergroup as the initial concept of quadripartitioned single-valued neutrosophic hyper-algebra and investigates some of its properties.

Chapter 17 studies NeutroSemigroups. NeutroSemigroups are alternatives to the classical semigroups where the closure law is allowed to be either partially well defined (with the degree of truth T) or partially indeterminate (with the degree of indeterminacy I) or partially outer defined (with the degree of falsehood F), and, associative axiom is equally allowed to be either partially true (with the degree of truth T) or partially indeterminate (with the degree of indeterminacy I) or partially outer defined (with the degree of falsehood F). Elementary properties of NeutroSemigroups with interesting examples are presented.

Florentin Smarandache
University of New Mexico, USA

Madeleine Al-Tahan
Lebanese International University, Lebanon

REFERENCES

Agboola, A. A. A. (2020a). Introduction to NeutroGroups. *International Journal of Neutrosophic Science*, *6*, 41–47. doi:10.5281/zenodo.3989823

Agboola, A. A. A. (2020b). Introduction to NeutroRings. *International Journal of Neutrosophic Science*, *7*, 62–73. doi:10.5281/zenodo.3991389

Agboola, A. A. A. (2020c). On Finite NeutroGroups of Type-NG. *International Journal of Neutrosophic Science*, *10*(2), 84–95. http://fs.unm.edu/IJNS/OnFiniteNeutroGroupsOfType-NG.pdf

Agboola, A. A. A. (2020d). On Finite and Infinite NeutroRings of Type-NR. *International Journal of Neutrosophic Science*, *11*(2), 87–99. http://fs.unm.edu/IJNS/OnFiniteAndInfiniteNeutroRings.pdf

Agboola, A. A. A. (2020e). Introduction to AntiGroups. *International Journal of Neutrosophic Science, 12*(2), 71-80. http://fs.unm.edu/IJNS/IntroductionAntiGroups.pdf

Agboola, A. A. A., Ibrahim, M. A., & Adeleke, E. O. (2020). Elementary Examination of NeutroAlgebras and AntiAlgebras viz-a-viz the Classical Number Systems. *International Journal of Neutrosophic Science*, *4*, 16–19. doi:10.5281/zenodo.3989530

Al-Tahan, . (2021a). NeutroOrderedAlgebra: Applications to Semigroups. *Neutrosophic Sets and Systems*, *39*, 133–147. doi:10.5281/zenodo.4444331

Al-Tahan, M. (2021b). NeutroOrderedAlgebra: Theory and Examples. In *3rd International Workshop on Advanced Topics in Dynamical Systems*. University of Kufa.

Al-Tahan, M., Davvaz, B., Smarandache, F., & Anis, O. (2021). On Some NeutroHyperstructures. *Symmetry*, *13*(535), 1–12. doi:10.3390ym13040535

Hamidi, M., & Smarandache, F. (2020). Neutro-BCK-Algebra. *International Journal of Neutrosophic Science*, *8*, 110–117. doi:10.5281/zenodo.3991437

IbrahimM. A.AgboolaA. A. A. (2020). Introduction to NeutroHyperGroups. *Neutrosophic Sets and Systems, 38*, 15-32. http://fs.unm.edu/NSS/IntroductionToNeutroHyperGroups2.pdf

Jiménez, D. S., Juan, A. V. M., Mara, E. R. U., & Hernández, N. B. (2021). NeutroAlgebra for the evaluation of barriers to migrants' access in Primary Health Care in Chile based on PROSPECTOR function. *Neutrosophic Sets and Systems*, *39*, 1–9. doi:10.5281/zenodo.4444189

KandasamyW. B. V.SmarandacheF.MapsF. C.MapsN. C.XiquanP. (2003). http://fs.unm.edu/NCMs.pdf

MohammadzadehRezaei. (2020). On NeutroNilpotentGroups. *Neutrosophic Sets and Systems, 38*, 33-40. http://fs.unm.edu/NSS/OnNeutroNilpotentGroups3.pdf

Rezaei, A., & Smarandache, F. (2020). On Neutro-BE-algebras and Anti-BE-algebras. *International Journal of Neutrosophic Science*, *4*, 8–15. doi:10.5281/zenodo.3989550

Rezaei, A., Smarandache, F., & Mirvakili, S. (2021). Applications of (Neutro/Anti)sophications to Semihypergroups. *Journal of Mathematics*, 1–7. doi:10.1155/2021/6649349

Smarandache, Rezaei, Agboola, Jun, Borzooei, Davvaz, Broumand, Saeid, Akram, Hamidi, & Mirvakilii. (2021). On NeutroQuadrupleGroups. *51st Annual Mathematics Conference*.

Smarandache, F. (2019). Introduction to NeutroAlgebraic Structures and AntiAlgebraic Structure. In *Advances of Standard and Nonstandard Neutrosophic Theories*. Pons Publishing House. http://fs.unm.edu/AdvancesOfStandardAndNonstandard.pdf

Smarandache, F. (2020a). NeutroAlgebra is a Generalization of Partial Algebra. *International Journal of Neutrosophic Science*, *2*, 8–17. doi:10.5281/zenodo.3989285

SmarandacheF. (2020b). Introduction to NeutroAlgebraic Structures and AntiAlgebraic Structures (revisited). *Neutrosophic Sets and Systems, 31*, 1-16. http://fs.unm.edu/NSS/NeutroAlgebraic-AntiAlgebraic-Structures.pdf

Smarandache, F. (2020c). Generalizations and Alternatives of Classical Algebraic Structures to NeutroAlgebraic Structures and AntiAlgebraic Structures. *Journal of Fuzzy Extension and Applications*, *1*(2), 85–87. http://fs.unm.edu/NeutroAlgebra-general.pdf

SmarandacheF. (2020d). Extension of HyperGraph to n-SuperHyperGraph and to Plithogenic n-SuperHyperGraph, and Extension of HyperAlgebra to n-ary (Classical-/Neutro-/Anti-)HyperAlgebra. *Neutrosophic Sets and Systems, 33*, 290-296. http://fs.unm.edu/NSS/n-SuperHyperGraph-n-HyperAlgebra.pdf

Smarandache, F. (2020e). Structure, NeutroStructure, and AntiStructure in Science. *International Journal of Neutrosophic Science, 13*(1), 28-33. http://fs.unm.edu/IJNS/NeutroStructure.pdf

Smarandache, F. (2021a). Universal NeutroAlgebra and Universal AntiAlgebra. In NeutroAlgebra Theory (vol. 1). Educational Publ.

Smarandache, F. (2021b). *NeutroGeometry & AntiGeometry are alternatives and generalizations of the Non-Euclidean Geometries* (Vol. 46). Neutrosophic Sets and Systems. doi:10.5281/zenodo.5553552

Smarandache, F., Rezaei, A., & Kim, H. S. (2020). A New Trend to Extensions of CI-algebras. *International Journal of Neutrosophic Science*, *5*(1), 8–15. doi:10.5281/zenodo.3788124

Chapter 1
NeutroAlgebra and AntiAlgebra Are Generalizations of Classical Algebras

Florentin Smarandache
https://orcid.org/0000-0002-5560-5926
University of New Mexico, USA

Madeleine Al-Tahan
Lebanese International University, Lebanon

ABSTRACT

Neutrosophy as a branch of philosophy has many applications in almost every field of science. It studies neutralities, which are found in almost every real-life problem. This chapter is a review of reproduction of the evolution from the foundation of paradoxism (1980s), then of neutrosophy (1995) and its derivatives (neutrosophic set/logic/probability/statistics), then neutrosophic algebraic structures (2003), and ending with the NeutroAlgebra and AntiAlgebra (2019) as well as the NeutroGeometry and AntiGeometry (2021) of the previous work.

FROM PARADOXISM TO NEUTROSOPHY

Paradoxism is an international movement in science and culture, founded by Florentin Smarandache in 1980s, based on excessive use of antitheses, oxymoron, contradictions, and paradoxes. During three decades (1980-2020) hundreds of

DOI: 10.4018/978-1-6684-3495-6.ch001

authors from tens of countries around the globe contributed papers to 15 international paradoxist anthologies.

In 1995, he extended the *paradoxism* (based on opposites) to a new branch of philosophy called ***neutrosophy*** (based on opposites and their neutral), that gave birth to many scientific branches, such as: neutrosophic logic, neutrosophic set, neutrosophic probability and statistics, neutrosophic algebraic structures, and so on with multiple applications in engineering, computer science, administrative work, medical research, etc.

Neutrosophy is an extension of Dialectics that have derived from the Yin-Yan Ancient Chinese Philosophy.

The **Neutrosophic Algebraic Structures** are based on neutrosophic numbers of the form a+bI, where a, b are real or complex, while I = literal indeterminacy with $I^2 = I$.

FROM CLASSICAL ALGEBRAIC STRUCTURES TO NEUTROALGEBRAIC STRUCTURES AND ANTIALGEBRAIC STRUCTURES

In 2019, Smarandache (Smarandache, 2019) generalized the classical Algebraic Structures to NeutroAlgebraic Structures (or **NeutroAlgebras**) {whose operations and axioms are partially true, partially indeterminate, and partially false} as extensions of Partial Algebra, and to AntiAlgebraic Structures (or **AntiAlgebras**) {whose operations and axioms are totally false} and on 2020 he continued to develop them (Smarandache, 2020a; Smarandache, 2020b; Smarandache, 2020c).

The NeutroAlgebras & AntiAlgebras are a *new field of research*, which is inspired from our real world.

In classical algebraic structures, all operations are 100% well-defined, and all axioms are 100% true, but in real life, in many cases these restrictions are too harsh, since in our world we have things that only partially verify some operations or some laws.

Using the process of *NeutroSophication* of a classical algebraic structure we produce a NeutroAlgebra, while the process of *AntiSophication* of a classical algebraic structure produces an AntiAlgebra.

NEUTROGEOMETRY AND ANTIGEOMETRY ARE ALTERNATIVES AND GENERALIZATIONS OF THE NON-EUCLIDEAN GEOMETRIES

While the Non-Euclidean Geometries resulted from the total negation of only one specific axiom (Euclid's Fifth Postulate), the AntiGeometry results from the total negation of any axiom and even of more axioms from any geometric axiomatic system (Euclid's, Hilbert's, etc.), and the NeutroGeometry results from the partial negation of one or more axioms [and no total negation of no axiom] from any geometric axiomatic system. Generally, instead of a classical geometric Axiom, one may take any classical geometric Theorem and transform it by NeutroSophication or AntiSophication into a NeutroTheorem or AntiTheorem respectively to construct a NeutroGeometry or AntiGeometry (Smarandache, 2021b). Therefore, the NeutroGeometry and AntiGeometry founded by Smarandache in 2021 are respectively alternatives and generalizations of the Non-Euclidean Geometries.

OPERATION, NEUTROOPERATION, ANTIOPERATION

When we define an operation on a given set, it does not automatically mean that the operation is well-defined. There are three possibilities:

1. The operation is well-defined (also called inner-defined) for all set's elements [degree of truth T = 1] (as in classical algebraic structures; this is a classical **Operation**). Neutrosophically we write: Operation(1,0,0).
2. The operation if well-defined for some elements [degree of truth T], indeterminate for other elements [degree of indeterminacy I], and outer-defined for the other elements [degree of falsehood F], where (T,I,F) is different from (1,0,0) and from (0,0,1) (this is a **NeutroOperation**). Neutrosophically we write: NeutroOperation(T,I,F).
3. The operation is outer-defined for all set's elements [degree of falsehood F = 1] (this is an **AntiOperation**). Neutrosophically we write: AntiOperation(0,0,1).

AXIOM, NEUTROAXIOM, ANTIAXIOM

Similarly for an axiom, defined on a given set, endowed with some operation(s). When we define an axiom on a given set, it does not automatically mean that the axiom is true for all set's elements. We have three possibilities again:

1. The axiom is true for all set's elements (totally true) [degree of truth T = 1] (as in classical algebraic structures; this is a classical **Axiom**). Neutrosophically we write: Axiom(1,0,0).
2. The axiom if true for some elements [degree of truth T], indeterminate for other elements [degree of indeterminacy I], and false for other elements [degree of falsehood F], where (T,I,F) is different from (1,0,0) and from (0,0,1) (this is **NeutroAxiom**). Neutrosophically we write NeutroAxiom(T,I,F).
3. The axiom is false for all set's elements [degree of falsehood F = 1](this is **AntiAxiom**). Neutrosophically we write AntiAxiom(0,0,1).

THEOREM, NEUTROTHEOREM, ANTITHEOREM

In any science, a classical Theorem, defined on a given space, is a statement that is 100% true (i.e., true for all elements of the space). To prove that a classical theorem is false, it is sufficient to get a single counterexample where the statement is false. Therefore, the classical sciences do not leave room for *partial truth* of a theorem (or a statement). But, in our world and in our everyday life, we have many more examples of statements that are only partially true, than statements that are totally true. The NeutroTheorem and AntiTheorem are generalizations and alternatives of the classical Theorem in any science.

Let's consider a theorem, stated on a given set, endowed with some operation(s). When we construct the theorem on a given set, it does not automatically mean that the theorem is true for all set's elements. We have three possibilities again:

1. The theorem is true for all set's elements [totally true] (as in classical algebraic structures; this is a classical **Theorem**). Neutrosophically we write: Theorem(1,0,0).
2. The theorem if true for some elements [degree of truth T], indeterminate for other elements [degree of indeterminacy I], and false for the other elements [degree of falsehood F], where (T,I,F) is different from (1,0,0) and from (0,0,1) (this is a **NeutroTheorem**). Neutrosophically we write: NeutroTheorem(T,I,F).
3. The theorem is false for all set's elements (this is an **AntiTheorem**). Neutrosophically we write: AntiTheorem(0,0,1).

And similarly for (Lemma, NeutroLemma, AntiLemma), (Consequence, NeutroConsequence, AntiConsequence), (Algorithm, NeutroAlgorithm, AntiAlgorithm), (Property, NeutroProperty, AntiProperty), etc.

ALGEBRA, NEUTROALGEBRA, ANTIALGEBRA

1. An algebraic structure who's all operations are well-defined and all axioms are totally true is called a classical Algebraic Structure (or **Algebra**).
2. An algebraic structure that has at least one NeutroOperation or one NeutroAxiom (and no AntiOperation and no AntiAxiom) is called a NeutroAlgebraic Structure (or **NeutroAlgebra**).
3. An algebraic structure that has at least one AntiOperation or one Anti Axiom is called an AntiAlgebraic Structure (or **AntiAlgebra**) ().

Therefore, a neutrosophic triplet is formed: <Algebra, NeutroAlgebra, AntiAlgebra>, where "Algebra" can be any classical algebraic structure, such as: a groupoid, semigroup, monoid, group, commutative group, ring, field, vector space, BCK-Algebra, BCI-Algebra, etc.

STRUCTURE, NEUTROSTRUCTURE, ANTISTRUCTURE IN ANY FIELD OF KNOWLEDGE

In general, by NeutroSophication, Smarandache extended any classical *Structure*, in no matter what field of knowledge, to a *NeutroStructure*, and by AntiSophication to an *AntiStructure*.

A classical Structure, in any field of knowledge, is composed of: a non-empty space, populated by some elements, and both (the space and all elements) are characterized by some relations among themselves (such as: operations, laws, axioms, properties, functions, theorems, lemmas, consequences, algorithms, charts, hierarchies, equations, inequalities, etc.), and by their attributes (size, weight, color, shape, location, etc.).

Of course, when analysing a structure, it counts with respect to what relations and what attributes we do it.

RELATION, NEUTRORELATION, ANTIRELATION

1. A classical **Relation** is a relation that is true for all elements of the set (degree of truth T = 1). Neutrosophically we write Relation(1,0,0).
2. A **NeutroRelation** is a relation that is true for some of the elements (degree of truth T), indeterminate for other elements (degree of indeterminacy I), and false for the other elements (degree of falsehood F). Neutrosophically we write Relation(T,I,F), where (T,I,F) is different from (1,0,0) and (0,0,1).

3. An **AntiRelation** is a relation that is false for all elements (degree of falsehood F = 1). Neutrosophically we write Relation(0,0,1).

ATTRIBUTE, NEUTROATTRIBUTE, ANTIATTRIBUTE

1. A classical **Attribute** is an attribute that is true for all elements of the set (degree of truth T = 1). Neutrosophically we write Attribute(1,0,0).
2. A **NeutroAttribute** is an attribute that is true for some of the elements (degree of truth T), indeterminate for other elements (degree of indeterminacy I), and false for the other elements (degree of falsehood F). Neutrosophically we write Attribute(T,I,F), where (T,I,F) is different from (1,0,0) and (0,0,1).
3. An **AntiAttribute** is an attribute that is false for all elements (degree of falsehood F = 1). Neutrosophically we write Attribute(0,0,1).

STRUCTURE, NEUTROSTRUCTURE, ANTISTRUCTURE

1. A classical **Structure** is a structure whose all elements are characterized by the same given Relationships and Attributes.
2. A **NeutroStructure** is a structure that has at least one NeutroRelation or one NeutroAttribute, and no AntiRelation and no AntiAttribute.
3. An **AntiStructure** is a structure that has at least one AntiRelation or one AntiAttribute.

ALMOST ALL STRUCTURES ARE NEUTROSTRUCTURES

The Classical Structures in science mostly exist in theoretical, abstract, perfect, homogeneous, idealistic spaces - because in our everyday life almost all structures are NeutroStructures, since they are neither perfect nor applying to the whole population, and not all elements of the space have the same relations and same attributes in the same degree (not all elements behave in the same way).

The indeterminacy and partiality, with respect to the space, to their elements, to their relations or to their attributes are not taken into consideration in the Classical Structures. But our Real World is full of structures with indeterminate (vague, unclear, conflicting, unknown, etc.) data and partialities.

There are exceptions to almost all laws, and the laws are perceived in different degrees by different people.

Examples of NeutroStructures From Our Real World

1. In the Christian society the marriage law is defined as the union between a male and a female (*degree of truth*).

But, in the last decades, this law has become less than 100% true, since persons of the same sex were allowed to marry as well (*degree of falsehood*).

On the other hand, there are transgender people (whose sex is indeterminate), and people who have changed the sex by surgical procedures, and these people (and their marriage) cannot be included in the first two categories (*degree of indeterminacy*).

Therefore, since we have a NeutroLaw (with respect to the Law of Marriage) we have a Christian NeutroStructure.

2. In India, the law of marriage is not the same for all citizen: Hindi religious men may marry only one wife, while the Muslims may marry up to four wives.
3. Not always the difference between good and bad may be clear, from a point of view a thing may be good, while from another point of view bad. There are things that are partially good, partially neutral, and partially bad.
4. The laws do not equally apply to all citizens, so they are NeutroLaws. Some laws apply to some degree to a category of citizens, and to a different degree to another category. Almost always there are exceptions to the law! As such, there is an American folkloric joke: All people are born equal, but some people are more equal than others!
    a. There are powerful people that are above the laws, and other people that benefit of immunity with respect to the laws.
    b. For example, in the court of law, privileged people benefit from better defense lawyers than the lower classes, so they may get a lighter sentence.
    c. Not all criminals go to jail, but only those caught and proven guilty in the court of law. Nor criminals that for reason of insanity cannot stand trail and do not go to jail since they cannot make a difference between right and wrong.
    d. Unfortunately, even innocent people went and might go to jail because of sometimes jurisdiction mistakes...
    e. The Hypocrisy and Double Standard are widely spread: some regulation applies to some people, but not to others!
5. Anti-Abortion Law does not apply to all pregnant women: the incest, rapes, and women whose life is threatened may get abortions.
6. Gun-Control Law does not apply to all citizen: the police, army, security, professional hunters are allowed to bear arms. Etc.

REFERENCES

Agboola, A. A. A. (2020a). Introduction to NeutroGroups. *International Journal of Neutrosophic Science*, *6*, 41–47.

Agboola, A. A. A. (2020b). Introduction to NeutroRings. *International Journal of Neutrosophic Science*, *7*, 62–73.

Agboola, A. A. A. (2020c). On Finite NeutroGroups of Type-NG. *International Journal of Neutrosophic Science*, *10*(2), 84–95. doi:10.5281/zenodo.4277243

Agboola, A. A. A. (2020d). On Finite and Infinite NeutroRings of Type-NR. *International Journal of Neutrosophic Science*, *11*(2), 87–99. doi:10.5281/zenodo.4276366

Agboola, A. A. A. (2020e). Introduction to AntiGroups. *International Journal of Neutrosophic Science, 12*(2), 71-80. http://fs.unm.edu/IJNS/IntroductionAntiGroups.pdf

Agboola, A. A. A., Ibrahim, M. A., & Adeleke, E. O. (2020). Elementary Examination of NeutroAlgebras and AntiAlgebras viz-a-viz the Classical Number Systems. *International Journal of Neutrosophic Science*, *4*, 16–19. doi:10.54216/IJNS.040102

Al-Tahan, M. (2021). NeutroOrderedAlgebra: Theory and Examples. In *3rd International Workshop on Advanced Topics in Dynamical Systems*. University of Kufa.

Al-Tahan, M., Davvaz, B., Smarandache, F., & Anis, O. (2021). On Some NeutroHyperstructures. *Symmetry*, *13*(535), 1–12. doi:10.3390ym13040535

Hamidi, M., & Smarandache, F. (2020). Neutro-BCK-Algebra. *International Journal of Neutrosophic Science*, *8*, 110–117. doi:10.54216/IJNS.080204

IbrahimM. A.AgboolaA. A. A. (2020). Introduction to NeutroHyperGroups. *Neutrosophic Sets and Systems, 38*, 15-32. http://fs.unm.edu/NSS/IntroductionToNeutroHyperGroups2.pdf doi:10.5281/zenodo.4300363

Jiménez, D. S., Juan, A. V. M., Mara, E. R. U., & Hernández, N. B. (2021). NeutroAlgebra for the evaluation of barriers to migrants' access in Primary Health Care in Chile based on PROSPECTOR function. *Neutrosophic Sets and Systems*, *39*, 1–9. doi:10.5281/zenodo.4444189

Madeleine Al-Tahan, F. (2021). NeutroOrderedAlgebra: Applications to Semigroups. *Neutrosophic Sets and Systems*, *39*, 133–147. doi:10.5281/zenodo.4444331

MohammadzadehRezaei. (2020). On NeutroNilpotentGroups. *Neutrosophic Sets and Systems, 38*, 33-40. http://fs.unm.edu/NSS/OnNeutroNilpotentGroups3.pdf doi:10.5281/zenodo.4300370

Rezaei, A., & Smarandache, F. (2020). On Neutro-BE-algebras and Anti-BE-algebras. *International Journal of Neutrosophic Science*, *4*, 8–15.

Rezaei, A., Smarandache, F., & Mirvakili, S. (2021). Applications of (Neutro/Anti)sophications to Semihypergroups. *Journal of Mathematics*, 1–7. doi:10.1155/2021/6649349

Smarandache, Rezaei, Agboola, Jun, Borzooei, Davvaz, Broumand Saeid, Akram, Hamidi, & Mirvakilii. (2021). On NeutroQuadrupleGroups. *51st Annual Mathematics Conference.*

Smarandache, F. (2019). Introduction to NeutroAlgebraic Structures and AntiAlgebraic Structures. In *Advances of Standard and Nonstandard Neutrosophic Theories*. Pons Publishing House. http://fs.unm.edu/AdvancesOfStandardAndNonstandard.pdf

Smarandache, F. (2020a). NeutroAlgebra is a Generalization of Partial Algebra. *International Journal of Neutrosophic Science*, *2*, 8–17. doi:10.54216/IJNS.020103

Smarandache, F. (2020b). Introduction to NeutroAlgebraic Structures and AntiAlgebraic Structures (revisited). *Neutrosophic Sets and Systems*, *31*, 1–16. doi:10.5281/zenodo.3638232

Smarandache, F. (2020c). Generalizations and Alternatives of Classical Algebraic Structures to NeutroAlgebraic Structures and AntiAlgebraic Structures, Journal of Fuzzy Extension and Applications (JFEA), J. Fuzzy. *Ext. Appl.*, *1*(2), 85–87. doi:10.22105/jfea.2020.248816.1008

Smarandache, F. (2020d). Extension of HyperGraph to n-SuperHyperGraph and to Plithogenic n-SuperHyperGraph, and Extension of HyperAlgebra to n-ary (Classical-/Neutro-/Anti-)HyperAlgebra. *Neutrosophic Sets and Systems*, *33*, 290–296. doi:10.5281/zenodo.3783103

Smarandache, F. (2020e). Structure, NeutroStructure, and AntiStructure in Science. *International Journal of Neutrosophic Science, 13*(1), 28-33. http://fs.unm.edu/IJNS/NeutroStructure.pdf

Smarandache, F. (2021a). Universal NeutroAlgebra and Universal AntiAlgebra. In NeutroAlgebra Theory (Vol. 1). Educational Publ.

SmarandacheF. (2021b). NeutroGeometry & AntiGeometry are alternatives and generalizations of the Non-Euclidean Geometries. *Neutrosophic Sets and Systems, 46*, 456-477. http://fs.unm.edu/NSS/NeutroGeometryAntiGeometry31.pdf doi:10.5281/zenodo.5553552

Smarandache, F., Rezaei, A., & Kim, H. S. (2020). A New Trend to Extensions of CI-algebras. *International Journal of Neutrosophic Science*, *5*(1), 8–15. doi:10.54216/IJNS.050101

Vasantha Kandasamy & Smarandache. (2003). *Fuzzy Cognitive Maps and Neutrosophic Cognitive Maps*. Phoenix. http://fs.unm.edu/NCMs.pdf

Chapter 2
An Introduction to NeutroOrderSemihypergroups

Madeleine Ali Al-Tahan
Lebanese International University, Lebanon

Osman Mohammad Anis
Lebanese International University, Lebanon

Ahmed A. Salama
https://orcid.org/0000-0003-2300-615X
Faculty of Science, Port Said University, Egypt

Rafael Rojas-Gualdron
Universidad de los Andes, Colombia

ABSTRACT

Neutroosphy is the study of neutralities founded by Smarandache in 1998. Starting with a partial order on a NeutroSemihypergroup, the authors get a NeutroHyperstructure. If it satisfies the conditions of NeutroOrder, it becomes a NeutroOrderSemihypergroup. This chapter introduces the field of NeutroOrderHyperstructures by considering a special class of it: NeutroOrderSemihypergroups. Some new definitions are provided, some properties of the new defined concepts are investigated, and illustrative examples are presented.

INTRODUCTION

Neutrosophy, the study of neutralities, is a new branch of Philosophy initiated by Smarandache (Smarandache, F., 1998) and it has connections with different domains

DOI: 10.4018/978-1-6684-3495-6.ch002

in Mathematics. One of these connections is Algebra. As a generalization of partial algebra (Smarandache, F., 2020), NeutroAlgebra was introduced whose operations and axioms are partially true, partially indeterminate, and partially false. On the other hand, algebraic hyperstructures are a generalization of algebraic structures and they were introduced in 1934 by Marty (Marty, F., 1934). As a combination between the two notions: algebraic hypertsructures and NeutroAlgebra, NeutroHyperstructure theory was introduced (Ibrahim, M.A. & Agboola, A.A.A, 2020) where Ibrahim and Agboola defined NeutroHypergroups and studied a special type of it.

An ordered algebraic structure is an algebraic structure associated with a partial order satisfying an additional property (Monotone property). As an extension of ordered algebraic structures, NeutroOrderAlgebra was introduced and applied to NeutroSemigroups (Al-Tahan, M., Smarandache, F., & Davvaz, B., 2021). New related notions were defined such as NeutroOrderSubsemigroup, NeutroOrderFilter, NeutroOrderStrongIsomorphism, etc. and some of their properties were investigated.

This chapter introduces NeutroOrderHyperstructure by considering a partial order on NeutroSemihypergroups. Such an order when it satisfies the property of NeutroOrder, the whole system becomes a NeutroOrderSemihypergroup. More precisely, the authors define (for the first time) NeutroOrderSemihypergroups and illustrate their new definition by non-trivial examples. Also, they proved that there are infinitely many NeutroOrderSemihypergroups by showing that there is an NOHS of cardinality n for every $n > 2$. Moreover, they study some subsets of this new structure and investigate some of their properties. Finally, they leave some ideas for future related research.

BACKGROUND

Hyperstructure theory is a generalization of classical algebraic structures and it was introduced in 1934 at the eighth Congress of Scandinavian Mathematicians by Marty (Marty, F. 1934). Marty generalized the notion of groups by defining hypergroups. The class of algebraic hyperstructures is larger than that of algebraic structures where the operation on two elements in the latter is again an element, whereas the hyperoperation of two elements in the first class is a non-void set. This theory has many applications in various fields of science like Cryptography, Automata theory, Fuzzy sets, Biology, Chemistry, Physics, Neutrosophy, Coding theory. Some of the applications can be found in the book "Hyperstructures and Their Representations" (Vougiouklis, T., 1994).

Let H be a non-empty set and $P^*(H)$ be the set of all non-empty subsets of H. Then a hyperoperation on H is a mapping $\bigcirc: H \times H \rightarrow P^*(H)$. The couple $(H,\bigcirc)$ is called a hypergroupoid.

A hypergroupoid $(H,\bigcirc)$ is called semihypergroup if it satisfies the associative axiom. i.e. $a\bigcirc(b\bigcirc c) = (a\bigcirc b)\bigcirc c$ for all $a,b,c \in H$. (Corsini, P., 1993)

Example 1. Let Q be the set of rational numbers and $*: Q \times Q \rightarrow P^*(Q)$ be defined as $q * t = Q$ for all $q,t \in Q$. Then $(Q,*)$ is a semihypergroup.

Al-Tahan et al. (Al-Tahan, M., Davvaz, B., Smarandache, F., & Anis, O., 2021) studied NeutroSemihypergroups and they defined some new related concepts such as NeutroSubsemihypergroups, NeutroHyperideals, etc. and they proved some related properties.

Definition 1. (Al-Tahan, M., Davvaz, B., Smarandache, F., & Anis, O., 2021) Let A be any non-empty set and "$\bigcirc: A \times A \rightarrow P^*(A)$" be a map. Then "$\bigcirc$" is called a NeutroHyperoperation on A if some (or all) of the following conditions hold in a way that

$$(T,I,F) \notin \{(1,0,0),(0,0,1)\}.$$

1. There exist $x,y \in A$ with $x\bigcirc y \subseteq A$. (This condition is called degree of truth, "T").
2. There exist $x,y \in A$ with $x\bigcirc y \subsetneq A$. (This condition is called degree of falsity, "F").
3. There exist $x,y \in A$ with $x\bigcirc y$ is indeterminate in A. (This condition is called degree of indeterminacy, "I").

The couple $(A,\bigcirc)$ is called a NeutroHypergroupoid.

Definition 2. (Al-Tahan, M., Davvaz, B., Smarandache, F., & Anis, O., 2021) Let A be any non-empty set and "$\bigcirc: A \times A \rightarrow P^*(A)$" be a map. Then "$\bigcirc$" is called an AntiHyperoperation on A if $x\bigcirc y \subsetneq A$ for all $x,y \in A$. The couple $(A,\bigcirc)$ is called an AntiHypergroupoid.

Example 2. Let Q,R be the set of rational and real numbers respectively and $\Delta,\nabla: Q \times Q \rightarrow P^*(Q)$ be maps on Q defined as $q\Delta t = R$ for all

$$q,t \in Q \, and \, q\nabla t = \begin{cases} R & if \, q = t, \\ Q & otherwise. \end{cases}$$

Then $(Q,\Delta),(Q,\nabla)$ are AntiHypergroupoid and NeutroHypergroupoid respectively.

Definition 3. (Al-Tahan, M., Davvaz, B., Smarandache, F., & Anis, O., 2021) Let A be any non-empty set and "$\bigcirc$: $A \times A \to P^*(A)$ be a map. Then "$\bigcirc$" is called a NeutroAssociative on A if some (or all) of the following conditions hold in a way that

$$(T,I,F) \notin \{(1,0,0),(0,0,1)\}.$$

1. There exist $x,y,z \in A$ with $x\bigcirc(y\bigcirc z) = (x\bigcirc y)\bigcirc z$. (This condition is called degree of truth, "T").
2. There exist $x,y,z \in A$ with $x\bigcirc(y\bigcirc z) \neq (x\bigcirc y)\bigcirc z$. (This condition is called degree of falsity, "F").
3. There exist $x,y,z \in A$ with $x\bigcirc(y\bigcirc z)$ is indeterminate or $(x\bigcirc y)\bigcirc z$ is indeterminate or we can not find if $x\bigcirc(y\bigcirc z)$ and $(x\bigcirc y)\bigcirc z$ are equal. (This condition is called degree of indeterminacy, "I").

Definition 4. (Al-Tahan, M., Davvaz, B., Smarandache, F., & Anis, O., 2021) Let A be any non-empty set and "$\bigcirc$: $A \times A \to P^*(A)$" be a map. Then "$\bigcirc$" is called an AntiAssociative on A if $x\bigcirc(y\bigcirc z) \neq (x\bigcirc y)\bigcirc z$ for all $x,y,z \in A$.

Definition 5. (Al-Tahan, M., Davvaz, B., Smarandache, F., & Anis, O., 2021) Let A be any non-empty set and "$\bigcirc$: $A \times A \to P^*(A)$" be a map. Then $(A,\bigcirc)$ is called NeutroSemihypergroup if "$\bigcirc$" is a NeutroAssociative but not an AntiHyperoperation.

Example 3. Let R be the set of real numbers and define "$\square$: $R \times R \to P^*(R)$" on R as follows.

$$x\square y=\begin{cases}\left[x,y\right] & if\; x<y,\\ \left[y,x\right] & if\; y<x,\\ 0 & if\; x=y=0,\\ \dfrac{1}{x} & otherwise.\end{cases}$$

Then $(R,\square)$ is a NeutroSemihypergroup. This is clear as:

$0\square(0\square 0)=0=(0\square 0)\square 0$ and $1\square(2\square 2)=[0.5,1]\neq[1,2]\square\{0.5\}=(1\square 2)\square 2.$

Example 4. Let $A=\{m,a,d\}$, ♥: $A\times A\rightarrow P^{*}(A)$, and define $(A,$♥$)$ by Table 1.

Table 1. The NeutroSemihypergroup $(A,$♥$)$

♥	m	a	d
m	m	m	m
a	m	{m,a}	d
d	m	d	d

Then $(A,$♥$)$ is a *NeutroSemihypergroup.*

Example 5. Let Z be the set of integers and define "$\oplus: Z^2\times Z^2\rightarrow P^{*}(Z^2)$" on Z^2 as follows. For all $m,n,p,q\in Z$,

$$(m,0)\oplus(0,0)=(0,0)\oplus(m,0)=\{(0,0),(m,0)\},$$

$$(0,n)\oplus(0,0)=(0,0)\oplus(0,n)=\{(0,0),(0,n)\},$$

and if

$$(n,p,q)\neq(0,0,0),(m,p,q)\neq(0,0,0)$$

$$(m,n) \oplus (p,q) = (m+p, n+q).$$

Then $(Z^2, \oplus)$ is a NeutroSemihypergroup.

Example 6. Let Z be the set of real numbers and define "$\bullet : Z^2 \times Z^2 \to P^*(Z^2)$" on Z^2 as follows. For all $m,n,p,q \in Z$,

$$(m,n) \bullet (1,1) = (1,1) \bullet (m,n) = \{(1,1),(m,n)\}$$

and

$$(m,n) \bullet (p,q) = (mp, nq) \text{if } (m,n) \neq (1,1),(p,q) \neq (1,1).$$

Then $(Z^2, \bullet)$ is a NeutroSemihypergroup.

Example 7. Let

$$A' = \{0,1,2,3,4\}, \heartsuit' : A' \times A' \to P^*(A'),$$

and define $(A', \heartsuit')$ by Table 2.

Table 2. The NeutroSemihypergroup $(A', \heartsuit')$

♥'	0	1	2	3	4
0	0	0	0	3	0
1	0	1	2	1	1
2	0	4	2	3	3
3	0	4	2	3	3
4	0	0	0	4	A'

Then $(A', \heartsuit')$ is a NeutroSemihypergroup.

Al-Tahan et al. (Al-Tahan, M., Davvaz, B., Smarandache, F., & Anis, O., 2021) defined the extension NeutroSemihypergroup as follows.

Let $\left(H,\bigcirc\right)$ be an NSH and J a non-empty set such that $H\cap J=\varnothing$ and $\left(H\bigcirc H\right)\cap J=\varnothing$. The extension $H\left[J\right]$ of H by J is given as $H\left[J\right]=H\cup J$ and the map "$\circledcirc : H\left[J\right]\rightarrow P^{*}\left(H\left[J\right]\right)$" is defined as follows:

$$x\circledcirc y=\begin{cases} x\bigcirc y & if\, x,y\in H, \\ H\cup J & otherwise. \end{cases}$$

Theorem 1. (Al-Tahan, M., Davvaz, B., Smarandache, F., & Anis, O., 2021) Let $\left(H,\bigcirc\right)$ be an NSH and J a non-empty set such that $H\cap J=\varnothing$ and $\left(H\bigcirc H\right)\cap J=\varnothing$. Then $\left(H\left[J\right],\circledcirc\right)$ is an NSH.

NEUTROORDERHYPERSTRUCTURES

Inspired by NeutroOrderAlgebra (Al-Tahan, M., Smarandache, F., & Davvaz, B., 2021) and NeutroHyperstructures (Al-Tahan, M., Davvaz, B., Smarandache, F., & Anis, O., 2021), the authors combine the two notions together to get NeutroOrderHyperstructure.

NeutroOrderSemihypergroups, Definition, and Examples

Definition 6. (Heidari, D. & Davvaz, B.,2011) Let $\left(H,\bigcirc\right)$ be a semihypergroup and "$\leq$" be a partial order on H. Then $\left(H,\bigcirc,\leq\right)$ is an ordered semihypergroup if $a\leq b\in H$ and $c\in H$ then $a\bigcirc c\leq b\bigcirc c$ and $c\bigcirc a\leq c\bigcirc b$.

Definition 7. (Al-Tahan, M., Smarandache, F., & Davvaz, B., 2021) Let $\left(A,\bigcirc\right)$ be a NeutroSemigroup and "$\leq$" be a partial order on A. Then $\left(A,\bigcirc,\leq\right)$ is a NeutroOrderSemigroup (NOSH) if some (or all) of the following conditions hold in a way that

$$\left(T,I,F\right)\notin\left\{\left(1,0,0\right),\left(0,0,1\right)\right\}.$$

1. There exist $x\leq y\in A$ with $x\neq y$ such that $z\bigcirc x\leq z\bigcirc y$ and $x\bigcirc z\leq y\bigcirc z$ for all z $\in A$. (This condition is called degree of truth, "T").
2. There exist $x\leq y\in A$ such that $z\bigcirc x\nleq z\bigcirc y$ or $x\bigcirc z\nleq y\bigcirc z$ for some z $\in A$. (This condition is called degree of falsity, "F").

3. There exist $x \leq y \in A$ with $z \bigcirc x$ or $z \bigcirc y$ or $x \bigcirc z$ or $y \bigcirc z$ are indeterminate in A, or the relation between $z \bigcirc x$ and $z \bigcirc y$, or the relation between $x \bigcirc z$ and $y \bigcirc z$ are indeterminate. (This condition is called degree of indeterminacy, "I").

Definition 8. Let $(A, \bigcirc)$ be a NeutroSemihypergroup and "$\leq$" be a partial order on A. Then $(A, \bigcirc, \leq)$ is a NeutroOrderSemihypergroup (NOSH) if some (or all) of the following conditions hold in a way that

$(T, I, F) \notin \{(1,0,0),(0,0,1)\}$.

1. There exist $x \leq y \in A$ with $x \neq y$ such that $z \bigcirc x \leq z \bigcirc y$ and $x \bigcirc z \leq y \bigcirc z$ for all z $\in A$. (This condition is called degree of truth, "T").
2. There exist $x \leq y \in A$ such that $z \bigcirc x \nleq z \bigcirc y$ or $x \bigcirc z \nleq y \bigcirc z$ for some z $\in A$. (This condition is called degree of falsity, "F").
3. There exist $x \leq y \in A$ with $z \bigcirc x$ or $z \bigcirc y$ or $x \bigcirc z$ or $y \bigcirc z$ are indeterminate in A, or the relation between $z \bigcirc x$ and $z \bigcirc y$, or the relation between $x \bigcirc z$ and $y \bigcirc z$ are indeterminate. (This condition is called degree of indeterminacy, "I").

Notation 1. Let $(\boldsymbol{A}, \bigcirc)$ be an NSH, $\boldsymbol{X}, \boldsymbol{Y} \neq \varnothing \subseteq \boldsymbol{A}$, and "$\leq$" a partial order on $\boldsymbol{A}$. By $\boldsymbol{X} \leq \boldsymbol{Y}$ we mean for every $\boldsymbol{x} \in \boldsymbol{X}$ there exists $\boldsymbol{y} \in \boldsymbol{Y}$ such that $\boldsymbol{x} \leq \boldsymbol{y}$.

Remark 1. Every NeutroOrderSemigroup is a NeutroOrderSemihypergroup.

Example 8. Let $(\boldsymbol{R}, \square)$ be the NeutroSemihypergroup in Example 3 and "$\leq_u$" be the usual order of real numbers. Then $(\boldsymbol{R}, \square, \leq_u)$ is an NOSH. This is clear as:

$0 \leq_u 1$ and

$$0 \square x = x \square 0 = \begin{cases} 0 & \text{if } x = 0, \\ [0, x] & \text{if } 0 < x \langle 1 \, or \, x \rangle 1, \\ [x, 0] & \text{if } x < 0, \\ [0, 1] & \text{if } x = 1. \end{cases} \leq_u 1 \square x = x \square 1 = \begin{cases} [0, 1] & \text{if } x = 0, \\ [1, x] & \text{if } x > 1, \\ [x, 1] & \text{otherwise}. \end{cases}$$

holds and $0 \leq_u 2$ but $0 \square 2 = [0, 2] \leq_u 2 \square 2 = \frac{1}{2}$ does not hold.

Example 9. Let $(A,♥)$ be the NeutroSemihypergroup in Example 4 and "$\leq_t$" be the partial order on A given as follows.

$$\leq_t = \{(a,m),(a,d),(a,a),(d,d),(d,m),(m,m)\}.$$

Then $(A,♥,\leq_t)$ is an NOSH. This is clear as:
$d \leq_t m$ and

$$d♥x = x♥d \leq_u m♥x = x♥m = m$$

holds for all $x \in A$ and $a \leq_t d$ but

$$a♥a = \{m,a\} \leq_u d♥a = d$$

does not hold.

Example 10. Let $(Z^2,\bullet)$ be the NeutroSemihypergroup in Example 6 and "$\leq_p$" be the partial order on Z^2 given as follows for all $(m,n),(p,q) \in Z^2$.

$$(m,n) \leq_p (p,q) \Leftrightarrow m \leq p \, and \, n \leq q.$$

Then $(Z^2,\bullet,\leq_p)$ is an NOSH. This is clear as:
$(0,0) \leq_p (1,1)$ and

$$(0,0)\bullet(m,n) = (m,n)\bullet(0,0) = \{(0,0)\} \leq_p \{(1,1),(m,n)\} = (1,1)\bullet(m.n) = (m.n)\bullet(1,1)$$

holds.

And $(0,0) \leq_p (2,3)$ and

$$(0,0)\bullet(-1,0) = \{(0,0)\} \leq_p (-2,0) = (2,3)\bullet(-1,0)$$

does not hold.

Example 11. Let $(A',♥')$ be the NeutroSemihypergroup in Example 7 and define "$\leq'$" on A' as follows.

$$\leq' = \{(0,0),(0,1),(0,3),(0,4),(1,1),(1,3),(1,4),(2,2),(3,3),(3,4),(4,4)\}.$$

Then $(A', \heartsuit', \leq')$ is an NOSH. This is clear as $0 \leq' 4$ and

$$0\heartsuit' x \leq' 4\heartsuit' x, x\heartsuit' 0 \leq' x\heartsuit' 4$$

holds for all $x \in A'$ and $0 \leq' 1$ but $0\heartsuit' 2 \leq' 1\heartsuit' 2$ does not hold.

Theorem 2. Let $(A_1, \bigcirc_1, \leq_1), (A_2, \bigcirc_2, \leq_2)$ be NOSHs. Then $(A_1 \times A_2, \bigcirc, \leq)$ is an NOSH. Here for all

$$(x, y), (z, w) \in A_1 \times A_2,$$

$$(x, y) \bigcirc (z, w) = \{(a, m) : a \in x \bigcirc_1 z \text{ and } m \in y \bigcirc_2 w\}$$

and $(x, y) \leq (z, w)$ is equivalent to $x \leq_1 z$ and $y \leq_2 w$.

Proof. Since $(A_1, \bigcirc_1), (A_2, \bigcirc_2)$ are NSHs, it follows from (Al-Tahan, M., Davvaz, B., Smarandache, F., & Anis, O., 2021) that $(A_1 \times A_2, \bigcirc)$ is an NSH. Having $(A_1, \bigcirc_1, \leq_1), (A_2, \bigcirc_2, \leq_2)$ NOSHs, we get:

1. Thereexist $x \leq y \in A_1$ with $x \neq y$ suchthat $z \bigcirc_1 x \leq z, \bigcirc_1 y$ and $x, \bigcirc_1 z \leq y, \bigcirc_1 z$ for all z $\in A_1$. (This condition is called degree of truth, "T").
2. There exist $x \leq y \in A_1$ such that $z \bigcirc_1 x \not\leq z \bigcirc_1 y$ or $x \bigcirc_1 z \not\leq y \bigcirc_1 z$ for some z $\in A_1$. (This condition is called degree of falsity, "F").
3. Thereexist $x \leq y \in A_1$ with $z \bigcirc_1 x$ or $z \bigcirc_1 y$ or $x \bigcirc_1 z$ or $y \bigcirc_1 z$ are indeterminate in A_1, or the relation between $z \bigcirc_1 x$ and $z \bigcirc_1 y$, or the relation between $x \bigcirc_1 z$ and $y \bigcirc_1 z$ are indeterminate. (This condition is called degree of indeterminacy, "I").
4. There exist $x' \leq y' \in A_2$ with $x' \neq y'$ 'such that $z' \bigcirc x' \leq z' \bigcirc y'$ and $x' \bigcirc z' \leq y' \bigcirc z'$ for all z' $\in A_2$. (This condition is called degree of truth, "T").
5. There exist $x' \leq y' \in A_2$ such that $z' \bigcirc x' \not\leq z' \bigcirc y'$ or $x' \bigcirc z' \not\leq y' \bigcirc z'$ for some z $\in A_2$. (This condition is called degree of falsity, "F").
6. There exist $x \leq y \in A_2$ with $z' \bigcirc_2 x'$ or $z' \bigcirc_2 y'$ or $x' \bigcirc_2 z'$ or $y' \bigcirc_2 z'$ are indeterminate in A_2, or the relation between $z' \bigcirc_2 x'$ and $z' \bigcirc_2 y'$, or the

relation between $x'\bigcirc_2 z'$ and $y'\bigcirc_2 z'$ are indeterminate. (This condition is called degree of indeterminacy, "I").

If Condition 3. or Condition 6. is (are) satisfied, we get "$\leq$" is a NeutroOrder on $\boldsymbol{A}_1 \times \boldsymbol{A}_2$ and we are done. Without loss of generality, suppose that Conditions 1., 2., 4., and 5. are satisfied. Then

1. There exist $(x,x') \leq (y,y') \in A_1 \times A_2$ with $(x,x') \neq (y,y')$ such that

$$(z,z')\bigcirc(x,x') \leq (z,z')\bigcirc(y,y') \text{ and } (x,x')\bigcirc(z,z') \leq (y,y')\bigcirc(z,z')$$

for all $(z,z') \in A_1 \times A_2$. (This condition is called degree of truth, "T").

2. There exist $(x,x') \leq (y,y') \in A_1 \times A_2$ such that

$$(z,z')\bigcirc(x,x') \nleq (z,z')\bigcirc(y,y') \text{ or } (x,x')\bigcirc(z,z') \nleq (y,y')\bigcirc(z,z')$$

for some $(z,z') \in A_1 \times A_2$. (This condition is called degree of falsity, "F").

Therefore, $(\boldsymbol{A}_1 \times \boldsymbol{A}_2, \bigcirc, \leq)$ is an NOSH.

Notation 1. The NOSH $(\boldsymbol{A}_1 \times \boldsymbol{A}_2, \bigcirc, \leq)$ in Theorem 2 is called the productional NOSH of $\boldsymbol{A}_1, \boldsymbol{A}_2$.

Theorem 3. Let $(\boldsymbol{H}, \bigcirc, \leq)$ be an NOSH and $\boldsymbol{J}$ a non-empty set such that $\boldsymbol{H} \cap \boldsymbol{J} = \varnothing$ and $(\boldsymbol{H}\bigcirc\boldsymbol{H}) \cap \boldsymbol{J} = \varnothing$. Then $(\boldsymbol{H}[\boldsymbol{J}], \circledcirc, \leq')$ is an NOSH. Here, "$\leq'$" is defined as follows: For all $\boldsymbol{x}, \boldsymbol{y} \in \boldsymbol{H}[\boldsymbol{J}]$,

$$\boldsymbol{x} \leq' \boldsymbol{y} \Leftrightarrow \begin{cases} \boldsymbol{x} \leq \boldsymbol{y} & \textit{if } \boldsymbol{x}, \boldsymbol{y} \in \boldsymbol{H}, \\ \boldsymbol{x} = \boldsymbol{y} & \textit{if } \boldsymbol{x}, \boldsymbol{y} \in \boldsymbol{J}. \end{cases}$$

Proof. Theorem 1 asserts that $(\boldsymbol{H}[\boldsymbol{J}], \circledcirc)$ is an NSH. Having $(\boldsymbol{H}, \bigcirc, \leq)$ an NOSH implies that:

1. There exist $x \leq y \in H$ with $x \neq y$ such that $z \bigcirc x \leq z \bigcirc y$ and $x \bigcirc z \leq y \bigcirc z$ for all z $\in H$. (This condition is called degree of truth, "T").
2. There exist $x \leq y \in H$ such that $z \bigcirc x \nleq z \bigcirc y$ or $x \bigcirc z \nleq y \bigcirc z$ for some z $\in H$. (This condition is called degree of falsity, "F").
3. There exist $x \leq y \in H$ with $z \bigcirc x$ or $z \bigcirc y$ or $x \bigcirc z$ or $y \bigcirc z$ are indeterminate in H, or the relation between $z \bigcirc x$ and $z \bigcirc y$, or the relation between $x \bigcirc z$ and $y \bigcirc z$ are indeterminate. (This condition is called degree of indeterminacy, "I").

If Condition 3. is satisfied, we are done. Suppose that Conditions 1. and 2. are satisfied. Condition 1. implies that there exist $x \leq' y \in H \subset H[J]$ with $x \neq y$ such that $z \circledcirc x \leq z \circledcirc y$ and $x \circledcirc z \leq y \circledcirc z$ for all $z \in H$. Let $j \in J$. Then

$$j \circledcirc x = j \circledcirc y = x \circledcirc j = y \circledcirc j = H \cup J .$$

Having $H \cup J \leq' H \cup J$ implies that there exist $x \leq' y \in H[J]$ with $x \neq y$ such that $z \circledcirc x \leq z \circledcirc y$ and $x \circledcirc z \leq y \circledcirc z$ for all $z \in H[J]$. Condition 2. implies that there exist $x \leq' y \in H \subset H[J]$ such that

$$z \circledcirc x = z \bigcirc x \nleq z \circledcirc y = z \bigcirc y$$

or

$$x \circledcirc z = x \bigcirc z \nleq y \circledcirc z = y \bigcirc z$$

for some $z \in H \subset H[J]$.

Therefore, $(H[J], \circledcirc, \leq')$ is an NOSH.

Example 12. Let $(A, \heartsuit, \leq_t)$ be the NOSH in Example 9 and $J = \{o\}$. Then $(A[J], \circledcirc, \leq'_t)$ is an NOSH with the following partial order and table.

$$\leq'_t = \{(a,m),(a,d),(a,a),(d,d),(d,m),(m,m),(o,o)\}.$$

Table 3. The NeutroSemihypergroup $\left(A\left[J\right],\circledcirc\right)$

$\circledcirc$	m	a	d	o
m	m	m	m	{m,a,d,o}
a	m	{m,a}	d	{m,a,d,o}
d	m	d	d	{m,a,d,o}
o	{m,a,d,o}	{m,a,d,o}	{m,a,d,o}	{m,a,d,o}

Theorem 4. Let $n \geq 3$ be an integer. Then there exist an NOHS of cardinality n.

Proof. Example 12 asserts that there exists an NOSH, $\left(A,\heartsuit,\leq_t\right)$, of cardinality 3. Let $k = n-3$, $m_i \notin A$ for $i = 1,\cdots,k$, and $J = \left\{m_1,\cdots,m_k\right\}$. Theorem 3 asserts that $\left(A\left[J\right],\circledcirc,\leq_t'\right)$ is an NOSH of cardinality equal to n.

Corollary 1. There are infinitely many NOSHs.

Proof. The proof follows from Theorem 4.

NeutroOrderSubhyperstructures

Definition 9. Let $\left(A,\bigcirc,\leq\right)$ be an NOSH and $M \neq \varnothing \subseteq A$. Then M is a NeutroOrderSubsemihypergroup if $\left(M,\bigcirc,\leq\right)$ is an NOSH and there exists $m \in M$ such that

$$\left(m\right] = \left\{a \in A : a \leq m\right\} \subseteq M.$$

Definition 10. Let $\left(A,\bigcirc,\leq\right)$ be an NOSH and $M \neq \varnothing \subseteq A$ a NeutroOrderSubsemihypergroup (NOSSH) of A. Then M is:

1. a NeutroOrderLeftHyperideal of A if there exists $m \in M$ such that $r\bigcirc m \subseteq M$ for all $r \in A$.
2. a NeutroOrderRightHyperideal of A if there exists $m \in M$ such that $m\bigcirc r \subseteq M$ for all $r \in A$.
3. a NeutroOrderHyperideal of A if there exists $m \in M$ such that $\left(m\bigcirc r\right)\cup\left(r\bigcirc m\right) \subseteq M$ for all $r \in A$.

Example 13. Let $(\boldsymbol{R},\square,\leq_u)$ be the NOSH in Example 8. Then $]-\infty,0]$ is an NOSSH of $\boldsymbol{R}$. This is clear as:

1. $]-\infty,0]$ is an NSH since $0\square(0\square 0)\neq(0\square 0)\square 0$ and $0\square(-2\square -2)\neq(0\square -2)\square -2$.
2. $-1\leq_u 0$ and for all $x\in\,]-\infty,0]$

$$-1\square x = x\square -1 = \begin{cases} -1 & if\ x=-1, \\ [x,-1] & if\ x<-1, \\ [-1,x] & otherwise. \end{cases} \leq_u 0\square x = x\square 0 = \begin{cases} 0 & if\ x=0, \\ [x,0] & otherwise. \end{cases}$$

holds and $-3\leq_u -1$ but

$$-3\square -3 = \frac{-1}{3}\leq_u [-3,-1] = -3\square -1$$

does not hold.

3. $(0] =]-\infty,0]\subseteq\,]-\infty,0]$.

Moreover, $]-\infty,0]$ is neither a NeutroOrderRightHyperideal nor a NeutroOrderLeftHyperideal of $\boldsymbol{R}$.

Example 14. Let $(\boldsymbol{Z}^2,\bullet,\leq_p)$ be the NOSH in Example 10. Then $\boldsymbol{I}=\{(m,n):m,n\leq 1\}$ is a NeutroOrderHyperideal of $\boldsymbol{Z}^2$.

$(\boldsymbol{I},\bullet,\leq_p)$ is an NOSH as:

$$(0,0)\bullet((0,0)(0,0))=\{(0,0)\}=((0,0)(0,0))\bullet(0,0),$$

$$(1,1)\bullet((-1,0)\bullet(1,0))\neq((1,1)(-1,0))\bullet(1,0),$$

$$(0,0)\leq_p(1,1)$$

and

$$(0,0)\bullet(m,n)=(m,n)\bullet(0,0)=\{(0,0)\}\leq_p\{(1,1),(m,n)\}=(1,1)\bullet(m.n)=(m.n)\bullet(1,1)$$

holds, $(0,0)\leq_p(0,1)$ and

$$(0,0)\bullet(0,-1)=\{(0,0)\}\leq_p(0,-1)=(0,1)\bullet(0,-1)$$

does not hold, $((1,1)]=I$, and

$$(m,n)\bullet(0,0)=(0,0)\bullet(m,n)=\{(0,0)\}\subseteq I$$

for all $(m,n)\in Z^2$.

Remark 2. Let $(A,\bigcirc,\leq)$ be an NOSH and M_1,M_2 be NeutroOrderSubsemihypergroups of A. Then $M_1\cap M_2$ is not necessary a NeutroOrderSubsemihypergroup of A.

Remark 2 is illustrated via the following example.

Example 15. Let $(A',\heartsuit',\leq')$ be the NOSH in Example 11. One can easily see that $M_1=\{0,1,2\}$ and $M_2=\{0,1,3\}$ are NeutroOrderSubsemihypergroups of A' but $M_1\cap M_2=\{0,1\}$ is not an NOSSH.

Theorem 5. Let $(v_1,\bigcirc_1,\leq_1),(A_2,\bigcirc_2,\leq_2)$ be NOSH and M_1,M_2 be NeutroOrderSubsemihypergroups of A_1,A_2 respectively. Then $M_1\times M_2$ is a NeutroOrderSubsemihypergroup of $A_1\times A_2$.

Proof. Theorem 2 asserts that $(M_1\times M_2,\bigcirc,\leq)$ is an NOSH. Since M_1,M_2 are NeutroOrderSubsemihypergroups of A_1,A_2 respectively, it follows that there exist $m_1\in M_1,m_2\in M_2$ such that $(m_1]\subseteq M_1$ and $(m_2]\subseteq M_2$. Thus,

$$((m_1,m_2)]=(m_1]\times(m_2]\subseteq M_1\times M_2.$$

Therefore, $M_1\times M_2$ is a NeutroOrderSubsemihypergroup of $A_1\times A_2$.

Theorem 6. Let $(A_1,\bigcirc_1,\leq_1),(A_2,\bigcirc_2,\leq_2)$ be NOSH and M_1,M_2 be NeutroOrderLeftHyperideals (NeutroOrderRightHyperideals or

NeutroOrderHyperideals) of A_1, A_2 respectively. Then $M_1 \times M_2$ is a NeutroOrderLeftHyperideal (NeutroOrderRightHyperideal or NeutroOrderHyperideal) of $A_1 \times A_2$.

Proof. The theorem is proved for the case M_1, M_2 are NeutroOrderRightHyperideals of A_1, A_2 respectively. The other two cases are done similarly. Let M_1, M_2 be NeutroOrderRightHyperideals of A_1, A_2 respectively. Theorem 5 asserts that $M_1 \times M_2$ is a NeutroOrderSubsemihypergroup of $A_1 \times A_2$. Having M_1, M_2 are NeutroOrderLeftHyperideals of A_1, A_2 respectively implies that there exist, $u \in M_1$, $v \in M_2$ such that $u \bigcirc_1 x \subseteq M_1$ and $v \bigcirc_2 y \subseteq M_2$ for all $x \in A_1, y \in A_2$. The latter implies that for all $(x, y) \in A_1 \times A_2$, there exist $(u, v) \in M_1 \times M_2$ such that

$$(u, v) \bigcirc (x, y) \subseteq M_1 \times M_2.$$

Example 16. Let $(R, \square, \leq_u)$ be the NOSH in Example 8. Then

$$I = \{(x, y, z) : x, y, z \leq 0\}$$

is a NeutroOrderHyperideal of the productional NOSH R^3.

FUTURE RESEARCH DIRECTIONS

The results in this chapter may be considered as a base for any possible study in the field of NeutroOrderHyperstructures. The authors suggest some research ideas in this direction.

1. Find all NeutroOrderSemihypergroups of small order (up to a certain type of equivalence).
2. Define other NeutroOrderHyperstructures such as NeutroOrderHypergroup, NeutroOrderPolygroup, NeutroOrderHyperring, etc.
3. Find applications of NeutroOrderHyperstructures in different fields.
4. Define AntiOrderAlgebra and AntiOrderHyperstructures.
5. Find a necessary and sufficient condition for a non-empty subset of the Cartesian product of NeutroOrderSemihypergroups to be a NeutroOrderSubemihypergroup (NeutroOrderHyperideal).
6. Find conditions for the Cartesian product of sets associated to maps and partial orders to be a NeutroOrderSemihypergroup.

CONCLUSION

This chapter introduced NeutroOrderSemihypergroups as a generalization of NeutroOrderSemigroups and as an extension of ordered semigroups. New definitions were introduced and were illustrated by non-trivial examples. Moreover, some properties of the new notions were investigated and the existence of infinitely many NeutroOrderSemihypergroups was shown.

FUNDING

This research received no specific grant from any funding agency in the public, commercial, or not-for-profit sectors.

ACKNOWLEDGMENT

The authors are grateful to the referees for their valuable comments and suggestions.

REFERENCES

Al-Tahan, M., Davvaz, B., Smarandache, F., & Anis, O. (2021). On some NeutroHyperstructures. *Symmetry*, *13*(4), 535. doi:10.3390ym13040535

Al-Tahan, M., Smarandache, F., & Davvaz, B. (2021). NeutroOrderedAlgebra: Applications to semigroups. *Neutrosophic Sets and System*, *39*, 133–147.

Corsini, P. (1993). Prolegomena of Hypergroup Theory. Udine Aviani Editore.

Heidari, D., & Davvaz, B. (2011). On ordered hyperstructures. *Politehn. Univ. Bucharest Sci. Bull. Ser. A Appl. Mathematics and Physics*, *73*(2), 85–96.

Ibrahim, M. A., & Agboola, A. A. A. (2020). Introduction to NeutroHyperGroups. *Neutrosophic Sets and System*, *38*, 15–32.

Marty, F. (1934). Sur une generalization de la notion de groupe. *Proceedings of the 8th Congress on Mathmatics, Scandinaves*, 45–49.

Smarandache, F. (1998). Neutrosophy. Neutrosophic Probability, Set, and Logic, 105.

Smarandache, F. (2020). NeutroAlgebra is a generalization of partial algebra. *International Journal of Neutrosophic Science*, *2*, 8–17. doi:10.54216/IJNS.020103

Smarandache, F. (2020). Introduction to NeutroAlgebraic structures and AntiAlgebraic structures (revisited). *Neutrosophic Sets and System*, *31*, 1–16.

Vougiouklis, T. (1994). *Hyperstructures and Their Representations*. Hadronic Press, Inc.

ADDITIONAL READING

Al-Tahan, M., & Davvaz, B. (2017). Algebraic Hyperstructures Associated to Biological Inheritance. *Mathematical Biosciences*, *285*, 112–118. doi:10.1016/j.mbs.2017.01.002 PMID:28108295

Al-Tahan, M., Davvaz, B., Smarandache, F., & Anis, O. (2021). Properties of Productional NeutroOrderedSemigroups. *Neutrosophic Sets and System*, *42*, 178–190.

Davvaz, B., & Leoreanu-Fotea, V. (2008). *Hyperring Theory and Applications*. International Academic Press.

Davvaz, B., & Omidi, S. (2016). Basic notions and properties of ordered semihyperrings. *Categ. Gen. Algebr. Struct. Appl.*, *4*(1), 43–62.

Fuchs, L. (1963). *Partially Ordered Algebraic Systems, Int. Ser. of Monographs on Pure and Appl. Math. 28*. Pergamon Press.

Hamidi, M., & Smarandache, F. (2020). Neutro-BCK-Algebra. *International Journal of Neutrosophic Science*, *8*, 110–117. doi:10.54216/IJNS.080204

Rezaei, A., & Smarandache, F. (2020). On Neutro-BE-algebras and Anti-BE-algebras. *International Journal of Neutrosophic Science*, *4*, 8–15.

Smarandache, F. (2020). Extension of HyperGraph to n-SuperHyperGraph and to Plithogenic n-SuperHyperGraph, and Extension of HyperAlgebra to n-ary (Classical-/Neutro-/Anti-) HyperAlgebra. *Neutrosophic Sets and Systems*, *33*, 290–296.

Chapter 3
On NeutroQuadrupleRings

Muritala Abiodun Ibrahim
Federal University of Agriculture, Abeokuta, Nigeria

Agboola Adesina Abdul Akeem
Federal University of Agriculture, Abeokuta, Nigeria

Zulaihat Hassan-Ibrahim
Auburn University, USA

Emmanuel O. Adeleke
Federal University of Agriculture, Abeokuta, Nigeria

ABSTRACT

In an attempt to generalize and create alternatives for classical algebraic structures, Smarandache in 2019 introduced the idea of treating binary operations on sets as partially well (totally outer) defined with axioms that are partially true (totally false) and even indeterminate for some elements of any given algebraic structure. The structures formed with these types of operations called Neutro(Anti) Operations are called Neutro(Anti) Structures and are named according to the axioms that are neutro(anti)-sophicated. In this chapter, the authors consider some properties of neutrosophic quadruple numbers with examples. In particular, they introduce and study the concept of NeutroQuadrupleRings and their substructures. Some basic definitions and a few important results are presented. It is shown that the intersection and union of any two NeutroQuadrupleSubrings of a particular class does not necessarily belong to the same class. Also, they give a necessary and sufficient condition for a neutrosophic quadruple ring (NQ(X), +, ·) to be a NeutroQuadrupleRing.

DOI: 10.4018/978-1-6684-3495-6.ch003

INTRODUCTION

Neutrosophic logic/set was introduced by Florentin Smarandache in 1995 as a generalization of fuzzy logic/set, intuitionistic fuzzy logic and soft logic/set. In neutrosophic logic, each proposition is char- acterized with the degree of truth in the set (T), the degree of indeterminacy in the set (I), and the degree of falsity in the set (F), where T, I, F are standard or non-standard subsets of]−0, 1+[, see (Smarandache, 1998; Smarandache, 2003; Smarandache, 2014). Neutrosophic set extends the notions of classical set and serves as a useful tool in the study of imprecise, incomplete and indeterminate information. The development of the theory of neutrosophy has brought about the introduction of neutrosophic quadruple numbers by Smarandache (2015) where he defined and presented arithmetic operations of addition, subtraction, multiplication, and scalar multiplication on neutrosophic sets of quadruple numbers. A neutrosophic quadruple number (*NQ*), is a number of the form: $NQ = a + bT + cI + dF$, where a, b, c, d are real or complex numbers, and the components T, I and F have their usual logical meanings. For each *NQ*, a is called the determinate part of NQ while b, c, and d, are called the indeterminate parts of NQ. A Preference Law, with respect to T, I and F, may be define on the set of neutrosophic quadruple numbers, for instance we may define the preference of these component as T < I < F or T > I > F, also other preferences may be developed. With reference to any of these Preference Laws, the Absorbance Law for the multiplication of T, I and F can be defined.

In (Agboola, 2020a), Akinleye et.al., considered and discussed neutrosophic quadruple algebraic structures. Later in (Agboola et al., 2017) Agboola et al., generalized the work of Akineye et.al., by discussing neutrosophic quadruple hyperstructure. Ever since then, several research papers on neutrosophic quadruple algebraic structures and algebraic hyperstructures have been published. For example, Vasantha K.,et.al., in (Vasantha Kandasamy, 2019)introduced for the first time the concept of neutrosophic quadruple vector spaces over the classical field R, C and Z_p. Ibrahim and Agboola in (Ibrahim & Agboola, 2020a) generalized the concept of neutrosophic hypervector spaces by presenting their neutrosophic quadruple forms. A comprehensive review of neutrosophic quadruple algebraic structures (hyperstructures) can be found in (Al-Tahan & Davvaz, 2020a; Al-Tahan & Davvaz, 2020b; Ibrahim & Agboola, 2020b; Jun et al., 2018; Jun et al., 2019; Li et al., 2019; Muhiuddin, 2019; Sahin & Kargın, 2019; Sahin & Kargın, 2019).

Recently, Florentin Smarandache in (Smarandache, 2019) presented new ways of studying and analyzing indeterminate and imprecise information. He called these concepts NeutroSophication and AntiSophication. Neutro- Sophication and AntiSophication are processes through which NeutroAlgebraicStructures and AntiAlgebraicStructures are respectively generated from any given classical Structure.

Since the introduction of these new ideas, several research papers on NeutroSophication and AntiSophication of algebraic structures (hyperstructures) such as groups, semigroups, vector spaces, BCI-algebras, BCK/BCI al- gebras, hypergroups and hyperrings (Akinleye et al., 2016; Ibrahim et al., 2020; Rezaei & Smarandache, 2020) have been published. The present paper is motivated by the work of Smarandache, et.al (Smarandache et al., 2021), that was presented at the 51st Annual Iranian Mathematics Conference 2021. The present paper is concerned with the study of some classes of Neu- troQuadrupleRings and their substructures. It is shown that the intersection and union of any two NeutroQuadrupleSubrings of a particular class does not necessarily belong to the same class. Also, we gave a necessary and sufficient condition for $\left(NQ\left(\mathrm{X}\right),+,\cdot\right)$ to be a NeutroQuadrupleRing.

PRELIMINARIES

In this section, we will give some definitions, examples and results that will be used in the sequel.

Definition 2.1: *A neutrosophic quadruple number is a number of the form (a, bT, cI, dF) where T, I, F* have their usual neutrosophic logic meanings and a, b, c, d ∈ R or C. The set NQ defined by

$$NQ = \{(a, bT, cI, dF): a, b, c, d \in \mathbb{R} \text{ or } \mathbb{C}\}$$

is called a neutrosophic quadruple set.

Definition 2.2: *(Agboola, 2020a) Suppose in an optimistic way we consider the prevalence order $T > I > F$. Then the combination of the usual Neutrosophic tools T, I, F are:*

$TI = IT = \max\{T, I\} = T,$

$TF = FT = \max\{T, F\} = T,$

$IF = FI = \max\{I, F\} = I,$

$TT = T^2 = T,$

$II = I^2 = I,$

$FF = F^2 = F.$

Analogously, suppose in a pessimistic way we consider the prevalence order $T < I < F$. Then we have:

$TI = IT = \max\{T, I\} = I$,

$TF = FT = \max\{T, F\} = F$,

$IF = FI = \max\{I, F\} = F$,

$TT = T^2 = T$,

$II = I^2 = I$,

$FF = F^2 = F$.

We shall adopt the pessimistic way in this work.

Theorem 2.3. (Agboola, 2020a) (NQ, +) is an abelian group.

Theorem 2.4. (Agboola, 2020a) (NQ, ·) is a commutative monoid.

Theorem 2.5. (Agboola, 2020a) (NQ, ·) is not a group.

Theorem 2.6. (Agboola, 2020a) (NQ, +, ·) is a commutative ring.

Definition 2.7. *(Smarandache, 2020b)*

1. A ClassicalOperation is an operation well-defined for all the set's elements.
2. A ClassicalHyperOperation is a hyper-operation well-defined for all the set's elements.
3. A NeutroOperation is an operation partially well-defined, partially indeterminate, and partially outer defined on the given set.
4. An AntiOperation is an operation that is outer defined for all set's elements.
5. A classical law/axiom defined on a nonempty set is a law/axiom that is totally true (i.e., true for all set's elements).
6. A NeutroLaw/NeutroAxiom defined on a nonempty set is a law/axiom that is true for some set's element [degree of truth (T)], indeterminate for other set's elements [degree of indeterminacy (I)], or false for the other set's elements [degree of falsehood (F)], where $T, I, F \in [0, 1]$, with $(T, I, F) = (1, 0, 0)$ that represents the classical axiom, and $(T, I, F) = (0, 0, 1)$ that represents the AntiAxiom.
7. An AntiLaw/AntiAxiom defined on a nonempty set is a law/axiom that is false for all set's elements.
8. A NeutroHyperOperation is a hyper-operation partially well-defined, partially indeterminate, and partially outer-defined on the given set.

9. An AntiHyperOperation is a hyper-operation outer-defined for all set's elements.
10. A NeutroAlgebra is an algebra that has at least one NeutroOperation or one NeutroAxiom (axiom that is true for some elements, indeterminate for other elements, and false for other elements).
11. An AntiAlgebra is an algebra endowed with at least one AntiOperation or at least one Anti- Axiom.

Theorem 2.8: *(Rezaei & Smarandache, 2020) Let* U *be a nonempty finite or infinite universe of discourse and let S be a finite or infinitre subset of* U. *If n classical operations (laws and axioms) are defined on S where* $n \geq 1$, *then there will be* $(2^n - 1)$ *NeutroAlgebras and* $(3^n - 2^n)$ *AntiAlgebras.*

OPERATION ON NEUTROSOPHIC QUADRUPLE NUMBERS

Remark 3.1: Let $NQ(X)$ be a neutrosophic quadruple set and let $x = (a, bT, cI, dF)$ and $y = (e, fT, gI, hF) \in NQ(X)$ we have that

$$x + y = (a, bT, cI, dF) + (e, fT, gI, hF) = (a + e, (b + f)T, (c + g)I, (d + h)F)$$

and

$$x - y = (a, bT, cI, dF) - (e, fT, gI, hF) = (a - e, (b - f)T, (c - g)I, (d - h)F)$$

are in $NQ(X)$.

Definition 3.2: *Let*

$$a = (a_1, a_2T, a_3I, a_4F), b = (b_1, b_2T, b_3I, b_4F) \in NQ(Z).$$

Then

$$a \cdot b = (a_1, a_2T, a_3I, a_4F) \cdot (b_1, b_2T, b_3I, b_4F)$$

$$= (a_1b_1, (a_1b_2 + a_2b_1 + a_2b_2)T, (a_1b_3 + a_2b_3 + a_3b_1 + a_3b_2 + a_3b_3)I,$$

$$(a_1b_4 + a_2b_4, a_3b_4 + a_4b_1 + a_4b_2 + a_4b_3 + a_4b_4)F).$$

$$\frac{a}{b} = \frac{(a_1, a_2T, a_3I, a_4F)}{(b_1, b_2T, b_3I, b_4F)} = (c_1, c_2T, c_3I, c_4F)$$

$\Leftrightarrow (a_1, a_2T, a_3I, a_4F) = (b_1, b_2T, b_3I, b_4F)(c_1, c_2T, c_3I, c_4F).$

From this, we obtain the following relationships:

$$a_1 = b_1c_1 \tag{1}$$

$$a_2 = b_1c_2 + (c_1 + c_2)b_2 \tag{2}$$

$$a_3 = b_1c_3 + b_2c_3 + (c_1 + c_2 + c_3)b_3 \tag{3}$$

$$a_4 = (b_1 + b_2 + b_3)c_4 + (c_1 + c_2 + c_3 + c_4)b_4. \tag{4}$$

It is clear from (1) that

$b_1 \mid a_1$.

By adding (1) and (2), we obtain

$\left(b_1 + b_2\right) \mid \left(a_1 + a_2\right)$.

Also by adding (1), (2) and (3) we obtain

$\left(b_1 + b_2 + b_3\right) \mid \left(a_1 + a_2 + a_3 + a_4\right)$.

Also by adding (1), (2), (3) and (4) we have that

$\left(b_1 + b_2 + b_3 + b_4\right) \mid \left(a_1 + a_2 + a_3 + a_4\right)$.

This observation can be employed in the study of some number theoretic properties in neutrosophic quadruple integers, i.e., $NQ\left(\mathbb{Z}\right)$. Based on this observation, we shall give the condition for divisibility in $NQ\left(\mathbb{Z}\right)$.

Proposition 3.3: *Let*

$NQ(\mathbb{Z}) = \{(a, bT, cI, dF): a, b, c, d \in Z\}$

be the set of neutrosophic quadruple integers. Let

$a = (a_1, a_2T, a_3I, a_4F), b = (b_1, b_2T, b_3I, b_4F) \in NQ(Z)$,

a|b if and only if

$a_1|b_1, (a_1 + a_2)|(b_1 + b_2), (a_1 + a_2 + a_3)|(b_1 + b_2 + b_3)$

and

$(a_1 + a_2 + a_3 + a_4)|(b_1 + b_2 + b_3 + b_4)$.

Proof. This follows easily from Definition 3.2.

Example 3.4: Let $NQ(\mathbb{Z})$ be a neutrosophic quadruple integer. Let $x, y \in NQ(\mathbb{Z})$.

1. 1. If $x = (8, 4T, 3I, 5F)$ and $y = (2, -6T, I, F)$, there exists $z = (a, bT, cI, dF) \in NQ(\mathbb{Z})$ such that

$x = yz$, since $2|8$, $-4|12$, $-3|15$ and $-2|20$.

So we have that

$(8, 4T, 3I, 5F) = (2, -6T, I, F)(4, -7T, -2I, -5F)$.

Hence $y|x$.

2. If $x = (8, 4T, 3I, 5F)$ and $y = (3, 2T, 3I, 2F)$, there exists no $z = (a, bT, cI, dF) \in NQ(Z)$ such that $x = yz$, since $3 \mid 8$, $5 \mid 12$ and $8 \mid 15$. Hence $y \mid x$.

Definition 3.5: *Let*

$NQ(R) = \{(a, bT, cI, dF): a, b, c, d \in R\}$

be the set of neutrosophic quadruple real number. Let

$a = (a_1, a_2T, a_3I, a_4F), b = (b_1, b_2T, b_3I, b_4F) \in NQ(\mathbb{R})$,

we say that $a \leq b$ if

$a_1 \leq b_1$ and $a_1 + a_2 + a_3 + a_4 \leq b_1 + b_2 + b_3 + b_4$.

Remark 3.6: It should be noted that some neutrosophic quadruple real numbers cannot be compared. So, this shows that the set of neutrosophic quadruple real numbers NQ(R) is not ordered.

For example, let

$(2, 3T, 4I, 2F), (3, -5T, 2I, 13F) \in NQ(\mathrm{R})$.

We can say that

$(2, 3T, 4I, \leq (3, -5T, 2I, 13F)$ or $(3, -5T, 2I, 13F) \geq (2, 3T, 4I, 2F)$ 2F)

since $2 < 3$ and $11 < 13$.

But, if we consider

$(3, 2T, -4I, 6F), (5, -5T, 2I, 3F) \in NQ(\mathrm{R})$.

We can neither say

$(3, 2T, -4I, 6F) \leq (5, -5T, 2I, 3F)$ nor $(3, 2T, -4I, 6F) \geq (5, -5T, 2I, 3F)$.

Since $3 < 5$ and $7 > 5$.

Thus, we say that

$(3, 2T, -4I, 6F), (5, -5T, 2I, 3F) \in NQ(\mathrm{R})$

are incomparable and hence $\leq$ is a partial order on $NQ(\mathrm{R})$.

Remark 3.7: From the second illustration given in remark 3.6 we can easily see that Trichotomy law fails in the set of neutrosophic quadruple real numbers NQ(R).

Definition 3.8: *Let*

$\mathrm{NQ(R)} = \{(a, bT, cI, dF): a, b, c, d \in \mathrm{R}\}$

be the set of neutrosophic quadruple real number. Let

a = (a1, a2T, a3I, a4F), θ = (0, 0T, 0I, 0F) ∈ NQ(R),

we say that a is

1. 1. *non-negative neutrosophic quadruple real number if* $a \geq \theta$, *i.e.,*

$a_1 \geq 0$, and $a_1 + a_2 + a_3 + a_4 \geq 0$.

2. 2. *a positive neutrosophic quadruple real number if* $a > \theta$, *i.e*

$a_1 > 0$, and $a_1 + a_2 + a_3 + a_4 > 0$.

Remark 3.9: If we collapse the neutrosophic quadruple real numbers or integers into neutrosophic real numbers $\mathbb{R}$ (I) or neutrosophic integers $\mathbb{Z}(I)$ respectively, we shall also see that both of them are partially ordered.

Proposition 3.10: Let $NQ(\mathbb{Z})$ be the set of neutrosophic quadruple integers.

Define on $NQ(\mathbb{Z})$ a relation ~ by

$(a_1, a_2T, a_3I, a_4F) \sim (b_1, b_2T, b_3I, b_4F)$

if

$a_1 + a_2 + a_3 + a_4 = b_1 + b_2 + b_3 + b_4$.

The relation ~ is an equivalence relation on the set $NQ(\mathbb{Z})$.

Proof: We can see that for every

$(a_1, a_2T, a_3I, a_4F), (b_1, b_2T, b_3I, b_4F), (c_1, c_2T, c_3I, c_4T) \in NQ(\mathbb{Z})$.

1. $(a_1, a_2T, a_3I, a_4F) \sim (a_1, a_2T, a_3I, a_4F)$.
2. If $(a_1, a_2T, a_3I, a_4F) \sim (b_1, b_2T, b_3I, b_4F)$ then $(b_1, b_2T, b_3I, b_4F) \sim (a_1, a_2T, a_3I, a_4F)$.
3. If $(a_1, a_2T, a_3I, a_4F) \sim (b_1, b_2T, b_3I, b_4F)$ and $(b_1, b_2T, b_3I, b_4F) \sim (c_1, c_2T, c_3I, c_4F)$ then $(a_1, a_2T, a_3I, a_4F) \sim (c_1, c_2T, c_3I, c_4F)$.

It follows that the relation ~ is reflexive, symmetric and transitive, hence it is an equivalence relation on the set $NQ(\mathbb{Z})$.

Let the equivalence class of ~ be given by

$\overrightarrow{(a_1, a_2T, a_3I, a_4F)} = \{(b_1, b_2T, b_3I, b_4F) \in NQ(\mathbb{Z}).: b_1 + b_2 + b_3 + b_4 = a_1 + a_2 + a_3 + a_4\}$.

If a = $a_1 + a_2 + a_3 + a_4$ we shall denote the equivalence class $\overrightarrow{(a_1, a_2T, a_3I, a_4F)}$ by [a], i.e the equivalence class of

(3, 2T, −4I, 6F), (5, −5T, 2I, 3F) ∈ NQ(Z)

will be denoted by (Akinleye et al., 2016) and (Agboola et al., 2017) respectively.

In general, the equivalence class of the relation ~ on NQ(Z) is the set

$[a] = \{(x, yT, zI, uF) \in NQ(Z): x + y + z + u = a;\ a \in Z\}$.

It can be seen that

$\bigcup_{a\in\mathbb{Z}}[a] = NQ(\mathbb{Z})$ and for any two a, b $\in \mathbb{Z}$, $[a] \cap [b] = \varnothing$.

This observations are recorded in the next proposition.

Proposition 3.11: *Let NQ(Z) be a neutrosophic quadruple integers, define on NQ(Z) the relation ~ as in Proposition 3.10 above. ~ is an equivalent relation and it partitions NQ(Z).*

Proof: The proof follows easily from Proposition 3.10.

Proposition 3.12: *Let (NQ(Z), +) be a neutrosophic quadruple group and let ~ be the equivalence relation defined in Proposition 3.11. The quotient (NQ(Z)/ ~, +) is an abelian group.*

Proof: The proof is straight forward.

More number theoretic properties such as division algorithm, prime numbers, congruency, g.c.d, e.t.c, in neutrosophic quadruple environment shall be deeply studied in our future work.

FORMULATION OF NEUTROQUADRUPLERING

Definition 4.1: *[NeutroSophication of the laws and axioms of the neutrosophic quadruple ring]*

(NQR1) There exist some duplets (x, y), (u, v), (p, q) ∈ NQ(R) such that x+y ∈ NQ(R) (**degree** of truth T) and [u + v = outer-defined/indeterminate (degree of indeterminacy I) or p + q ∈/ NQ(R)] (degree of falsehood F) [NeutroClosure law of addition].

(NQR2) There exist some triplets (x, y, z), (p, q, r), (u, v, w) ∈ NQ(R) such that x +(y + z) = (x + y) + z (degree of truth T) and [[p + (q + r)] or [(p + q) + r] = outer-defined/indeterminate (degree of indeterminacy I) or u + (v + w) = (u + v) + w] (degree of falsehood F) [NeutroAxiom of associativity (NeutroAssociativity)].

(NQR3) There exists an element e ∈ NQ(R) such that x + e = x + e = x and [[x + e] or [e + x] = outer-defined/indeterminate or x + e = x = e + x] for at least one x ∈ NQ(R) [NeutroAxiom of existence of neutral element (NeutroNeutralElement)].

(NQR4) There exists −x ∈ NQ(R) such that x+(−x) = (−x)+x = e and [[−x+x] or [x+(−x)] = outer- defined/indeterminate or −x + x = e = x + (−x)] for at least one x ∈ NQ(R) [NeutroAxiom of existence of inverse element (NeutroInverseElement)].

(NQR5) There exist some duplets (x, y), (u, v), (p, q) ∈ NQ(R) such that x + y = y + x and [[p + q] or [q + p] = outer-defined/indeterminate (degree of indeterminacy I) or u + v =v + u] (degree of falsehood F)[NeutroAxiom of commutativity(NeutroCommutativity)].

(NQR6) There exist some duplets (x, y), (p, q), (u, v) ∈ NQ(R) such that x · y ∈ NQ(R) (degree of truth T) and [u · v = outer-defined/indeterminate (degree of indeterminacy I) or p · q ∉ NQ(R)] (degree of falsehood F) NeutroClosure law of multiplication].

(NQR7) There exist some triplets (x, y, z), (p, q, r), (u, v, w) ∈ NQ(R) such that x · (y · z) = (x · y) · z (degree of truth T) and [[p · (q · r)] or [(p · q) · r] = outer-defined/indeterminate (degree of indeterminacy I) or u · (v · w) = (u · v)· w] (degree of falsehood F) [NeutroAxiom of associativity (NeutroAssociativity)].

(NQR8) There exist some triplets (x, y, z), (p, q, r), (u, v, w) ∈ NQ(R) such that x·(y +z) = (x·y)+(x·z) (degree of truth T) and [[p · (q + r)] or [(p · q) + (p · r)]= outer-defined/indeterminate (degree of indeterminacy I) or u · (v + w) = (u · v + (u · w)] (degree of falsehood F) [NeutroAxiom of left distributivity (NeutroLeftDistributivity)].

(NQR9) There exist some triplets (x, y, z), (p, q, r), (u, v, w) ∈ NQ(R) such that (y +z)·x = (y ·x)+(z ·x) (degree of truth T) and [[(v + w) · u] or [(v · u) + (w ·

u)]= outer-defined/indeterminate (degree of indeterminacy I) or (v + w) · u = (v · u) + (w · u)] (degree of falsehood F) [NeutroAxiom of right distributivity (NeutroRightDistributivity)].

(NQR10) There exist some duplets (x, y), (p, q), (u, v) ∈ NQ(R) such that x · y = y · x (degree of truth T) and [[p · q] or [q · p] = outer-defined/indeterminate (degree of indeterminacy I) or u · v = v· u] (degree of falsehood F) [NeutroAxiom of commutativity (NeutroCommutativity)].

Definition 4.2: *[AntiSophication of the laws and axioms of the neutrosophic quadruple ring]*

(AQR1) For all duplets (x, y) ∈ NQ(R), x + y ∉ NQ(R) [AntiClosure law of addition].

(AQR2) For all the triplets (x, y, z) ∈ NQ(R), x + (y + z) ≠ (x + y) + z [AntiAxiom of associativity (AntiAssociativity)].

(AQR3) There doest not exist an element e ∈ NQ(R) such that x + e = x + e = x ∀ x ∈ NQ(R) [AntiAxiom of existence of neutral element (AntiNeutralElement)].

(AQR4) There does not exist −x ∈ NQ(R) such that x +(−x) = (−x)+x = e ∀x ∈ NQ(R) [AntiAxiom of existence of inverse element (AntiInverseElement)].

(AQR5) For all the duplets (x, y) ∈ NQ(R), x + y ≠ y + x [AntiAxiom of commutativity (AntiCommutativity)].

(AQR6) For all the duplets (x, y) ∈ NQ(R), x · y ∉ NQ(R) [AntiClosure law of multiplication].

(AQR7) For all the triplets (x, y, z) ∈ NQ(R), x · (y · z) ≠ (x · y) · z [AntiAxiom of associativity (Anti-Associativity)].

(AQR8) For all the triplets (x, y, z) ∈ NQ(R), x·(y +z) ≠ (x· y)+(x· z) [AntiAxiom of left distributivity (AntiLeft Distributivity)].

(AQR9) For all the triplets (x, y, z) ∈ NQ(R), (y+z)·x ≠ (y·x)+(z·x) [AntiAxiom of right distributivity (AntiRightDistributivity)]

(AQR10) For all the duplets (x, y) ∈ NQ(R), x · y ≠ y · x [AntiAxiom of commutativity (AntiCommutativity)].

Definition 4.3: *A NeutroQuadrupleRing NQ(R) is an alternative to the neutrosophic quadruple ring Q(R) that has at least one NeutroLaw or at least one of NQR1, NQR2, NQR3, NQR4, NQR5, NQR6, NQR7, NQR8 and NQR9 with no AntiLaw or AntiAxiom.*

Definition 4.4: *NeutroCommutativeQuadrupleRing NQ(R) is an alternative to the commutative neu- trosophic quadruple ring Q(R) that has at least one NeutroLaw or at least one of NQR1, NQR2, NQR3, NQR4, NQR5, NQR6, NQR7, NQR8, NQR9 and NQR10 with no AntiLaw or AntiAxiom.*

Definition 4.5: *An AntiQuadrupleRing AQ(R) is an alternative to the neutrosophic quadruple ring Q(R) that has at least one AntiLaw or at least one of AQR1, AQR2, AQR3, AQR4, AQR5, AQR6, AQR7, AQR8 and AQR9.*

Definition 4.6: *AntiCommutativeQuadrupleRing NQ(R) is an alternative to the commutative neu- trosophic quadruple ring Q(R) that has at least one AntiLaw or at least one of AQR1, AQR2, AQR3, AQR4, AQR5, AQR6, AQR7, AQR8, AQR9 and AQR10.*

Definition 4.7: *Let NQ(R) be a NeutroQuadrupleRing.*

1. NQ(R) is called finite of order n if the cardinality of NQ(R) is n that is o(NR) = n. Otherwise, NQ(R) is called an infinite NeutroQuadrupleRing and we write $o(NQ(R)) = \infty$.
2. NQ(R) is called a NeutroQuadrupleRing with unity if there exists a multiplicative unit element $u \in NQ(R)$ such that $ux = xu = x$ for at least one $x \in NQ(R)$.
3. NQ(R) is called a NeutroQuadrupleRing of characteristic n, if there exists a least positive integer n such that $nx = e$ for at least one $x \in NQ(R)$. If no such n exists, then NQ(R) is called a NeutroQuadrupleRing of characteristic zero.
4. An element $x \in NQ(R)$ is called a multiplicative inverse element if there exists at least one $y \in NQ(R)$ such that $xy = yx = u$ where u is the multiplicative unity element in NQ(R).
5. An element $x \in NQ(R)$ is called idempotent if $x^2 = x$.
6. An element $x \in NQ(R)$ is called a nilpotent element if for the least positive integer n, we have $x^n = e$.
7. An element $e = x \in NQ(R)$ is called a zero divisor element if there exists an element $e = y \in NQ(R)$ such that $xy = e$ or $yx = e$.

Proposition 4.8: *Let* (*NQ*(*R*), +, ·) *be a neutrosophic quadruple ring.*

1. *If NQ(R) is a commutative neutrosophic quadruple ring, then;*
   a. *There are 1023 types of NeutroQuadrupleCommutativeRings.*
   b. *There are 58025 types of AntiQuadrupleCommutativeRings.*
2. *If NQ(R) is not a commutative neutrosophic quadruple ring, then;*
   a. *There are 511 types of NeutroQuadrupleRings.*
   b. *There are 19171 types of AntiQuadrupleRings.*

Proof: The proof follows easily from Theorem 2.8.

It should be noted that the type/class where all the Laws and Axioms are NeutroSophicated or AntiSophicated are called the trivial type/class.

In what follows we shall give an example of AntiQuadrupleRings and study some classes of Neutro- QuadrupleRing through several examples.

Example 4.9: Let $Z_2 = \{0, 1\}$, the neutrosophic quadruple set $NQ(Z_2)$ is the set given as follows:

$$NQ(Z_2) = \left\{\theta, \vec{1}, \vec{2}, \vec{3}, \vec{4}, \vec{5}, \vec{6}, \vec{7}, \vec{8}, \vec{9}, \overrightarrow{10}, \overrightarrow{11}, \overrightarrow{12}, \overrightarrow{13}, \overrightarrow{14}, \overrightarrow{15}\right\}$$

where

$\theta = (0, 0T, 0I, 0F)$, $\vec{1} = (0, 0T, 0I, F)$, $\vec{2} = (0, 0T, I, 0F)$, $\vec{3} = (0, 0T, I, F)$,

$\vec{4} = (0, T, 0I, 0F)$, $\vec{5} = (0, T, 0I, F)$, $\vec{6} = (0, T, I, 0F)$, $\vec{7} = (0, T, I, F)$,

$\vec{8} = (1, 0T, 0I, 0F)$, $\vec{9} = (1, 0T, 0I, F)$, $\overrightarrow{10} = (1, 0T, I, 0F)$, $\overrightarrow{11} = (1, 0T, I, F)$,

$\overrightarrow{12} = (1, T, 0I, 0F)$, $\overrightarrow{13} = (1, T, 0I, F)$, $\overrightarrow{14} = (1, T, I, 0F)$ and $\overrightarrow{15} = (1, T, I, F)$.

For all

$(a_1, a_2T, a_3I, a_4F), (b_1, b_2T, b_3I, b_4F) \in NQ(Z_2)$

define the operation $\star'$ by

$(a_1, a_2T, a_3I, a_4F) \star' (b_1, b_2T, b_3I, b_4F) = (a_1\star b_1, (a_1\star b_2 + a_2\star b_1 + a_2\star b_2)T,$

$(a_1\star b_3 + a_2\star b_3 + a_3\star b_1 + a_3\star b_2 + a_3\star b_3)I,$

$(a_1\star b_4 + a_2\star b_4 + a_3\star b_4 + a_4\star b_1 + a_4\star b_2 + a_4\star b_3 + a_4\star b_4)F).$

If we define on Z_2 the binary operations $+$ and $\star$, for all $a, b \in Z_2$ by

$a + b = a +_2 b$ and $a\star b = min(a, b) + 1$.

Then the triple (NQ(Z2), +, $\star'$) is a [8, 9]– AntiQuadrupleRing. This can be deduced from Table 1 below.

Study of Some Classes of NeutroQuadrupleRings

In this section we shall study some classes of NeutroQuadrupleRings, ($NQ(R)$, $\oplus$, $\odot$) and present some of their basic properties with several examples. It should be noted that, the classes of NeutroQuadrupleRings in this paper are named according to the axioms or laws that are neutrosophicated. For example, a class will be called [7,8,9]-NeutroQuadrupleRings if axioms 7, 8 and 9 are neutrosophicated.

Example 4.10: Let $Z_2 = \{0, 1\}$, the neutrosophic quadruple set $NQ(Z_2)$ is the set given as follows:

$$NQ(Z_2) = \left\{\vec{\theta}, \vec{1}, \vec{2}, \vec{3}, \vec{4}, \vec{5}, \vec{6}, \vec{7}, \vec{8}, \vec{9}, \vec{10}, \vec{11}, \vec{12}, \vec{13}, \vec{14}, \vec{15}\right\}.$$

1. 1. Define on Z_2 the binary operations + and , for all $a, b \in Z_2$ by

$$a + b = a +_2 b \text{ and } a \star b = = \begin{cases} ab + 1, a = b = 1 \\ a + b, otherwise \end{cases}.$$

Define the operation $\star'$ as in Example 4.9 above, then the triple (NQ(Z2), +, $\star^1$) is a [8, 9]- NeutroQuadrupleRing.

To see this we consider the Tables 1 and 2.

a. That $\left(NQ\left(\mathbb{Z}_2\right), +\right)$ is a netrosophic abelian group is straight forward.
b. It can be seen from Table 2 that $\left(NQ\left(\mathbb{Z}\right), \star'\right)$ is a neutrosophic quadruplee commutative semigroup.
c. c. $\star^1$ is NeutroDistributive over +

$$\vec{9} \star' (\vec{8} + \vec{7}) = \vec{\theta} \neq \vec{15} = \vec{9} \star' \vec{8} + \vec{9} \star \vec{7} \text{ and } \vec{7} \star' (\vec{8} + \vec{1}) = \vec{15} = \vec{7} \star' \vec{8} + \vec{7} \star \vec{1}.$$

Hence, we can conclude that ($NQ(Z_2)$, +, $\star^1$) is a [8, 9]– NeutroQuadrupleRing.

2. 2. Define on Z_2 the binary operations + and $\star$, for all a, b $\in Z_2$ by

Table 1. Cayley table for the binary operation ★[1]

$\star^1$	θ	$\vec{1}$	$\vec{2}$	$\vec{3}$	$\vec{4}$	$\vec{5}$	$\vec{6}$	$\vec{7}$	$\vec{8}$	$\vec{9}$	$\vec{10}$	$\vec{11}$	$\vec{12}$	$\vec{13}$	$\vec{14}$	$\vec{15}$
θ	$\vec{15}$	$\vec{15}$	$\vec{15}$	$\vec{15}$	$\vec{15}$	$\vec{15}$	$\vec{15}$	$\vec{15}$	$\vec{15}$	$\vec{15}$	$\vec{15}$	$\vec{15}$	$\vec{15}$	$\vec{15}$	$\vec{15}$	$\vec{15}$
$\vec{1}$	$\vec{15}$	$\vec{14}$	$\vec{14}$	$\vec{15}$	$\vec{15}$	$\vec{15}$	$\vec{15}$	$\vec{14}$	$\vec{14}$	$\vec{15}$	$\vec{15}$	$\vec{14}$	$\vec{15}$	$\vec{14}$	$\vec{14}$	$\vec{15}$
$\vec{2}$	$\vec{15}$	$\vec{14}$	$\vec{13}$	$\vec{12}$	$\vec{13}$	$\vec{12}$	$\vec{15}$	$\vec{14}$	$\vec{13}$	$\vec{12}$	$\vec{15}$	$\vec{14}$	$\vec{15}$	$\vec{14}$	$\vec{13}$	$\vec{12}$
$\vec{3}$	$\vec{15}$	$\vec{15}$	$\vec{12}$	$\vec{12}$	$\vec{12}$	$\vec{13}$	$\vec{12}$	$\vec{15}$	$\vec{12}$	$\vec{12}$	$\vec{15}$	$\vec{15}$	$\vec{15}$	$\vec{15}$	$\vec{12}$	$\vec{12}$
$\vec{4}$	$\vec{15}$	$\vec{14}$	$\vec{13}$	$\vec{12}$	$\vec{11}$	$\vec{10}$	$\vec{9}$	$\vec{8}$	$\vec{11}$	$\vec{10}$	$\vec{9}$	$\vec{8}$	$\vec{15}$	$\vec{14}$	$\vec{13}$	$\vec{13}$
$\vec{5}$	$\vec{15}$	$\vec{15}$	$\vec{12}$	$\vec{13}$	$\vec{10}$	$\vec{10}$	$\vec{9}$	$\vec{9}$	$\vec{10}$	$\vec{10}$	$\vec{9}$	$\vec{9}$	$\vec{15}$	$\vec{15}$	$\vec{12}$	$\vec{12}$
$\vec{6}$	$\vec{15}$	$\vec{15}$	$\vec{15}$	$\vec{12}$	$\vec{9}$	$\vec{9}$	$\vec{9}$	$\vec{9}$	$\vec{9}$	$\vec{9}$	$\vec{9}$	$\vec{9}$	$\vec{15}$	$\vec{15}$	$\vec{15}$	$\vec{15}$
$\vec{7}$	$\vec{15}$	$\vec{14}$	$\vec{14}$	$\vec{15}$	$\vec{8}$	$\vec{9}$	$\vec{9}$	$\vec{8}$	$\vec{8}$	$\vec{9}$	$\vec{9}$	$\vec{8}$	$\vec{15}$	$\vec{14}$	$\vec{14}$	$\vec{15}$
$\vec{8}$	$\vec{15}$	$\vec{14}$	$\vec{13}$	$\vec{12}$	$\vec{11}$	$\vec{10}$	$\vec{9}$	$\vec{8}$	$\vec{7}$	$\vec{7}$	$\vec{5}$	$\vec{4}$	$\vec{3}$	$\vec{2}$	$\vec{1}$	$\vec{\theta}$
$\vec{9}$	$\vec{15}$	$\vec{15}$	$\vec{12}$	$\vec{12}$	$\vec{10}$	$\vec{10}$	$\vec{9}$	$\vec{9}$	$\vec{7}$	$\vec{6}$	$\vec{5}$	$\vec{5}$	$\vec{3}$	$\vec{3}$	$\vec{\theta}$	$\vec{\theta}$
$\vec{10}$	$\vec{15}$	$\vec{15}$	$\vec{15}$	$\vec{15}$	$\vec{9}$	$\vec{9}$	$\vec{9}$	$\vec{9}$	$\vec{5}$	$\vec{5}$	$\vec{5}$	$\vec{5}$	$\vec{3}$	$\vec{3}$	$\vec{3}$	$\vec{3}$
$\vec{11}$	$\vec{15}$	$\vec{14}$	$\vec{14}$	$\vec{15}$	$\vec{8}$	$\vec{9}$	$\vec{9}$	$\vec{8}$	$\vec{4}$	$\vec{5}$	$\vec{5}$	$\vec{4}$	$\vec{3}$	$\vec{2}$	$\vec{2}$	$\vec{3}$
$\vec{12}$	$\vec{15}$	$\vec{15}$	$\vec{15}$	$\vec{15}$	$\vec{15}$	$\vec{15}$	$\vec{15}$	$\vec{3}$	$\vec{3}$	$\vec{3}$	$\vec{3}$	$\vec{3}$	$\vec{3}$	$\vec{3}$	$\vec{3}$	$\vec{3}$
$\vec{13}$	$\vec{15}$	$\vec{14}$	$\vec{14}$	$\vec{15}$	$\vec{14}$	$\vec{15}$	$\vec{15}$	$\vec{14}$	$\vec{2}$	$\vec{3}$	$\vec{3}$	$\vec{2}$	$\vec{3}$	$\vec{2}$	$\vec{2}$	$\vec{3}$
$\vec{14}$	$\vec{15}$	$\vec{14}$	$\vec{13}$	$\vec{12}$	$\vec{13}$	$\vec{12}$	$\vec{15}$	$\vec{14}$	$\vec{1}$	$\vec{\theta}$	$\vec{3}$	$\vec{2}$	$\vec{3}$	$\vec{2}$	$\vec{1}$	$\vec{\theta}$
$\vec{15}$	$\vec{15}$	$\vec{15}$	$\vec{12}$	$\vec{12}$	$\vec{13}$	$\vec{12}$	$\vec{15}$	$\vec{15}$	$\vec{\theta}$	$\vec{\theta}$	$\vec{3}$	$\vec{3}$	$\vec{3}$	$\vec{3}$	$\vec{\theta}$	$\vec{\theta}$

Table 2. Cayley table for the binary operation ★[1]

$\star^1$	θ	$\vec{1}$	$\vec{2}$	$\vec{3}$	$\vec{4}$	$\vec{5}$	$\vec{6}$	$\vec{7}$	$\vec{8}$	$\vec{9}$	$\vec{10}$	$\vec{11}$	$\vec{12}$	$\vec{13}$	$\vec{14}$	$\vec{15}$
θ	θ	θ	$\vec{3}$	$\vec{3}$	$\vec{3}$	$\vec{3}$	θ	θ	$\vec{15}$	$\vec{15}$	$\vec{12}$	$\vec{12}$	$\vec{12}$	$\vec{12}$	$\vec{15}$	$\vec{15}$
$\vec{1}$	θ	θ	$\vec{3}$	$\vec{3}$	$\vec{3}$	$\vec{3}$	θ	θ	$\vec{15}$	$\vec{15}$	$\vec{12}$	$\vec{12}$	$\vec{12}$	$\vec{12}$	$\vec{15}$	$\vec{15}$
$\vec{2}$	$\vec{3}$	$\vec{3}$	θ	θ	θ	θ	$\vec{3}$	$\vec{3}$	$\vec{12}$	$\vec{12}$	$\vec{15}$	$\vec{15}$	$\vec{15}$	$\vec{15}$	$\vec{12}$	$\vec{12}$
$\vec{3}$	$\vec{3}$	$\vec{3}$	θ	θ	θ	θ	$\vec{3}$	$\vec{3}$	$\vec{12}$	$\vec{12}$	$\vec{15}$	$\vec{15}$	$\vec{15}$	$\vec{15}$	$\vec{12}$	$\vec{12}$
$\vec{4}$	$\vec{3}$	$\vec{3}$	θ	θ	θ	θ	$\vec{3}$	$\vec{3}$	$\vec{12}$	$\vec{12}$	$\vec{15}$	$\vec{15}$	$\vec{15}$	$\vec{15}$	$\vec{12}$	$\vec{12}$
$\vec{5}$	$\vec{3}$	$\vec{3}$	θ	θ	θ	θ	$\vec{3}$	$\vec{3}$	$\vec{12}$	$\vec{12}$	$\vec{15}$	$\vec{15}$	$\vec{15}$	$\vec{15}$	$\vec{12}$	$\vec{12}$
$\vec{6}$	θ	θ	$\vec{3}$	$\vec{3}$	$\vec{3}$	$\vec{3}$	θ	θ	$\vec{15}$	$\vec{15}$	$\vec{12}$	$\vec{12}$	$\vec{12}$	$\vec{12}$	$\vec{15}$	$\vec{15}$
$\vec{7}$	θ	θ	$\vec{3}$	$\vec{3}$	$\vec{3}$	$\vec{3}$	θ	θ	$\vec{15}$	$\vec{15}$	$\vec{12}$	$\vec{12}$	$\vec{12}$	$\vec{12}$	$\vec{15}$	$\vec{15}$
$\vec{8}$	$\vec{15}$	$\vec{15}$	$\vec{12}$	$\vec{12}$	$\vec{12}$	$\vec{12}$	$\vec{15}$	$\vec{15}$	θ	θ	$\vec{3}$	$\vec{3}$	$\vec{3}$	$\vec{3}$	θ	θ
$\vec{9}$	$\vec{15}$	$\vec{15}$	$\vec{12}$	$\vec{12}$	$\vec{12}$	$\vec{12}$	$\vec{15}$	$\vec{15}$	θ	θ	$\vec{3}$	$\vec{3}$	$\vec{3}$	$\vec{3}$	θ	θ
$\vec{10}$	$\vec{12}$	$\vec{12}$	$\vec{15}$	$\vec{15}$	$\vec{15}$	$\vec{15}$	$\vec{12}$	$\vec{12}$	$\vec{3}$	$\vec{3}$	θ	θ	θ	θ	$\vec{3}$	$\vec{3}$
$\vec{11}$	$\vec{12}$	$\vec{12}$	$\vec{15}$	$\vec{15}$	$\vec{15}$	$\vec{15}$	$\vec{12}$	$\vec{12}$	$\vec{3}$	$\vec{3}$	θ	θ	θ	θ	$\vec{3}$	$\vec{3}$
$\vec{12}$	$\vec{12}$	$\vec{12}$	$\vec{15}$	$\vec{15}$	$\vec{15}$	$\vec{15}$	$\vec{12}$	$\vec{12}$	$\vec{3}$	$\vec{3}$	θ	θ	θ	θ	$\vec{3}$	$\vec{3}$
$\vec{13}$	$\vec{12}$	$\vec{12}$	$\vec{15}$	$\vec{15}$	$\vec{15}$	$\vec{15}$	$\vec{12}$	$\vec{12}$	$\vec{3}$	$\vec{3}$	θ	θ	θ	θ	$\vec{3}$	$\vec{3}$
$\vec{14}$	$\vec{15}$	$\vec{15}$	$\vec{12}$	$\vec{12}$	$\vec{12}$	$\vec{12}$	$\vec{15}$	$\vec{15}$	θ	θ	$\vec{3}$	$\vec{3}$	$\vec{3}$	$\vec{3}$	θ	θ
$\vec{15}$	$\vec{15}$	$\vec{15}$	$\vec{12}$	$\vec{12}$	$\vec{12}$	$\vec{12}$	$\vec{15}$	$\vec{15}$	θ	θ	$\vec{3}$	$\vec{3}$	$\vec{3}$	$\vec{3}$	θ	θ

$$a+b=a+_2b \text{ and } a\star b= =\begin{cases} a+b, a=b \\ a, otherwise \end{cases}.$$

Define the operation $\star'$ as in Example 4.9 above, then the triple ($NQ(Z_2)$, +, $\star^1$) is a [7,8, 10]- NeutroQuadrupleRing.

To see this we consider the Table 3.

Table 3. Cayley table for the binary operation $\star^1$

$\star^1$	θ	$\vec{1}$	$\vec{2}$	$\vec{3}$	$\vec{4}$	$\vec{5}$	$\vec{6}$	$\vec{7}$	$\vec{8}$	$\vec{9}$	$\overrightarrow{10}$	$\overrightarrow{11}$	$\overrightarrow{12}$	$\overrightarrow{13}$	$\overrightarrow{14}$	$\overrightarrow{15}$
θ	θ	θ	θ	θ	θ	θ	θ	θ	θ	θ	θ	θ	θ	θ	θ	θ
$\vec{1}$	θ	$\vec{1}$	$\vec{1}$	θ	$\vec{1}$	θ	θ	$\vec{1}$	$\vec{1}$	θ	θ	$\vec{1}$	θ	$\vec{1}$	$\vec{1}$	θ
$\vec{2}$	$\vec{3}$	$\vec{2}$	$\vec{1}$	θ	$\vec{1}$	θ	$\vec{3}$	$\vec{2}$	$\vec{1}$	θ	$\vec{3}$	$\vec{2}$	$\vec{3}$	$\vec{2}$	$\vec{1}$	θ
$\vec{3}$	$\vec{3}$	$\vec{3}$	θ	θ	θ	θ	$\vec{3}$	$\vec{3}$	θ	θ	$\vec{3}$	$\vec{3}$	$\vec{3}$	$\vec{3}$	θ	θ
$\vec{4}$	$\vec{3}$	$\vec{2}$	$\vec{1}$	θ	$\vec{7}$	$\vec{6}$	$\vec{5}$	$\vec{4}$	$\vec{7}$	$\vec{6}$	$\vec{5}$	$\vec{4}$	$\vec{3}$	$\vec{2}$	$\vec{1}$	θ
$\vec{5}$	$\vec{3}$	$\vec{3}$	θ	θ	$\vec{6}$	$\vec{6}$	$\vec{5}$	$\vec{5}$	$\vec{6}$	$\vec{6}$	$\vec{5}$	$\vec{5}$	$\vec{3}$	$\vec{3}$	θ	θ
$\vec{6}$	θ	θ	θ	θ	$\vec{6}$	$\vec{6}$	$\vec{6}$	$\vec{6}$	$\vec{6}$	$\vec{6}$	$\vec{6}$	$\vec{6}$	θ	θ	θ	θ
$\vec{7}$	θ	$\vec{1}$	$\vec{1}$	θ	$\vec{7}$	$\vec{6}$	$\vec{6}$	$\vec{7}$	$\vec{7}$	$\vec{6}$	$\vec{6}$	$\vec{7}$	θ	$\vec{1}$	$\vec{1}$	θ
$\vec{8}$	$\overrightarrow{15}$	$\overrightarrow{14}$	$\overrightarrow{13}$	$\overrightarrow{12}$	$\overrightarrow{11}$	$\overrightarrow{10}$	$\vec{9}$	$\vec{8}$	$\vec{7}$	$\vec{6}$	$\vec{5}$	$\vec{4}$	$\vec{3}$	$\vec{2}$	$\vec{1}$	θ
$\vec{9}$	$\overrightarrow{15}$	$\overrightarrow{15}$	$\overrightarrow{12}$	$\overrightarrow{12}$	$\overrightarrow{10}$	$\overrightarrow{10}$	$\vec{9}$	$\vec{9}$	$\vec{6}$	$\vec{6}$	$\vec{6}$	$\vec{6}$	θ	θ	θ	θ
$\overrightarrow{10}$	$\overrightarrow{12}$	$\overrightarrow{12}$	$\overrightarrow{12}$	$\overrightarrow{12}$	$\overrightarrow{10}$	$\overrightarrow{10}$	$\overrightarrow{10}$	$\overrightarrow{10}$	$\vec{6}$	$\vec{6}$	$\vec{6}$	$\vec{6}$	θ	θ	θ	θ
$\overrightarrow{11}$	$\overrightarrow{12}$	$\overrightarrow{13}$	$\overrightarrow{13}$	$\overrightarrow{12}$	$\overrightarrow{11}$	$\overrightarrow{10}$	$\overrightarrow{10}$	$\overrightarrow{11}$	$\vec{7}$	$\vec{6}$	$\vec{6}$	$\vec{7}$	θ	$\vec{1}$	$\vec{1}$	θ
$\overrightarrow{12}$	$\overrightarrow{12}$	$\overrightarrow{12}$	$\overrightarrow{12}$	$\overrightarrow{12}$	$\overrightarrow{12}$	$\overrightarrow{12}$	$\overrightarrow{12}$	$\overrightarrow{12}$	θ	θ	θ	θ	θ	θ	θ	θ
$\overrightarrow{13}$	$\overrightarrow{12}$	$\overrightarrow{13}$	$\overrightarrow{13}$	$\overrightarrow{12}$	$\overrightarrow{13}$	$\overrightarrow{12}$	$\overrightarrow{12}$	$\overrightarrow{13}$	$\vec{1}$	θ	θ	$\vec{1}$	θ	$\vec{1}$	$\vec{1}$	θ
$\overrightarrow{14}$	$\overrightarrow{15}$	$\overrightarrow{14}$	$\overrightarrow{13}$	$\overrightarrow{12}$	$\overrightarrow{13}$	$\overrightarrow{12}$	$\overrightarrow{15}$	$\overrightarrow{14}$	$\vec{1}$	θ	$\vec{3}$	$\vec{2}$	$\vec{3}$	$\vec{2}$	$\vec{1}$	θ
$\overrightarrow{15}$	$\overrightarrow{15}$	$\overrightarrow{15}$	$\overrightarrow{12}$	$\overrightarrow{12}$	$\overrightarrow{12}$	$\overrightarrow{12}$	$\overrightarrow{15}$	$\overrightarrow{15}$	θ	θ	$\vec{3}$	$\vec{3}$	$\vec{3}$	$\vec{3}$	θ	θ

a. That $\left(NQ\left(\mathbb{Z}_2\right),+\right)$ is a netrosophic abelian group is straight forward.

b. It can be seen from Table 3 that $\left(NQ\left(\mathbb{Z}\right),\star'\right)$ is a neutrosophic quadruple groupoid.

c. c. For NeutroAssociative of $\star'$ consider the triples

$$\left(\theta,\overrightarrow{15},\overrightarrow{14}\right),\left(\overrightarrow{15},\overrightarrow{15},\overrightarrow{15}\right)\in NQ\left(\mathbb{Z}_2\right).$$

Then we have that

$$\theta \star' \left(\overrightarrow{15} \star' \overrightarrow{14}\right) = \left(\theta \star' \overrightarrow{15}\right) \star' \overrightarrow{14} \text{and } \left(\overrightarrow{15} \star' (\overrightarrow{15} \star' \overrightarrow{15}\right) \neq \left(\overrightarrow{15} \star' \overrightarrow{15}) \star' \overrightarrow{15}\right).$$

d. For distributive of $\star'$ over $+$. We note that ;

d. i. $\star'$ is NeutroLeftDistributive over $+$. To see this, consider the triples

$$\left(\theta, \overrightarrow{15}, \theta\right), \left(\overrightarrow{15}, \overrightarrow{15}, \theta\right) \in NQ\left(\mathbb{Z}_2\right).$$

$$\left(\theta \star' (\overrightarrow{15} + \theta\right) = \theta = \theta \star' \overrightarrow{15} + \theta \star' \theta \text{ and}$$

$$\overrightarrow{15} \star' \left(\overrightarrow{15} + \theta\right) = \theta \neq \overrightarrow{15} = \overrightarrow{15} \star' \overrightarrow{15} + \overrightarrow{15} \star' \theta.$$

ii. $\star'$ is right distributive over + for all triple $\left(x, y, z\right) \in NQ\left(\mathbb{Z}_2\right)$.

e. For NeutroCommutativity of $\star^1$ consider the duplet $\left(\theta, \overrightarrow{15}\right) \in NQ\left(\mathbb{Z}_2\right)$ we have that

$$\theta \star' \overrightarrow{15} = \theta \neq \overrightarrow{15} = \overrightarrow{15} \star' \theta$$

but the duplet (x, x) for all $x \in NQ\left(\mathbb{Z}_2\right)$ satisfies commutativity.

Hence, we can conclude that ($NQ(Z_2)$, +, $\star^1$) is a [7,8, 10]–NeutroQuadrupleRing.

Eample 4.11: Let *NQ*(Z) be neutrosophic quadruple integers and define on *NQ*(Z) the binary operations ⊕ and ⨀ as $a \oplus b = ab + a$, $a \odot b = ab$, for all $a = (a_1, a_2T, a_3I, a_4F)$, $b = (b_1, b_2T, b_3I, b_4F) \in NQ(Z)$. Then (*NQ*(Z), ⊕, ⨀) is a [2, 5, 8, 9]–NeutroQuadrupleRing.

Example 4.12: Let $NQ(Z^+)$ be neutrosophic quadruple positive integers and define on $NQ(Z^+)$ the binary operations of ordinary "+" and multiplication ".". Then for $\theta = (0, 0T, 0I, 0F)$ and for all $p, q, r, s \in -2Z^+$ the triple ($NQ(Z^+) \cup \{\theta, (p, qT, rI, sF)\}$, +, ·) is a [4]– NeutroQuadrupleRing.

Example 4.13: Let *NQ*(N) be neutrosophic quadruple natural numbers. Let $\theta = (0, 0T, 0I, 0F)$, defined for all $a = (a_1, a_2T, a_3I, a_4F)$, $b = (b_1, b_2T, b_3I, b_4F) \in NQ(N) \cup \{\theta, -(x, yT, zI, kF)\}$ with $x, y, z, k \in 3Z^+$, the binary operations of ordinary addition + and multiplication ⨀ by $a \odot b = a^2 b$.

Then ($NQ(N) \cup \{\theta, -(x, yT, zI, kF)\}$, +, ⨀) is a [4, 7]– NeutroQuadrupleRing.

Example 4.14: Let $NQ(\mathrm{R})$ be neutrosophic quadruple real numbers and define on $NQ(\mathrm{R})$ the binary operations $\oplus$ and 0 as $a \oplus b = ab + a$, $a \odot b = a^2 b$, for all $a = (a_1, a_2T, a_3I, a_4F)$, $b = (b_1, b_2T, b_3I, b_4F) \in NQ(\mathrm{R})$. Then $(NQ(\mathrm{R}), \oplus, \odot)$ is a [2, 5, 7, 8, 9]– NeutroQuadrupleRing.

Example 4.15: Let $NQ(\mathrm{Q})$ be neutrosophic quadruple rational numbers and define on $NQ(\mathrm{Q})$ the binary operations $\oplus$ and 0 as $a \oplus b = a + b$, $a \odot b = a \div b$, for all $a, b \in NQ(\mathrm{Q})$. Then $(NQ(\mathrm{Q}), \oplus, \odot)$ is a [7, 8, 9]– NeutroQuadrupleRing.

Example 4.16: Let $M_n(NQ(\mathrm{Z}))$ be the collection of neutrosophic quadruple $n \times n$ matrices hav- ing elements of $NQ(\mathrm{Z})$ as entries. For all $A, B \in M_n(NQ(\mathrm{Z}))$, define on $M_n(NQ(\mathrm{Z}))$ the binary operations $\oplus$ and 0 by $A \oplus B = AB + B$, $A \odot B = AB$. Then $(M_n(NQ(\mathrm{Z}), \oplus, \odot)$ is a [2, 5, 8, 9]–Non- CommutativeNeutroQuadrupleRing.

Example 4.17: Let $NQ(\mathrm{Z})$ be a neutrosophic quadruple integer, let + be the usual addition and $\star'$ be a binary operation defined on $NQ(\mathrm{Z})$ for all

$$(a_1, a_2T, a_3I, a_4F), (b_1, b_2T, b_3I, b_4F) \in NQ(\mathrm{Z})$$

by

$$(a_1, a_2T, a_3I, a_4F) \star^1 (b_1, b_2T, b_3I, b_4F) = (a_1 \star b_1, (a_1 \star b_2 + a_2 \star b_1 + a_2 \star b_2)T,$$

$$(a_1 \star b_3 + a_2 \star b_3 + a_3 \star b_1 + a_3 \star b_2 + a_3 \star b_3)I,$$

$$(a_1 \star b_4 + a_2 \star b_4 + a_3 \star b_4 + a_4 \star b_1 + a_4 \star b_2 + a_4 \star b_3 + a_4 \star b_4)F).$$

Where

$$a_i \star b_i = \max\left(a_i, b_i\right) \textit{for all } a_i, b_i \in \mathbb{Z}.$$

Then $(NQ(\mathrm{Z}), +, \star^1)$ is a [7, 8, 9]–NeutroQuadrupleRing.

1. It can be easily shown that $(NQ(\mathrm{Z}), +)$ is a neutrosophic quadruple abelian group.
2. It is easy to see that $(NQ(\mathrm{Z}), {}^1)$ is a neutrosophic quadruple commutative groupoid.
3. To show that $\star^1$ is NeutroAssociative.

We only need to find some triplets $((a_1, a_2T, a_3I.a_4F), (b_1, b_2T, b_3I, b_4F), (c_1, c_2T, c_3I, c_4F)), ((u_1, u_2T, u_3I.u_4F), (v_1, v_2T, v_3I, v_4F), (k_1, k_2T, k_3I, k_4F)) \in NQ(\mathrm{Z})$, such that $NQR7$ holds.

Let

$(a_1, a_2T, a_3I, a_4F) = (2, 3T, 4I, 5F), (b_1, b_2T, b_3I, b_4F) = (4, 3T, -2I,$ $3F)$

and $(c_1, c_2T, c_3I, c_4F) = (5, 8T, -7I, 6F)$. Then we have that

$(2, 3T, 4I, 5F) \star' ((4, 3T, -2I, 3F) \star' (5, 8T, -7I, 6F))$

$= (-1, 18T, 11I, 77F) =(2, 12T, 2I,$ $33F)$

$= ((2, 3T, 4I, 5F) \star' (4, 3T, -2I, 3F)) \star' (5, 8T, -7I, 6F).$

But if we let the triplets

$((u_1, u_2T, u_3I.u_4F), (v_1, v_2T, v_3I, v_4F), (k_1, k_2T, k_3I, k_4F))$

be such that

$u_1 = u_2 = \cdots = v_1 = v_2 = \cdots = k_1 = \cdots = k_4,$

then associative property holds.
For instance

$((4, 4T, 4I, \star^1 (4, 4T, 4I, 4F))\star^1 (4, 4T, 4I, 4F)$ $4F)$

$= (1, 3T, 8I, 28F) = (4, 4T, 4I, 4F)\star^1((4, 4T, 4I, 4F)\star^1 (4, 4T, 4I, 4F)).$

Hence, *NQR*7 holds.

4. To show that is NeutroDistributive over +.

We only need to find some triplets in *NQ*(Z) such that *NQR*8 and *NQR*9 hold. Since $\star^1$ is commutative it suffices to show that one of *NQR*8 or *NQR*9 holds.
Consider the triplets

$((2, 3T, 4I, 5F), (4, 3T, -2I, 3F), (5, 8T, -7I, 6F))$

$((3, 3T, 3I, 3F), (2, 2T, 2I, 2F), (1, T, I, F)) \in NQ(Z).$

We have that

(2, 3T, 4I, 5F)$\star^1$ ((4, 3T, −2I, 3F) + (5, 8T, −7I, 6F))

= (6, 22T, 14I, 40F) = (3, 13T, 9I, 30F) =

(2, 3T, 4I, 5F)$\star^1$ (4, 3T, −2I, 3F) + (2, 3T, 4I, 5F) $\star^1$ (5, 8T, −7I, 6F).

And

(3, 3T, 3I, 3F)$\star^1$ ((2, 2T, 2I, 2F) + (1, T, I, F)) = (0, 0T, 0I, 0F)

= (3, 3T, 3I, 3F)+(2, 2T, 2I, 2F)$\star^1$ (3, 3T, 3I, 3F)+(1, T, I, F).

Hence (NQ(Z), +, $\star^1$) is a [7, 8, 9]− NeutroQuadrupleRing.

It can be seen from all the examples given above that the triple (NQ(X), +, $\star^1$) is NeutroQuadrup- leRing if (X, +, $\star$) is a NeutroQuadrupleRing, where X is any of N, Zn, Z+, Z, Q, R. This observation is recorded in next proposition.

Proposition 4.18: *Let NQ*(X) *be a neutrosophic quadruple set.* (*NQ*(X), +,$\star$) *is NeutroQuadrupleRing if and only if* (X, +,$\star$) *is a NeutroQuadrupleRing.*

Proof: The proof follows easily from all the examples given above.

Example 4.19: Let $\theta = (0, 0T, 0I, 0F)$ denote additive neutral element.--

1. In Example 4.9 we have that;
   a. $\{\vec{8},\vec{9},\vec{14},\vec{15}\}$ are zero divisor elements.
   b. $\{\vec{15}\}$ is nilpotent element.
2. In 1 and 2 of Example 4.10,
   a. $\{\vec{1},\vec{2},\vec{3},\vec{4},\vec{5},\vec{6},\vec{7},\vec{8},\vec{9},\vec{10},\vec{11},\vec{12},\vec{13},\vec{14},\vec{15}\}$ are zero divisor elements.
   b. $\{\vec{1}, \vec{6}, \vec{7}\}$ are idempotent elements in 2 of Example 4.10 .
   c. Nilpotent elements in ;
      i. 1 of Example 4.10 are $\{\theta,\vec{1},\vec{2},\vec{3},\vec{4},\vec{5},\vec{6},\vec{7},\vec{8},\vec{9},\vec{10},\vec{11},\vec{12},\vec{13},\vec{14},\vec{15}\}$
      ii. 2 of Example 4.10 are $\{\theta,\vec{3},\vec{12},\vec{15}\}$
3. In Example 4.17 we have that;
   a. {(3, 3*T,* 3*I,* 3*F*)} is nilpotent element.
   b. {(3, 3*T,* 3*I,* 3*F*), (*x, yT, uI, vF*)} are zero divisor elements, where $x, y, u, v \leq 3$ and $\theta = (x, yT, uI, vI)$.

Remark 4.20: The following should be noted.

1. From Examples 4.9, 4.10, and 4.17 we can see that $\theta\star^1 x = x\star^1\theta = \theta$ is not true for some elements in some classes of AntiQuadrupleRings and NeutroQuadrupleRings.
2. Also, for some z in some classes of AntiQuadrupleRings and NeutroQuadrupleRings, it is not true that $z^n \star^1 z^m = z^{n+m}$ and $(z^n)^m = (z^m)^n = z^{nm}$.

From Table 1 of Example 4.9 we can see that

$$\overrightarrow{10}^2 \star' \overrightarrow{10}^3 = \overrightarrow{10} \neq \vec{9} = \overrightarrow{10}^5 \text{ and } \left(\overrightarrow{11}^2\right)^3 = \vec{8} \neq \vec{4} = \overrightarrow{11}^6 \neq \vec{7} = \left(\overrightarrow{11}^3\right)^2$$

Table 3 of Example 4.9 shows that

$$\vec{3}^1 \star' \vec{3}^2 = \vec{3} \neq \theta = \vec{3}^3$$

Also, from Example 4.17, if we consider the element $(3, 3T, 3I, 3F) \in NQ(Z)$. We will have that

$$(3, 3T, 3I, 3F)^2 \star^1 (3, 3T, 3I, 3F)^3 = (-3, -9T, -15I, -21F)$$

$$= (0, 0T, 0I, 0F) = (3, 3T, 3I, 3F)^5$$

and

$$[(3, 3T, 3I, 3F)^2]^3 = (-3, -9T, -15I, -21F) = (0, 0T, 0I, 0F) = (3, 3T, 3I, 3F)^6.$$

These observations are recorded in the next proposition.

Proposition 4.21: *Let $(NQ(R), +, \star^1)$ be a class/type of NeutroQuadrupleRing/AntiQuadrupleRing in which $\star^1$ is NeutroAssociative and let $\theta = (0, 0T, 0I, 0F)$ be the additive neutral element in $NQ(R)$. Then the properties:*

1. $\theta\star^1 z = z \star^1 \theta = \theta$,
2. $z^n\star^1 z^m = z^{n+m}$ *and* $(z^n)^m = (z^m)^n = z^{nm}$*, fail to hold for some* $z \in NQ(R)$ *and* $n, m \in Z$.

Proof: The proof follows easily from Remark 4.20.

Proposition 4.22: *Let*

$(NQ(R_1), +, \star_1), (NQ(R_2), +, \star_2)$

be any two NeutroQuadrupleRings of the same type/class. Define the binary operations $\oplus$ *and* $\odot$ *for all*

$((a_1, b_1T, c_1I, d_1F), (a_2, b_2T, c_2I, d_2F)), ((s_1, t_1T, u_1I, v_1F), (s_2, t_2T, u_2I, v_2F)) \in NQ(R_1)\times NQ(R_2)$

as follows;

$((a_1, b_1T, c_1I, d_1F), (a_2, b_2T, c_2I, d_2F)) \oplus ((s_1, t_1T, u_1I, v_1F), (s_2, t_2T, u_2I, v_2F))$

$=((a_1 + s_1, (b_1 + t_1)T, (c_1 + u_1)I, (d_1 + v_1)F), (a_2 + s_2, (b_2 + t_2)T, (c_2 + u_2)I, (d_2 + v_2)F))$

and

$((a_1, b_1T, c_1I, d_1F), (a_2, b_2T, c_2I, d_2F)) \odot ((s_1, t_1T, u_1I, v_1F), (s_2, t_2T, u_2I, v_2F))$

$=((a_1, b_1T, c_1I, d_1F) \star_1 (s_1, t_1T, u_1I, v_1F), (a_2, b_2T, c_2I, d_2F) \star_2 (s_2, t_2T, u_2I, v_2F)).$

Then $(NQ(R_1) \times NQ(R_2), \oplus, \odot)$ *gives a NeutroQuadrupleRing which belongs to the same class has*

$(NQ(R_1), +, \star_1)$ *and* $(NQ(R_2), +, \star_2)$.

Remark 4.23: It should be noted that;

1. If NQR1 in Proposition 4.22 is NeutroQuadripleRing and NQR2 is a neutrosophic quadruple ring. The Cartesian product will give a NeutroQuadrupleRing which belongs to the class of NQR1.
2. If NQR1 and NQR2 of Proposition 4.22 are any two NeutroQuadripleRings of different classes. The Cartesian product will give a NeutroQuadrupleRing which belongs to either the class of NQR1 or NQR2 or other classes other than the class of NQR1 and NQR2

Suppose we take $(NQR1, +_1, \star_1)$ and $(NQR2, +_2, \star_2)$ to be NeutroQuadrupleRings of type [a, b] and [a, b, c] respectively, where a, b and c are the neutrosophic quadruple ring axioms that are NeutroSophicated. Then the Cartesian product $(NQR1 \times NQR2, +, \star^1)$ will be a Neu- troQuadrupleRing of type [a, b, c], since all the axioms a, b and c will be NeutroSophicated in $(NQR1 \times NQR2, +, \star^1)$.

Also, if we (NQR1, $+_1$, $\star_1$) and (NQR2, $+_2$, $\star_2$) to be NeutroQuadrupleRings of type [a, b] and [c, e] respectively. Then the Cartesian product (NQR1 × NQR2, +, $\star^1$) will be a Neutro- QuadrupleRing of type [a, b, c, e], which is a class other than the class of (NQR1, $+_1$, $\star_1$) and (NQR2, $+_2$, $\star_2$).

Definition 4.24: *Let (NQ(R), ⊕, ⨀) be a NeutroQuadripleRing and let NQ(U) be nonempty subset of NQ(R).*

1. *NQ(U) is called a NeutroQuadrupleSubring of NQ(R) if (NQ(U), ⊕, ⨀) is also a NeutroQuadrupleRing.*
2. *NQ(U) is called a QuasiNeutroQuadrupleSubring of NQ(R) if (NQ(U), ⊕, ⨀) is a NeutroQuadrupleRing of the type/class different from the type/class of the parent NeutroQuadrupleRing (NQ(R), ⊕, ⨀).*

Example 4.25: Let ($NQ(Z_2)$, +, $\star^1$) be the NeutroQuadrupleRing in (2) of Example 4.10 and let

$$NQ(U_1) = \{\theta, \vec{1}, \vec{2}, \vec{3}\}, NQ(U_2) = \{\theta, \vec{1}, \overrightarrow{14}, \overrightarrow{15}\}$$

and

$$NQ(U_3) = \{\theta, \vec{1}, \vec{2}, \vec{3}, \vec{4}, \vec{5}, \vec{6}, \vec{7}\}$$

be subsets of $NQ(Z_2)$. Then

$(NQ(U_1), +, \star^1)$, $(NQ(U_2), +, \star^1)$

and $(NQ(U_3), +, \star^1)$ are NeutroQuadrupleRings which belong to the same class as $NQ(Z_2)$ and consequently NeutroQuadrupleSubrings of $NQ(Z_2)$.

This is true, since:

1. $(NQ(U_1), +)$, $(NQ(U_2), +)$ and $(NQ(U_3), +)$ are neutrosophic abelian groups. This can be seen from table 4, 5 and 6 below.
2. $(NQ(U_1), \star^1)$, $(NQ(U_2), \star^1)$ and $(NQ(U_3), \star^1)$ are neutrosophic commutative groupoids as can be seen from table 7, 8 and 9 below..
3. *NQR*7, *NQR*8 and *NQR*9 can be confirmed from table 7, 8 and 9 below.

Now, if we consider the following:

1. $NQ(U_1) \cup NQ(U_2) = \{\theta, \vec{1}, \vec{2}, \vec{3}, \overrightarrow{14}, \overrightarrow{15}\}$

2. $NQ(U_2) \cup NQ(U_3) = \{\theta, \vec{1}, \vec{2}, \vec{3}, \vec{4}, \vec{5}, \vec{6}, \vec{7}, \overrightarrow{14}, \overrightarrow{15}\}$
3. $NQ(U_1) \cap NQ(U_2) = \{\theta, \vec{1}\}$.

It can be shown that

(NQ(U1) ∪ NQ(U2), +, $\star^1$) and (NQ(U2) ∪ NQ(U3), +, $\star^1$)

is a NeutroQuadrupleRing of type other than the type of NQ(Z2). Hence

(NQ(U1) ∪ NQ(U2), +, $\star^1$)

is a QuasiNeutroQuadrupleSubring of NQ(Z2). It can also be verified that

(NQ(U1) ∩ NQ(U2), +, $\star^1$)

is neutrosophic quadruple ring and not a NuetroQuadrupleSubring of NQ(Z2).

Proposition 4.26: *Let NQ(R) be any class of NeutroQuadrupleRing, let NQR_1 and NQR_2 be any two NeutroQuadrupleSubrings of NQR. Then*

1. *NQK ∪ NQW is not necessarily a NeutroQuadrupleSubring of NQR.*
2. *NQK ∩ NQW is not necessarily a NeutroQuadrupleSubring of NQR.*

Table 4. Cayley table for the binary operation +

+	θ	$\vec{1}$	$\vec{2}$	$\vec{3}$
θ	θ	$\vec{1}$	$\vec{2}$	$\vec{3}$
$\vec{1}$	$\vec{1}$	θ	$\vec{3}$	$\vec{2}$
$\vec{2}$	$\vec{2}$	$\vec{3}$	θ	$\vec{1}$
$\vec{3}$	$\vec{3}$	$\vec{3}$	$\vec{1}$	θ

Table 5. Cayley table for the binary operation +

+	θ	$\vec{1}$	$\overrightarrow{14}$	$\overrightarrow{15}$
θ	θ	$\vec{1}$	$\overrightarrow{14}$	$\overrightarrow{15}$
$\vec{1}$	$\vec{1}$	θ	$\overrightarrow{15}$	$\overrightarrow{14}$
$\overrightarrow{14}$	$\overrightarrow{14}$	$\overrightarrow{15}$	θ	$\vec{1}$
$\overrightarrow{15}$	$\overrightarrow{15}$	$\overrightarrow{14}$	$\vec{1}$	θ

Table 6. Cayley table for the binary operation +

+	θ	$\vec{1}$	$\vec{2}$	$\vec{3}$	$\vec{4}$	$\vec{5}$	$\vec{6}$	$\vec{7}$
θ	θ	$\vec{1}$	$\vec{2}$	$\vec{3}$	$\vec{4}$	$\vec{5}$	$\vec{6}$	$\vec{7}$
$\vec{1}$	$\vec{1}$	θ	$\vec{3}$	$\vec{2}$	$\vec{5}$	$\vec{4}$	$\vec{7}$	$\vec{6}$
$\vec{2}$	$\vec{2}$	$\vec{3}$	θ	$\vec{1}$	$\vec{6}$	$\vec{7}$	$\vec{4}$	$\vec{5}$
$\vec{3}$	$\vec{3}$	$\vec{2}$	$\vec{1}$	θ	$\vec{7}$	$\vec{6}$	$\vec{5}$	$\vec{4}$
$\vec{4}$	$\vec{4}$	$\vec{5}$	$\vec{6}$	$\vec{7}$	θ	$\vec{1}$	$\vec{2}$	$\vec{3}$
$\vec{5}$	$\vec{5}$	$\vec{4}$	$\vec{7}$	$\vec{6}$	$\vec{1}$	θ	$\vec{3}$	$\vec{2}$
$\vec{6}$	$\vec{6}$	$\vec{7}$	$\vec{4}$	$\vec{5}$	$\vec{2}$	$\vec{3}$	θ	$\vec{1}$
$\vec{7}$	$\vec{7}$	$\vec{6}$	$\vec{5}$	$\vec{4}$	$\vec{3}$	$\vec{2}$	$\vec{1}$	θ

Table 7. Cayley table for the binary operation $\star^1$

$\star^1$	θ	$\vec{1}$	$\vec{2}$	$\vec{3}$
θ	θ	θ	$\vec{3}$	$\vec{3}$
$\vec{1}$	θ	θ	$\vec{3}$	$\vec{3}$
$\vec{2}$	$\vec{3}$	$\vec{3}$	θ	θ
$\vec{3}$	$\vec{3}$	$\vec{3}$	θ	θ

Table 8. Cayley table for the binary operation $\star^1$

+	θ	$\vec{1}$	$\overrightarrow{14}$	$\overrightarrow{15}$
θ	θ	θ	$\overrightarrow{15}$	$\overrightarrow{15}$
$\vec{1}$	θ	θ	$\overrightarrow{15}$	$\overrightarrow{15}$
$\overrightarrow{14}$	$\overrightarrow{15}$	$\overrightarrow{15}$	θ	θ
$\overrightarrow{15}$	$\overrightarrow{15}$	$\overrightarrow{15}$	θ	θ

Table 9. Cayley table for the binary operation $\star^1$

+	θ	$\vec{1}$	$\vec{2}$	$\vec{3}$	$\vec{4}$	$\vec{5}$	$\vec{6}$	$\vec{7}$
θ	θ	$\vec{3}$	$\vec{3}$	$\vec{3}$	$\vec{3}$	$\vec{3}$	θ	θ
$\vec{1}$	θ	θ	$\vec{3}$	$\vec{3}$	$\vec{3}$	$\vec{3}$	θ	θ
$\vec{2}$	$\vec{3}$	$\vec{3}$	θ	θ	θ	θ	$\vec{3}$	$\vec{3}$
$\vec{3}$	$\vec{3}$	$\vec{3}$	θ	θ	θ	θ	$\vec{3}$	$\vec{3}$
$\vec{4}$	$\vec{3}$	$\vec{3}$	θ	θ	θ	θ	$\vec{3}$	$\vec{3}$
$\vec{5}$	$\vec{3}$	$\vec{3}$	θ	θ	θ	θ	$\vec{3}$	$\vec{3}$
$\vec{6}$	θ	θ	$\vec{3}$	$\vec{3}$	$\vec{3}$	$\vec{3}$	θ	θ
$\vec{7}$	θ	θ	$\vec{3}$	$\vec{3}$	$\vec{3}$	$\vec{3}$	θ	θ

Proof: The proofs to 1 and 2 follow easily from the illustrations given in example 4.25 above.

CONCLUSION

In this paper, we have introduced some number theoretic properties of neutrosophic quadruple numbers. We presented and studied some basic properties of NeutroQuadrupleRing and its substructures. It was shown that the intersection and union of any two NeutroQuadrupleSubrings of a particular class do not necessarily belong to the same class. Also, we gave a necessary and sufficient condition for neutrosophic quadruple ring (NQ(X), +, ·) to be a NeutroQuadrupleRing. In our future work, we hope to study some other number theoretic properties of neutrosophic quadruple, like g.c.d, division algorithm, e.t.c. Also we hope to investigate some classes of Neutro(Anti) QuadrupleRings independently.

REFERENCES

Agboola, A. A. A. (2020a). Introduction to AntiRings. *Neutrosophic Sets and Systems*, *36*, 293–307.

Agboola, A.A.A. (2020b). On Finite and Infinite NeutroRings of Type-NR [8,9]. *International Journal of Neutrosophic Science, 11*(2), 87-99.

Agboola, A. A. A. (2020c). Introduction to NeutroGroups. *International Journal of Neutrosophic Science*, *6*(1), 41–47.

Agboola, A. A. A. (2020d). Introduction to NeutroRings. *International Journal of Neutrosophic Science*, *7*(2), 62–73.

Agboola, A.A.A., Davvaz, B., & Smarandache, F. (2017). Neutrosophic quadruple algebraic hyperstructures. Ann. Fuzzy Math. Inform., 14, 29–42.

Akinleye, S. A., Smarandache, F., & Agboola, A. A. A. (2016). On neutrosophic quadruple algebraic structures. *Neutrosophic Sets System.*, *12*, 122–126.

Al-Tahan, & Davvaz. (2020a). *On Some Properties of Neutrosophic Quadruple Hv-rings* (Vol. 36). Neutrosophic Sets and Systems.

Al-Tahan, & Davvaz. (2020b). On neutrosophic quadruple Hv-groups and their properties. *Southeast Asian Bulletin of Mathematics*, *44*, 613–625.

Ibrahim, M. A., & Agboola, A. A. A. (2020a). Introduction to NeutroVector Space I. *Neutrosophic Science and System*, *36*, 329–350.

Ibrahim, M. A., & Agboola, A. A. A. (2020b). Introduction to NeutroHyperGroups. *Neutrosophic Sets and Systems*, *38*, 15–32.

Ibrahim, M. A., Agboola, A. A. A., Adeleke, E. O., & Akinleye, S. A. (2020). On Neutrosophic Quadruple Hypervector Spaces. *International Journal of Neutrosophic Science*, *4*(1), 20–35. doi:10.54216/IJNS.040103

Jun, Y., Song, S. Z., Smarandache, F., & Bordbar, H. (2018). Neutrosophic quadruple BCK/BCI-algebras. *Axioms*, *7*(2), 41. doi:10.3390/axioms7020041

Jun, Y. B., Song, S. Z., & Kim, S. J. (2019). Neutrosophic Quadruple BCI-Positive Implicative Ideals. *Mathematics*, *7*(5), 385. doi:10.3390/math7050385

Li, Q., Ma, Y., Zhang, X., & Zhang, J. (2019). Neutrosophic Extended Triplet Group Based on Neutrosophic Quadruple Numbers. *Symmetry*, *11*(5), 696. doi:10.3390ym11050696

Muhiuddin, G. (2019). Neutrosophic quadruple ideals in neutrosophic quadruple BCI- algebras. *Neutrosophic Sets and Systems*, *25*, 161–173.

Rezaei, A., & Smarandache, F. (2020). On Neutro-BE-algebras and Anti-BE-algebras (revisited). *International Journal of Neutrosophic Science*, *4*(1), 8–15.

Sahin, M., & Kargın, A. (2019). Neutrosophic triplet groups based on set valued neutrosophic quadruple numbers. *Neutrosophic Set and Systems, 30*, 122-131.

Sahin, M., & Kargın, A. (2019). Single valued neutrosophic quadruple graphs. *Asian Journal of Mathematics and Computer Research*, *26*(4), 243–250.

Smarandache, F. (1998). *Neutrosophic Probability, Set and Logic.* Am Res Press.

Smarandache, F. (2003). *A Unifying Field in Logics: Neutrosophic Logic, Neutrosophy, Neutrosophic Set, Neutrosophic Probability*. American Research Press.

Smarandache, F. (2014). *Introduction to Neutrosophic Statistics*. Sitech and Education Publishing.

Smarandache, F. (2015). Neutrosophic quadruple numbers, refined neutrosophic quadruple numbers, absorbance law, and the multiplication of neutrosophic quadruple numbers. *Neutrosophic Sets and Systems*, *10*, 96–98.

Smarandache, F. (2019). Introduction to NeutroAlgebraic Structures. In *Advances of Standard and Nonstandard Neutro- sophic Theories* (pp. 240–265). Pons Publishing House Brussels.

Smarandache, F. (2020a). Introduction to NeutroAlgebraic Structures and AntiAlgebraic Structures (revisited). *Neutrosophic Sets and Systems*, *31*, 1–16. doi:10.5281/zenodo.3638232

Smarandache, F. (2020b). NeutroAlgebra is a Generalization of Partial Algebra. *International Journal of Neutrosophic Science, 2*(1), 8-17.

Smarandache, F., Razaei, A., Agboola, A. A. A., Jun, Y. B., Borzooei, R. A., & Davvaz, B. (2021). On NeutroQuadrupleGroups. *51st Annual Mathematics Conference.*

Vasantha Kandasamy, W. B. (2019). Neutrosophic Quadruple Vector Spaces and Their Properties. *Mathematics*, *7*, 758–765. doi:10.3390/math7080758

Chapter 4 Introduction to Antihypergroups

Muritala Abiodun Ibrahim
Federal University of Agriculture, Abeokuta, Nigeria

Agboola Adesina Abdul Akeem
Federal University of Agriculture, Abeokuta, Nigeria

Zulaihat Hassan-Ibrahim
Auburn University, USA

Emmanuel O. Adeleke
Federal University of Agriculture, Abeokuta, Nigeria

ABSTRACT

The process through which AntiAlgebraicStructures can be generated from any given classical structure is called AntiSophication. In this chapter, the authors introduce the concept of AntiHyperStructure and present some of its basic properties. It has been shown in our previous works that there are 19 types/classes of AntiHyperGroups. The chapter is devoted to the study of a particular class of these 19 classes, called a [2, 3] AntiHyperGroup, that is, a class in which associativity and reproductive axioms are totally false. Several examples and properties of [2, 3] AntiHyperGroup are presented. It is shown that the union of [2, 3] AntiSubhypergroups is a [2, 3] AntiHyperGroup and their intersection is not a [2, 3] AntiSubhypergroup. Also, it is shown that the Cartesian product of a [2, 3] AntiHyperGroup with any other class of AntiHyperGroup is not in general a [2, 3] AntiHyperGroup but its product with a classical Hypergroup/[2, 3]NeutroHyperGroup is a [2, 3] AntiHyperGroup. Further, it is shown that a [2, 3] AntiHyperGroup factored by a [2, 3] AntiSubhypergroup is a [2, 3] AntiHyperGroup.

DOI: 10.4018/978-1-6684-3495-6.ch004

INTRODUCTION

Recently, Florentin Smarandache in (Smarandache, 2020a), presented new tools for handling indeterminate, incomplete and imprecise information/processes. He called these ideas NeutroSophication and AntiSophication. AntiSophication and NeutroSophication are processes through which AntiAlgebraicStructures and Neu-troAlgebraicStructures are respectively generated from any given classical Structure. In (Rezaei & Smarandache, 2020), he recalled, improved and extended several definitions and properties of these new structures. In 2015, Agboola and Davvaz in (Agboola, 2013; Agboola & Akinleye, 2015), examined Hypergroups, canonical hypergroups and Hyperrings in the neutrosophic environment. Their work established the connections between neutrosophic set and algebraic hyper-structures, and it opened up new areas of research in the theory of neutrosophic set. Since then, several neutrosophic algebraic structures have been studied and many results have been obtained and published. In (Ibrahim & Agboola, 2020a), Neutrosophic Hypernear-ring was examined by Ibrahim and Agboola and some of its properties were presented.

The concept of NeutroAlgebra and AntiAlgebra was examined by Agboola et al. in (Agboola et al., 2020) viz-a-viz the classical number systems $\mathbb{N}$; $\mathbb{Z}$; $\mathbb{Q}$; $\mathbb{R}$; and $\mathbb{C}$. The idea introduced by Smarandanche in (Smarandache, 2020a) motivated a lot of neutrosophic researchers and as a result, many Neutro(Anti) Structures have been studied and presented. For example, In (Agboola, 2020a; Agboola, 2020b; Agboola, 2020c; Agboola et al., 2020), Agboola formally presented the notions of Finite NeutroGroups of Type-NG, finite and infinite NeutroRings of Type-NR, NeutroGroups, NeutroRings, NeutroIdeal, NeutroQuotientRings, and he established several properties of these structures and their substructures for the classes considered. Also in (10), Al-Tahan, M. et al., studied NeutroOrderedAlgebra and considered its applications to Semigroups. Rezaei and Smarandache in (Ibrahim & Agboola, 2020c) introduced the concept of Neutro-BE-algebras and Anti-BE-algebras and in (9, 13, 15) Agboola and Ibrahim introduced the concept of NeutroVectorSpaces, AntiRings and NeutroHyperGroups. In (Ibrahim & Agboola, 2020b) they presented applications of NeutroHyperGroup in biological and chemical sciences. The present paper will be concerned with the introduction of AntiHyperGroup and presentation of its basic properties.

PRELIMINARIES

In this section, we will give some definitions, examples and results that will be used in the sequel.

Definition 2.1: *(Davvaz & Leoreanu-Fotea, 2007) Let H be a non-empty set and* $\circ : H \times H \to P^*\left(H\right)$ *be a hyperoperation. The couple (H, ∘) is called a hypergroupoid. For any two non-empty subsets A and B of H and* $x \in H$ *; we define*

$$A \circ B = \bigcup_{a \in A, b \in B} a \circ b, A \circ x = A \circ \left\{x\right\} and\, x \circ B = \left\{x\right\} \circ B.$$

Definition 2.2: *(Davvaz & Leoreanu-Fotea, 2007) A hypergroupoid (H, ∘) is called a semihypergroup if for all a, b, c of H we have*

$$\left(a \circ b\right) \circ c = a \circ \left(b \circ c\right)$$

which means

$$\bigcup_{u \in a \circ b} u \circ c = \bigcup_{v \in b \circ c} a \circ v.$$

A hypergroupoid (H, ∘) is called a quasihypergroup if for all a of H we have $a \circ H = H \circ a = H$. *This condition is also called the reproduction axiom.*

Definition 2.3: *(Davvaz & Leoreanu-Fotea, 2007) A hypergroupoid (H, ∘) which is both a semihypergroup and a quasi- hypergroup is called a hypergroup.*

Definition 2.4: *(Davvaz & Leoreanu-Fotea, 2007) Let (H, ∘) and (H`, ∘`) be two hypergroupoids. A map* $\phi : H \to H'$ *is called*

1. *an inclusion homomorphism if for all x, y of H; we have* $\phi\left(x \circ y\right) \subseteq \phi\left(x\right) \circ' \phi(y)$.
2. *a good homomorphism if for all x, y of H, we have* $\phi\left(x \circ y\right) = \phi\left(x\right) \circ' \phi(y)$.

Definition 2.5: *(Davvaz & Leoreanu-Fotea, 2007) Let H be a non-empty set and let + be a hyperoperation on H. The couple (H; +) is called a canonical hypergroup if the following conditions hold:*

1. x + y = y + x; for all x, y in H;
2. *x + (y + z) = (x + y) + z; for all* $x, y, z \in H$ *;*
3. *there exists a neutral element* $0 \in H$ *such that x + 0 = {x} = 0 + x; for all* $x \in H$ *;*
4. *for every* $x \in H$ *there exists a unique element* $-x \in H$ *such that* $0 \in x + \left(-x\right) \cap \left(-x\right) + x$;

5. $z \in x + y$ *implies* $y \in -x + z$ *and* $x \in z - y$ *for all* $x, y, z \in H$.

Definition 2.6: *(Smarandache, 2019)*

1. (Classical) Operation is an operation well-defined for all the set's elements.
2. (Classical) HyperOperation is a hyper-operation well-defined for all the set's elements.
3. (Neutro) Operation is an operation partially well-defined, partially indeterminate, and partially outer defined on the given set.
4. (Anti) Operation is an operation that is outer defined for all set's elements.
5. A classical law/axiom defined on a nonempty set is a law/axiom that is totally true (i.e., true for all set's elements).
6. A NeutroLaw/NeutroAxiom defined on a nonempty set is a law/axiom that is true for some set's element [degree of truth (T)], indeterminate for other set's elements [degree of indeterminacy.
7. *(I)], or false for the other set's elements [degree of falsehood (F)], where* $T, I, F \in [0,1]$ *with* $(T, I, F) \neq (1,0,0)$ *that represents the classical axiom, and with* $(T, I, F) \neq (1,0,0)$ *that represents the AntiAxiom.*
8. An AntiLaw/AntiAxiom defined on a nonempty set is a law/axiom that is false for all set's elements.
9. NeutroHyperOperation is a hyper-operation partially well-defined, partially indeterminate, and partially outer-defined on a given set.
10. AntiHyperOperation is a hyper-operation outer-defined for all set's elements.
11. A NeutroAlgebra is an algebra that has at least one NeutroOperation or one NeutroAxiom (axiom that is true for some elements, indeterminate for some elements, and false for other elements).
12. An AntiAlgebra is an algebra endowed with at least one AntiOperation or at least one Anti-Axiom.

Theorem 2.7: (Ibrahim & Agboola, 2020c) Let U be a nonempty finite or infinite universe of discourse and let S be a finite or infinite subset of U: If n classical operations (laws and axioms) are defined on S where n>1; then there will be $(2^n - 1)$ NeutroAlgebras and $(3^n - 2^n)$ AntiAlgebras.

Definition 2.8: *(Ibrahim & Agboola, 2020b)[NeutroSophication of law and axioms of the classical hypergroup] Let H be a non-empty set and* $\circ : H \times H \rightarrow P^*(H)$ *be a hyperoperation.*

The NeutroSophication of the Law (degree of well-defined, degree of indeterminacy, degree of outer-defined).

(NH1) There exist some duplets $(u,v),(x,y) \in H$ such that $u \circ v \subseteq H$ and [$x \circ y$ = indeterminate or $x \circ y \subsetneq H$].

The NeutroSophication of the Axioms (degree of truth, degree of indeterminacy, degree of false-hood)

(NH2) There exist some triplets

$$(u,v,w),(x,y,z),(p,q,r) \in H$$

such that

$$u \circ (v \circ w) = (u \circ v) \circ w$$

and

[$x \circ (y \circ z) or (x \circ y) \circ z$ = indeterminate or $(p \circ q) \circ r \neq p \circ (q \circ r)$].

(NH3) There exist some $u,v,x \in H$ such that

$u \circ H = H \circ u = H$ and [$v \circ H$ or $H \circ v$ = indeterminate or $x \circ H \neq H (H \circ x \neq H)$].

Definition 2.9: *(Ibrahim & Agboola, 2020b)[AntiSophication of the law and axioms of the classical hypergroup] Let H be a non-empty set and* $\circ : H \times H \to P^*(H)$ *be a hyperoperation.*

The AntiSophication of the Law (totally outer-defined).

(AH1) $u \circ v \subsetneq H$ *for all* $u,v \in H$:

The AntiSophication of the Axioms (totally false)

(AH2) $u \circ (v \circ w) \neq (u \circ v) \circ w$ *for all* $(u,v,w) \in H$:
(AH3) For all $v \in H; v \circ H \neq H$ *and* $H \circ v \neq H$.

Definition 2.10: *(Ibrahim & Agboola, 2020b) A NeutroHypergroup* $(NH, \circ)$ *is an alternative to the classical hypergroup* $(H, \circ)$ *that has a NeutroLaw or at least one of NH2 and NH3 with no Antilaw or AntiAxiom.*

Definition 2.11: *(Ibrahim & Agboola, 2020b) An AntiHypergroup $(AH,\circ)$ is an alternative to the classical hypergroup $(H,\circ)$ that has an AntiLaw or at least one of AH2 and AH3.*

Example 2.12: (Ibrahim & Agboola, 2020b) Let $V=\{u,v,w,s,t,z\}$ be a universe of discourse and let $NG=\{v,w,s,z\}$ be a subset of V: Define on N G the binary Operation as shown in the Table 1.

Table 1.

o	v	w	s	z
v	s	t	z	v
w	z	u	v	s
s	v	w	s	z
z	z	u	v	s

It can be shown from the table that $(NG,\circ)$ is a trivial NeutroGroup. The subset $NK=\{v,s\}$ of N G is also a trivial NeutroGroup and hence a NeutroSubgroup of NG.

Now, consider the NeutroSubgroup $NK=\{v,s\}$. Define on NG a hyperoperation $\star_{NH}$ as follows:

$$x \star_{NH} y = \begin{cases} x\circ NK\circ y=\{x\circ z\circ y : z\in NK\}, if\ x=y \\ \{x, x\circ y\}, if\ x\neq y. \end{cases}$$

Hence $(NG,\star_{NH})$ is a trivial NeutroHyperGroup.

Theorem 2.13: (Ibrahim & Agboola, 2020b) Let $(H,\circ)$ be a classical hypergroup. Then,

1. there are 7 classes of NeutroHypergroup.
2. there are 19 classes of AntiHypergroup.

Proof: The proof follows easily from Theorem 2.7 .

Theorem 2.13 shows that there are 19 classes of AntiHypergroup.

The case where AH1 - AH3 hold is called the trivial case. Examples of this case are presented below.

Example 2.14: Let

$$AV = \{a, b, c, d, e, u, v, w, s, t, z\}$$

be a universe of discourse and let $AH = \{u, v, w\}$ be a subset of AV. Define on AH the AntiHyperoperation $\star$ as shown in the Table 2.

Table 2.

$\star$	u	v	w
u	{v,s}	{w,t}	{z,b}
v	{w,t}	{s,z}	{a,c}
w	{z,b}	{a,c}	{d,v}

Since for all $u, v \in AH, u \star v$ is outer defined, we say that $(AH, \star)$ is a trivial AntiHypergroup.

Example 2.15: Let

$$AK = \{0,1,2,3,4,5,6,7,8,9\}$$

be a universe of discourse and let $AP = \{0,1,2\}$ be a subset of AK: Define on AP the AntiHyperoperation $\star$ as shown in the Table 3.

Table 3.

$\star$	0	1	2
0	3	4	7
1	4	{3,5}	{6,8}
2	7	{6,8}	{9,1}

For all $x, y \in AK$, we can see from the table that $x \star y$ is outer defined. Hence, (AK, $\star$) is a trivial AntiHypergroup.

1. Study of a Class of AntiHyperGroup

Let $(AG,\star)$ be an AntiHyperGroup as defined in Section 2. Here, we shall consider a particular class of AntiHyperGroup where

1. (AG, $\star$) is a classical hypergroupoid, i.e AH1 fails.
2. The hypergroupoid (AG, $\star$) is AntiSemiHyperGroup, i.e., AH2 holds.
3. The hypergroupoid (AG, $\star$) is AntiQuasiHyperGroup, i.e., AH3 holds.

We will refer to this class of AntiHyperGroup as (2, 3)-AntiHyperGroup (i.e., Axioms 2 and 3 are AntiAxioms).

Example 3.1: Let AG = $\mathbb{R}-\{0\}$ be the set of non-zero real number. Define for all $r_1, r_2 \in AG$ the AntiHyperoperation $\star$ as

$$r_1 \star r_2 = \begin{cases} \{r_1^2+1, r_1 - r_2\}, if\, r_1 \neq r_2 \\ \{r_1^2-2, 1\}, if\, r_1 = r_2. \end{cases}$$

The couple (AG, $\star$) is a (2,3) – AntiHyperGroup.

Proof: To see that (AG, $\star$) is a (2,3) - AntiHgypergroup, we proceed thus;

That $\star$ is closed as follows from the definition of $\star$

For all $r_1, r_2, r_3 \in AG$, we want to show that $r_1 \star (r_2 \star r_3) \neq (r_1 \star r_2) \star r_3$. We consider the following cases:

CASE1: When all the r_i are distinct.

$$r_1 \star (r_2 \star r_3) = r_1 \star \{r_2^2+1, r_2 - r_3\} = \{r_1 \star (r_2^2+1), r_1 \star (r_2 - r_3)\}$$
$$= \{r_1^2+1, r_1 - r_2^2 - 1, r_1 - r_2 + r_3\}.$$

Now consider

$$(r_1 \star r_2) \star r_3 = \{r_1^2+1, r_1 - r_2\} \star r_3 = \{(r_1^2+1) \star r_3, (r_1 - r_2) \star r_3\}$$
$$= \{(r_1^2+1)^2+1, r_1^2 - r_3 + 1, (r_1 - r_2)^2 + 1, r_1 - r_2 - r_3\}.$$

Hence, $r_1 \star (r_2 \star r_3) \neq (r_1 \star r_2) \star r_3$.

CASE 2: When the r_i are not distinct

$$r_1 \star (r_2 \star r_3) = r_1 \star \{r_2^2 - 2, 1\} = \{r_1 \star (r_2^2 - 2), r_1 \star 1\} = \{r_1^2 + 1, r_1^2 - r_2^2 + 2, r_1 - 1\}.$$

Now consider

$$(r_1 \star r_2) \star r_3 = \{r_1^2 - 2, 1\} \star r_3 = \{(r_1^2 - 2) \star r_3, 1 \star r_3\} = \{(r_1^2 - 2)^2 + 1, r_1^2 - 2 - r_3, 2, 1 - r_3\}.$$

Hence, $r_1 \star (r_2 \star r_3) \neq (r_1 \star r_2) \star r_3$.

Accordingly $\star$ is AntiAssociative .

4. 4. For AntiQuasiHypergroup. For all

$$x \in AG, x \star AG \neq AG \star x \neq AG.$$

To see this, consider

$$\begin{aligned} x \star AG &= x \star \{h : h \in AG \, and \, h \neq x\} = \{x \star h : h \in AG \, and \, h \neq x\} \\ &= \{\{x^2 + 1, x - h\} : h \in AG \, and \, h \neq x\} \neq AG \end{aligned}.$$

$$x \star AG = \{x \star h : h \in AG \, and \, h = x\} = \{x^2 - 2, 1\} \neq AG.$$

Similarly, it can be show that $AG \star x \neq AG$. Hence, (AG, $\star$) is a [2,3] - AntiHgypergroup.

Example 3.2: Let AG be a set of real numbers defined for all $x, y \in AG$ the AntiHyperoperation $\circ$ by

$$x \circ y = \begin{cases} \left\{ \dfrac{x+y}{2}, x/y \right\}, if\ x \neq y, x, y \neq 0 \\ \left\{ \dfrac{1}{x^2+1} \right\}, if\ otherwise. \end{cases}$$

Then (AG, $\circ$) is a (2,3) – AntiHyperGroup.

Proof: It follows same approach as Example 3.1.

Example 3.3. Let AG be a set of real numbers. Define for all $x, y \in AG$ an AntiHyperoperation $\circ$ given by

$$x \circ y = \left\{ x^2 + x + 1, x^2 - 1 \right\}.$$

Then (AG, $\circ$) is a (2,3) – AntiHyperGroup.

Proof:

1. For $\circ$ is closed by definition.
2. For AntiAssociativity,

$$x \circ (y \circ z) = x \circ \left\{ y^2 + y + 1, y^2 - 1 \right\} = \left\{ x \circ (y^2 + y + 1), x \circ (y^2 - 1) \right\} = \left\{ x^2 + x + 1, x^2 - 1 \right\}.$$

And

$$(x \circ y) \circ z = \left\{ x^2 + x + 1, x^2 - 1 \right\} \circ z = \left\{ (x^2 + x + 1) \circ z, (x^2 - 1) \circ z \right\}$$
$$= \left\{ (x^2 + x + 1)^2 + x^2 + x + 2, (x^2 - 1)^2 + x^2 \right\}.$$

This implies $x \circ (y \circ z) \neq (x \circ y) \circ z$.

3. For AntiQuasiHypergroup. For all

$$x \in AG, x \circ AG \neq AG \circ x \neq AG.$$

To see this, consider

$$x \circ AG = x \circ \{y : y \in AG\} = \{x \circ y : y \in AG\} = \{x^2 + x + 1, x^2 - 1\} \neq AG.$$

Similarly, it can be shown that $AG \circ x \neq AG$. Hence, (AG, $\circ$) is a (2,3) – AntiHyperGroup.

Example 3.4: Let AG = $\mathbb{R}$ be the set of non-zero real numbers. Define for all $r_1, r_2 \in AG$ the AntiHyperoperation $\star$ as

$$r_1 \star r_2 = \{r_1^2 + n, r_1^2 - n\}.$$

The couple (AG, $\star$) is a (2,3) – AntiHyperGroup.

Proof: The proof follows similar approach as Example 3.3.

Example 3.5: Let $AG = \mathbb{Z}$. Define for all $x, y \in AG$ the AntiHyperoperation $\circ$ given by

$$x \circ y = \{2 + x^2, 2 - x^2\}.$$

Then (AG, $\circ$) is a (2,3) – AntiHyperGroup.

Proof: This follows similar approach as the proof of Example 3.3.

Example 3.6: Let $AG = V_K$ be a vector space over a field K. For all $u, v \in AG$ define the AntiHyperoperation $\circ$ by

$$u \circ v = \{k(u^2 + 1) : k \in K\}.$$

Then (AG, $\circ$) is a (2,3) – AntiHyperGroup.

Proposition 3.7: Let $(AR_1, \star_1)\, and\, (AR_2, \star_2)$ be any two (2,3)–AntiHyperGroups. Let

$$AR_1 \times AR_2 = \{(x, y) : x \in AR_1\, and\, y \in AR_2\}.$$

For

$$x = (h_1, k_1), y = (h_2, k_2) \in AR_1 \times AR_2$$

define:

$x \star y = \left(\left(h_1 \star_1 h_2\right), \left(k_1 \star_2 k_2\right)\right).$

Then $\left(AR_1 \times AR_2, \star\right) is\, a \left(2,3\right)$ AntiHyperGroup.

Proof:

1. 1. $x = \left(h_1, k_1\right), y = \left(h_2, k_2\right) \in AR_1 \times AR_2$ then

$$x \star y = \left(h_1, k_1\right) \star \left(h_2, k_2\right) = \left(\left(h_1 \star_1 h_2\right), \left(k_1 \star_2 k_2\right)\right) \subseteq AR_1 \times AR_2$$

$\because \left(h_1 \star_1 h_2\right)) \subseteq AR_1$ and $\left(k_1 \star_2 k_2\right)) \subseteq AR_2$

Hence ($AR_1 \times AR_2, \star$) is a Hypergroupoid.

2. Then for any triplets

$$\left(h_1, b_1\right), \left(h_2, b_2\right), \left(h_3, b_3\right) \in AR_1 \times AR_2$$

we have that

$$\left(\left(h_1, b_1\right) \star \left(h_2, b_2\right)\right) \star \left(h_3, b_3\right) = \left(\left(h_1 \star_1 h_2\right), \left(b_1 \star_2 b_2\right)\right) \star \left(h_3, b_3\right)$$

$$= \left(\left(h_1 \star_1 h_2\right) \star_1 h_3, \left(b_1 \star_2 b_2\right) \star_2 b_3\right)$$

$$\neq \left(h_1 \star_1 \left(h_2 \star_1 h_3\right) b_1 \star_2 \left(b_2 \star_2 b_3\right)\right) \left(\because AH1\, holds\, in\, AR_1\, and\, AR_2\right)$$

$$= \left(h_1, b_1\right) \star \left(\left(h_2 \star_1 h_3\right), \left(b_2 \star_2 b_3\right)\right)$$

$$= \left(h_1, b_1\right) \star \left(\left(h_2, b_2\right) \star \left(h_3, b_3\right)\right).$$

Hence AH2 holds in $AR_1 \times AR_2$

3. 3. For all $(h,q) \in holds\, in\, AR_1\, and\, AR_2$ we have that

$$(h,q) \star AR_1 \times AR_2 = (h,q) \star \{(h_1,q_1) : h_1 \in AR_1, q_1 \in AR_2\}$$

$$= \{(h,q) \star (h_1,q_1) : h_1 \in AR_1, q_1 \in AR_2\}$$

$$= \{(h \star_1 h_1),(q \star_2 q_1) : h_1 \in AR_1, q_1 \in AR_2\}$$

$$= (h \star_1 AR_1, q \star_2 AR_2)$$

$$\neq (AR_1 \star_1 h, AR_2 \star_2 q)(\because AH3\, holds\, in\, AR_1\, and\, AR_2)$$

$$= AR_1 \times AR_2 \star (h,q).$$

Accordingly, $(AR_1 \times AR_2, \star)\, is\, a\, (2,3)$ AntiHyperGroup.

Proposition 3.8: Let

$$(AR_1, \star_1)\, be\, a\, [2,3]\, AntiHyperGroup\, and\, let\, (AR_2, \star_2)$$

be any other type of AntiHyperGroups. Let

$$AR_1 \times AR_2 = \{(x,y) : x \in AR_1\, and\, y \in AR_2\}.$$

For

$$x = (h_1, k_1), y = (h_2, k_2) \in AR_1 \times AR_2$$

define

$$x \star y = ((h_1 \star_1 h_2),(k_1 \star_2 k_2)).$$

Then

$((AR_1 \times AR_2, \star)\, is\, not\, in\, gerneral\, a\, (2,3)$ AntiHyperGroup.

Proof: To proof this we consider the following cases:

1. Case 1. Let $(AR_1, \star_1)$ be (2, 3) AntiHyperGroup and let $(AR_2, \star_2)$ be a trivial AntiHyperGroup. Then $(AR_1 \times AR_2, \star)$ is a trivial AntiHyperGroup.
2. Case 2. Let $(AR_1, \star_1)$ be (2, 3) AntiHyperGroup and let $(AR_2, \star_2)$ be a non trivial classes of AntiHy-perGroup in which AH1 holds.
   a. Then $(AR_1 \times AR_2, \star)$ is a trivial AntiHyperGroup.
3. Case 3. Let $(AR_1, \star_1)$ be (2, 3) AntiHyperGroup and let $(AR_2, \star_2)$ be an AntiHyperGroup from the classes of AntiHyperGroup in which AH1 fails to holds.
   a. Then $(AR_1 \times AR_2, \star)$ is a (2; 3) AntiHyperGroup.

Hence, considering cases 1 to 3, we have the proof.

Proposition 3.9: Let

$$(AR_1, \star_1)\, be\, a\, [2,3]\, AntiHyperGroup\, and\, let\, (AR_2, \star_2)$$

be a classical Hypergroup/[2,3]-NeutroHypergrup. Let

$$AR_1 \times AR_2 = \{(x,y) : x \in AR_1\, and\, y \in AR_2\}.$$

For

$$x = (h_1, k_1), y = (h_2, k_2) \in AR_1 \times AR_2$$

define

$$x \star y = ((h_1 \star_1 h_2), (k_1 \star_2 k_2)).$$

Then

$((AR_1 \times AR_2, \star)$ *is not in gerneral a* $(2,3)$ AntiHyperGroup.

Proof: The proof follows similar approach as the proof of proposition 3.7.

Definition 3.10: Let AR be a (2, 3) AntiHyperGroup, a non-empty subset AK of AR is called a (2, 3) AntiSubhypergroup of AR if AK is itself a (2, 3) AntiHyperGroup.

Example 3.11: Let $(AR, \circ)$ be the (2, 3) AntiHyperGroup defined in Example 3.5 and let AK; AW be the set of even and odd integers respectively. Then (AK, $\circ$) and (AW, $\circ$) are (2, 3) AntiSubhypergroups of AG.

Example 3.12: Let (AG, $\star$) be the (2, 3) AntiHyperGroups in Example 3.4 and let AK be the set of positive real numbers. $r_1, r_2 \in AK$ define $\star$ as

$$r_1 \star r_2 = \{r_1^2 + n, r_1^2 - n\}, for\, r_{i's} > n \in \mathbb{N}.$$

Then the couple (AK, $\star$) is a (2; 3) AntiSubhypergroup of AG.

Proposition 3.13: Let AK; AW be any (2,3) AntiSubhypergoup of a (2,3) AntiHyperGroup AG. Then $AK \cup AW$ is a [2,3] AntiSubhypergroup of AG.

This proposition will be illustrated in the example given below.

Example 3.14: Let AK; AW be any (2,3) AntiSubhypergoup of a (2,3) AntiHyperGroup AG=Z defined in 3.11 $AK \cup AW = AG$. Hence $(AK \cup AW, \circ)$ is a (2,3) – AntiSubhypergroup of AG .

Remark 3.15: It should be noted that the intersection of (2,3) AntiSubhypergroups is not necessarily a (2,3) AntiSubhypergroup. For instance in example 3.11 $AK \cap AW = \varnothing$.

Definition 3.16: Let (AW, $\star$) be any (2,3)-AntiSubhypergoup of a (2,3) AntiHyperGroup (AR, $\star$). The quotient AR / AW is defined by the set $\{g \star AW : g \in AR\}$.

Proposition 3.17: Let (AR, $\star$) be any (2,3)-AntiHypergoup and let (AW, $\star$) be a (2,3) AntiSubhyperGroup of AR. For

$$a \star AW, b \star AW \in AR / AW$$

with $a, b \in AR$, let $\circ$ be a hyperopration defined on AR/AW by

$$a \star AW \circ b \star AW = \{c \star d \mid c \in a \star AW, d \in b \star AW\}.$$

Then (AR/AW, $\circ$) is a [2.3] AntiHypergroup which we call a (2,3)-AntiQuotientHyperGroup.

Proof: By the definition of '$\circ'$ it is easy to see that (AR/AW, $\circ$) is a hypergroupoid.

$$\left(a \star AW \circ b \star AW\right) \circ c \star AW = \left\{x \star y \mid x \in a \star AW, y \in b \star AW\right\} \circ c \star AW$$

$$= \left\{\left(x \star y\right) \star z \mid x \in a \star AW, y \in b \star AW, z \in c \star AW\right\}$$

$$\neq \left\{x \star \left(y \star z\right) \mid x \in a \star AW, y \in b \star AW, z \in c \star AW\right\}\left(\because AH2\, holds\, in\, AR\right).$$

$$= a \star AW \circ \left\{\left(y \star z\right) \mid y \in b \star AW, z \in c \star AW\right\}$$

$$= a \star AW \circ \left(b \star AW \circ c \star AW\right).$$

Finally for all $a \star AW \in AR / AW$, we have

$$a \star AW \circ AR / AW = \left\{c \star d \mid c \in a \star AW, d \in AR / AW\right\}$$

$$\neq AR / AW\}\left(\because AH3\, holds\, in\, AR\right)$$

Similarly we can show that

$$AR / AW \circ a \star AW \neq AR / AW.$$

Hence

$\left(\left(AR / AW, \circ\right)\right) is\, a \left(2,3\right)$ AntihyperGroup.

CONCLUSION

In this paper, we have introduced the concept of AntiHyperGroups. Specifically, a class of An-tiHyperGroup called (2,3) AntiHyperGroup was studied and some of its basic properties were presented. It was shown that the intersection of (2,3) AntiSubhypergrpoups is not a (2, 3) Anti-Subhypergroup and their union is a (2, 3) AntiHyperGroup. Also, It was shown that the Cartesian product of a (2, 3) AntiHyperGroup and any other class of AntiHyperGroup is not a (2, 3) Anti-HyperGroup in general, but its product with a classical Hypergroup or a (2, 3) NeutroHyperGroup is a (2, 3) AntiHyperGroup. Further, it was shown that a (2, 3) AntiHyperGroup factored by a (2, 3) AntiSubhypergroup is a (2, 3) AntiHyperGroup. In our future work, we hope to investigate other classes of AntiHyperGroups and study the relationship between them.

REFERENCES

Agboola, A. A A. (2013). *Introduction to Neutrosophic Hypergroups. ROMAI J.*, *9*(2), 1–10.

Agboola, A. A. A. (2014). On Neutrosophic Canonical Hypergroups and Neutrosophic Hyperrings. *Neutro-sophic Sets and Systems.*, *2*, 34–41.

Agboola, A. A. A. (2020a). On Finite NeutroGroups of Type-NG. *International Journal of Neutrosophic Science*, *10*(2), 84–95.

Agboola, A. A. A. (2020b). On Finite and In nite NeutroRings of Type-NR. *International Journal of Neutrosophic Science*, *11*(2), 87–99.

Agboola, A. A. A. (2020c). Introduction to NeutroGroups. *International Journal of Neutrosophic Science*, *6*(1), 41–47.

Agboola, A. A. A. (2020d). Introduction to NeutroRings. *International Journal of Neutrosophic Science*, *7*(2), 62–73.

Agboola, A. A. A. (2020e). Introduction to AntiRings. *Neutrosophic Science and System*, *36*, 293–307.

Agboola, A. A. A., & Akinleye, S. A. (2015). Neutrosophic Hypervector Spaces. *ROMAI Journal*, *11*, 1–16.

Agboola, A.A.A., Ibrahim, M.A., & Adeleke, E.O. (2020). Elementary Examination of NeutroAlgebras and AntiAlgebras viz-a-viz the Classical Number Systems. International Journal of Neutrosophic Science, 4(1), 16–19.

Al-Tahan, M., Smarandache, F., & Davvaz, B. (2021). NeutroOrderedAlgebra: Applications to Semigroups. *Neutrosophic Sets and Systems*, *39*, 133–147.

Davvaz, B. (2012). Chemical Hyperalgebra: Dismutation Reactions. *MATCH Commun. Math.Comput.Chem.*, *67*, 55–63.

Davvaz, B., & Leoreanu-Fotea, V. (2007). *Hyperring Theory and Applications*. International Academic Press.

Ibrahim, M. A., & Agboola, A. A. A. (2020a). Introduction to NeutroVector Space I. *Neutrosophic Science and System*, *36*, 329–350.

Ibrahim, M. A., & Agboola, A. A. A. (2020b). Introduction to Neutrosophic Hypernear-rings. *International Journal of Neutro-sophic Science*, *10*(1), 9–22.

Ibrahim, M. A., & Agboola, A. A. A. (2020c). Introduction to NeutroHyperGroups. *Neutrosophic Sets and Systems*, *38*, 15–32.

Rezaei, A., & Smarandache, F. (2020). On Neutro-BE-algebras and Anti-BE-algebras (revisited). *International Journal of Neutrosophic Science*, *4*(1), 8–15.

Smarandache, F. (2019). Introduction to NeutroAlgebraic Structures. In *Advances of Standard and Nonstandard Neutro-sophic Theories* (pp. 240–265). Pons Publishing House.

Smarandache, F. (2020a). Introduction to NeutroAlgebraic Structures and AntiAlgebraic Structures (revisited). *Neutrosophic Sets and Systems*, *31*, 1–16. doi:10.5281/zenodo.3638232

Smarandache, F. (2020b). NeutroAlgebra is a Generalization of Partial Algebra. *International Journal of Neutrosophic Science*, *2*(1), 8-17.

Chapter 5
Neutro Boolean Algebra:
An Extension of Classical Boolean Algebra

Somen Debnath
https://orcid.org/0000-0003-3585-0868
Umakanta Academy, India

ABSTRACT

The main focus of the chapter is to generalize classical Boolean algebra by introducing the neutro Boolean algebra. In many real-life situations, researchers experience some evidence where the classical algebraic structure failed to address the issues. This leads to a new field of research, known as neutro algebraic structure and the anti-algebraic structure proposed in 2019. Like this, neutro Boolean algebraic structure is introduced to take care of the issues where the Huntington postulates of Boolean algebra are partially true and partially false. More precisely, in neutro Boolean algebra, the outcome of any given two elements under logical operations has three possibilities, such as appurtenance, non-appurtenance, or indeterminate. Finally, some fundamental definitions, theorems, and properties based on neutro Boolean algebra are proposed.

INTRODUCTION

In 1998 (Smarandache, 1998) Smarandache introduced a new branch of philosophy called neutrosophy. After that, he has been proposed neutrosophy as a multiple-valued logic (Smarandache, 2002). It brings a new idea, axiom, theorem, postulate, or attribute that may be partly true, partly false, and indeterminate. So, it provides a

DOI: 10.4018/978-1-6684-3495-6.ch005

new level of thinking to the researchers to present the indeterminate, inconsistent, incomplete information logically. After that, neutrosophic set (NS), pointed out in (Smarandache, 2005), is defined as a part of neutrosophy and it is taken care of the issues that cannot be explained with the help of fuzzy set (Zadeh, 1965) and intuitionistic fuzzy set (Atanassov, 1986). The non-restriction condition in NS makes it a powerful and flexible tool in solving various problems involving uncertainty, ambiguity, vagueness, redundancy, ignorance, etc. arising from social science, philosophy, science and technology, computer science, artificial intelligence, etc.

It is well known that, in classical algebra, the basic compositional laws (or axioms) such as commutative, associative, distributive, etc defined on a set are always well defined. But, in the context of scientific knowledge, the classical algebraic structure alone can't address all possible issues that we encounter in various practical applications. There are some notions of knowledge where the axioms defined on a set are partially true (partially false) or totally false. To handle these, neutro algebraic and anti algebraic structures are introduced (Smarandache, 2019). In (Smarandache, 2019), Smarandache revisited the notion of neutro algebra and anti algebra. In the context of the new algebraic structure, an idea or an axiom or a proposition or a theorem, etc say denoted by $\langle X \rangle$. Then, its neutral notion is denoted by $\langle neut\, X \rangle$ and its opposite notion is denoted by $\langle anti\, X \rangle$. Thus, in neutro algebra, there exists at least one neutro operation that is true for some elements, indeterminate for other elements, false for other elements. Some other research works inspired by neutro algebra and anti algebra are presented in the form: neutro algebra as an extension of partial algebra(Smarandache, 2020). Agboola et al. investigate the neutro and anti algebras in the viz-a-viz classical number systems i.e real numbers (Agboola, Ibrahim & Adeleke, 2020).The notion of AntiGroups is presented in(Agboola, 2021). The formal presentation of NeutroGroups and study their properties are described in (Agboola, 2020). Smarandache introduced the neutro quadruple numbers and their laws (Smarandache, 2015). NeutroQuadrupleGroups are introduced in(Smarandache, Rezaei, Agboola, Jun, Borzooei, Davvaz, Borumand Saeid, Akram, Hamidi & Mirvakili, 2021), Ibrahim et al. (Ibrahim & Agboola, 2020) defined the NeutroHyperGroups. By introducing the NeutroAbelianGroup, Neutrosemigroup, and NeutroDistributivity, NeutroRings and their properties are developed in (Agboola, 2020). Moreover, an application on neutro algebra based on PROSPECTOR function is proposed in (Jiménez, Mayorga, Ubilla & Hernández, 2021). Hamidi et al. (Smarandache & Hamidi, 2020) investigated the neutro-BCK-algebra. The two new notions such as neutro-CI-algebras and anti-CI-algebras are proposed in (Smarandache, Rezaei & Kim, 2020). The neutro nilpotent groups and their properties are studied in(Mohammadzadeh & Rezaei, 2020), etc.

George Boole is the pioneer of introducing the modern symbolic algebra of logic that is known as Boolean algebra. There is a likeness among set theory, logic, and Boolean algebra in the sense that the algebra of sets and the algebra of statements are almost the same. Several researchers defined Boolean algebra in different ways. In this work, we mainly focus on the definition of Boolean algebra that is given by Huntington (Huntington, 1904). It is based on Mathematical two-valued logic where the values of the variables are 0(for false) and 1(for true). For any two elements or variables in a Boolean algebra, logicians only can perform logical operations like conjunction $(\wedge)$, disjunction $(\vee)$, negation $(\sim)$, etc. Mathematicians also relate the notion of Boolean algebra practically like on and off, pass and fail, friend and enemy, head and tail(unbiased coin), open and closed, logical **And** and **OR,** or many more instances that are inspired from our real world. Later on, Shannon (Shannon, 1949) developed the switching circuit as a practical implication of Boolean algebra. To get insight into the core concept of Boolean algebra, the author point out some works given in (Barnett, 2011; Muller, 1954; Hanf, 1975; Hailperin, 1981).

In the current study, the author's main aim is to develop a new notion known as neutro Boolean algebra (NBA), as an extension of classical Boolean algebra (CBA). Motivating from Huntington's postulates for Boolean algebra, in this chapter Huntington's postulates for NBA are newly introduced. In NBA, decision-makers are not sure whether the operations defined on a set under the NBA environment are well-defined or not. After developing the neutro Boolean algebraic structure, the author give some fundamental theorems and properties with proper justification.

To validate the present study, the author gives a comparative analysis in Table 1.

SOME FUNDAMENTAL CONCEPTS

In this section, the author discusses some basic definitions of neutro algebra, anti algebra, partial algebra, and Boolean algebra.

Definition 2.1 (Smarandache, 2020)

An algebra that has at least one partial operation, and all axioms are classical (i.e true for all elements) is called a partial algebra.

Definition 2.2 (Smarandache, 2019)

Let Ω be a universal set and Γ be a non-empty space in Ω. Suppose $\langle \mathrm{K} \rangle$ be an object (attribute, axiom, concept, theorem, proposition, etc) defined on Γ. Then, the set Γ can be split into three exhaustive regions [two opposite $\langle \mathrm{K} \rangle$ and $\langle anti\, \mathrm{K} \rangle$, and one neutral $\langle neut\, \mathrm{K} \rangle$ between them]. A neutro algebra is regarded as an algebra that has at least one neutro operation or one neutro axiom i.e. an axiom that is true

Table 1. A comparative study between the proposed concept and the existing concepts

Algebra of Sets	Algebra of Logic	Boolean Algebra	Neutro Boolean Algebra
The algebra of sets is mainly concerned with some fundamental laws and identities associated with sets and set-theoretic operations that are established for executing calculations. Some of them given as $X \cup Y = Y \cup X$ $X \cap Y = Y \cap X$ $X \cup \varnothing = X, X \cap U = X$ $X \cup X^c = U, X \cap X^c = \varnothing$ etc. In the classical set, just like 0 and 1, we use $\varnothing$ and U respectively. Similarly, the operators $\cup$ and $\cap$ are used instead of + and $\times$ etc. Thus, the algebra of sets is nothing but the modern way of expressing human thoughts.	Computation and reasoning are the two important aspects of mathematical logic. It is a symbolic re-presentation of a statement or proposition. In logical algebra, our main purpose is to use the notion of algebraic operations logically to justify the validity of the proposed statement with proper reasoning. We use symbols like p, q, r, s, t, etc. to represent a simple statement. Various logical connectives are used to make compound statements. $p \vee q = q \vee p$ $p \wedge q = q \wedge p$ $\sim p$ (not p) $p \rightarrow q$ (If p then q) $p \leftrightarrow q$ (p if and only if q)	In Boolean algebra, the values of the variables are either true (1) or false (0) i.e. they are binary variables. Here we perform logistic operations on variables. The values of the variables are called the truth values of a variable. For n variables, there are exactly 2^n truth values. Therefore, it is an essential tool for framing computer programming. It is a mathematical structure that is denoted by a sextuple $\langle B,+,.,',0,1\rangle$. In Boolean algebra, $a + a = a, a.a = a$ $a + 1 = a, a.0 = 0$ $1' = 0, 0' = 1$	If the axioms, propositions, or theorems in a Boolean algebra are not well-defined i.e., they are partly true and partly false or totally false, then it allows us an opportunity to investigate another algebraic structure known as neutro Boolean algebraic structure or neutro Boolean algebra. To address the complex interpretation of binary variables, the neutro Boolean algebra(NBA) is introduced. So to realize the importance and practical implication of the neutro Boolean algebra, the present study is adopted. In an NBA, there is no guarantee to make a belief that the variables are binary. For a detailed study, see Section 3.

for some elements, indeterminate for other elements, and false for other elements. In neutro algebra, there is no anti operation or anti axiom. Neutro algebra is a generalization of partial algebra.

Definition 2.3 (Smarandache, 2019)

Algebra or an algebraic structure that has at least one anti operation or anti axiom is called an anti algebra or an anti algebraic structure.

Thus, $\langle algebra, neutro\, algebra, anti\, algebra \rangle$ formed a neutrosophic triplet structure that is capable to handle the indeterminacy, uncertainty, ambiguity, vagueness, or inconsistency that researchers address in various human knowledge.

Definition 2.4 (Ibrahim & Agboola, 2020)

Let X be a non-empty finite or infinite set and A be a subset of X. If p classical operations (laws or axioms) are defined on A where $p \geq 1$, then there will be $\left(2^p - 1\right)$ neutro algebras and $\left(3^p - 2^p\right)$ anti algebras.

Definition 2.5 (Huntington, 1904)

A Boolean algebra is a mathematical system consisting of a non-empty set B together with two binary operations $'+'$ $\left(\vee\right)$ and $'\bullet'$ $\left(\wedge\right)$ and one unary operation ' / ' defined on the set B such that the following postulates(Huntington, 1904) are satisfied:

P$_{1:}$ For every $a, b \in$ B, $a + b = b + a$ and $a \bullet b = b \bullet a$ (commutative laws)

P$_{2:}$ For every $a, b, c \in$ B, $a \bullet \left(b + c\right) = a \bullet b + a \bullet c$ and $a + \left(b \bullet c\right) = \left(a + b\right) \bullet \left(a + c\right)$ (Distributive laws)

P$_{3:}$ There exist two distinct elements $0, 1 \in$ B such that $a + 0 = 0 + a = a$ and $a \bullet 1 = 1 \bullet a = a$ (Existence of identity)

P$_{4:}$ For every $a \in$ B, there exists an element $a^/$ such that $a + a^/ = a^/ + a = 1$ and $a \bullet a^/ = a^/ \bullet a = 0$ (Existence of complement)

Thus, the Boolean algebra is generally denoted by 6-tuple< B, +,•, / , 0, 1>.

Examples

Let $\Gamma = \left\{\varnothing, S, X, X^/\right\}$, where X be any non-empty subset of a universal set S. Then the system $\left\langle \Gamma, \cup, \cap \right\rangle$ represents a Boolean algebra.

Let D_{21} be the set of all natural numbers which are divisors of 21, i.e., $D_{21} = \left\{1, 3, 7, 21\right\}$. Then, for any $x, y \in D_{21}$, suppose $x + y = lcm(x, y)$; $x.y = \gcd(x, y)$ and $x^/ = \dfrac{21}{x}$, $\left\langle D_{21}, +, ., /, 1, 21 \right\rangle$ is a Boolean algebra.

Note: For D_n, if n is divisible by any square integer greater than 1, then $\left\langle D_n, +, ., /, 1, n \right\rangle$ is not a Boolean algebra.

HUNTINGTON POSTULATES FOR NEUTRO BOOLEAN ALGEBRA AND PROPERTIES

In this section, first, the author gives the notion of neutro Boolean algebra based on Huntington postulates and then describe its properties.

Definition 3.1

Let NB a non-empty set together with two binary operations $\oplus$ and $\odot$ and one unary operation$'$ defined on NB. Then the system containing NB is said to be a neutro Boolean algebra(NBA) if the following postulates hold:

NB$_1$: The binary operations $\oplus$ and $\odot$ are said to be neutro commutative if there exists at least one couple $\left(x, y\right) \in$ NB such that $x \oplus y = y \oplus x$ and $x \odot y = y \odot x$.

And there exists at least one couple $\left(r, s\right) \in$ NB such that $x \oplus y \neq y \oplus x$ and $x \odot y \neq y \odot x$.

NB$_2$: $\oplus$ and $\odot$ are said to be neutro distributive if there exists at least one triplet $\left(x, y, z\right) \in$ NB such that

$$a \odot \left(b \oplus c\right) = a \odot b + a \odot c$$

and

$$a \oplus \left(b \odot c\right) = \left(a \oplus b\right) \odot \left(a \oplus c\right).$$

And there exists at least one triplet $\left(r, s, t\right) \in$ NB such that

$$r \odot \left(s \oplus t\right) \neq r \odot s \oplus r \odot t$$

and

$$a \oplus \left(b \odot c\right) \neq \left(a \oplus b\right) \odot \left(a \oplus c\right).$$

NB$_3$: There exists at least one element $x \in$ NB for which the neutro identity elements $0_I, 1_I \in$ NB such that $x \oplus 0_I = x$ and $x \odot 1_I = x$.

And there exists at least one element $r \in$ NB for which the neutro identity elements $0_I, 1_I \in$ NB such that $r \oplus 0_I \neq r$ and $r \odot 1_I \neq r$.

NB$_4$: There exists at least one element $a \in$ NB for which a unique element $a' \in$ NB such that $a \oplus a' = 1_I$ and $a \odot a' = 0_I$.

And there exist at least one element r for which two elements $r_1, r_2 \in$ NB such that $r \oplus r_1 = r \oplus r_2 = 1_I$ and $r \odot r_1 = r \odot r_2 = 0_I$

Therefore, the neutro Boolean algebra is usually denoted by the 6-tuple < NB, $\oplus$, $\odot$, ', 0_I, 1_I>.

Example 3.1.1: Let NB= $\{1,2,3,6,9,10,15,45,90\}$ represents the set of some positive divisors of 90. We consider the following operators on NB as follows: a$\oplus$ b=LCM (a, b), a$\odot$b=HCF(a, b), a' = $\frac{90}{a}$, where $a,b \in$NB.

Now, NB$_1$: 2 $\oplus$3=6 $\in$NB, 2 $\odot$3=1$\in$NB. But, 6$\oplus$9=18$\notin$ NB, 10$\odot$ 15=5$\notin$ NB.

Therefore, $\oplus$ and $\odot$ are neutro binary operations.

NB$_2$: $2\oplus(3\odot 6)$=6=$(2\oplus 3)\odot(2\oplus 6)$, $2\odot(3\oplus 6) = 2 = (2\odot 3)\oplus(2\odot 6)$

Again, $6\oplus(9\odot 90) = 6\oplus 9 = 18 \notin$ NB

Therefore, operations are neutro distributive.

NB$_3$: $2\oplus 1 = 2, 2\odot 90 = 2$. Again, $6\oplus 3 = 6$, $2\odot$ 10=2

Therefore, there exists neutro identity elements under the neutro binary operations.

NB$_4$: $2\oplus 2' = 2\oplus \frac{90}{2} = 2\oplus 45 = 90 \neq 1$.

Again,

$$2\odot 2' = 2\odot \frac{90}{2} = 2\odot 45 = 1 \neq 90$$

$$6\oplus 6' = 6\oplus \frac{90}{6} = 6\oplus 15 = 3$$

Therefore, a' exists for some $a \in$NB.

Thus, $\langle NB, \oplus, \odot, ' \rangle$ is a NBA.

Definition 3.2: An NBA with only single element is called a trivial NBA or a degenerate NBA.

Definition 3.3: The set NB= $\{0_I, 1_I\}$ together with two neutro binary operations $\oplus$ and $\odot$ and one neutro Boolean complement ' is called a two-element NBA.

Definition 3.4: Let X be a NBA and $a, b \in X$ such that $a \odot b = a \ or \ a \odot b = b$ i.e., $a \tilde{\leq} b \ or \ a \tilde{\geq} b$. The relation $\tilde{\leq}$ or $\tilde{\geq}$ is called a partial order relation in a NBA.

Theorem 3.2 (Principle of Neutro Duality): Let NB be any neutro Boolean algebra and A be a proposition derived from the axioms of NB. Then, the dual of A sometimes derivable from the axioms of NB and sometimes not derivable (or indeterminate) from the axioms of NB. That is, by interchanging the operations $\oplus$ and $\odot$ and the elements 0_I and 1_I in a neutro Boolean algebra, then its postulates are partly valid and partly invalid.

Theorem 3.3: Let < NB, $\oplus$, $\odot$, ' ,0_I,1_I> be a NBA. Then,

The identity elements 0 and 1 are unique for some $x \in$ NB and for some $y \in$ NB the elements 0 and 1 are not unique (or indeterminate or does not exists)

For some $x \in$ NB there exists a unique element $x' \in$ NB and for some $y \in$ NB, there exists $y_1', y_2' \in$ NB such that $y_1' \neq y_2'$.

For some $x \in$ NB, $(x')' = x$ or $(x')' \neq x$.

$0_I' = 1_I$ or $0_I' \neq 1_I$ and $1_I' = 0_I$ or $1_I' \neq 0_I$

For some $x \in$ NB, $x \oplus 1_I = 1_I$ or $x \oplus 1_I \neq 1_I$ and $x \odot 0_I = 0_I$ or $x \odot 0_I \neq 0_I$

For some $x \in$ NB, $x \oplus x = x$ or $x \oplus x \neq x$ and $x \odot x = x$ or $x \odot x \neq x$

For some couple $(x, y) \in$ NB, $x \odot (x \oplus y) = x$ or $x \odot (x \oplus y) \neq x$ and $x \oplus (x \odot y) = x$ or $x \oplus (x \odot y) \neq x$.

For some triplet $(x, y, z) \in$ NB, $x \oplus (y \oplus z) = (x \oplus y) \oplus z$ or $x \oplus (y \oplus z) \neq (x \oplus y) \oplus z$ and $x \odot (y \odot z) = (x \odot y) \odot z$ or $x \odot (y \odot z) \neq (x \odot y) \odot z$.

For some couple $(x, y) \in$ NB, $(x \oplus y)' = x' \odot y'$ or $(x \oplus y)' \neq x' \odot y'$ and $(x \odot y)' = x' \oplus y'$ or $(x \odot y)' \neq x' \oplus y'$

Proof

1. Let 0_{I_1} and 0_{I_2} be two zero elements in NB. Then, there exists an element $x \in NB$ such that $x \oplus 0_{I_1} = x \oplus 0_{I_2} = x$. Thus, $0_{I_1} = 0_{I_2}$.

And for an element $y \in NB$, there does not exist a zero element or the zero elements are not unique or indeterminate.

By the neutro duality principle, we claim the same conclusion for unit element 1_I in NB.

2. For $x \in NB$, there exists a unique element $x^{'}$ such that $x \oplus x^{'} = 1_I$ and $x \odot x^{'} = 0_I$.

And for $y \in \mathrm{NB}$, there exists two elements $y_1^{/}$ and $y_2^{/}$ such that $y_1^{/} \neq y_2^{/}$ i.e., the complement of y is not defined.

3. For any $x \in \mathrm{NB}$,

$$x^{/} \oplus x = 1_I \text{ or} \neq 1_I \text{ and } x^{/} \odot x = 0_I \text{ or } \neq 0_I$$

This gives us $\left(x^{/}\right)^{/} = x \text{ or } \neq x$.

4. For some $x \in \mathrm{NB}$, we have $x \oplus 0_I = x$ and $x \odot 1_I = x$. By replacing x by 0_I and 1_I, it can be written as $1_I \oplus 0_I = 1_I$ and $0_I \odot 1_I = 0_I$. It is evident that, $1^{'}_I = 0^{'}$ and $0_I^{/} = 1_I$.

And for some $y \in$ NB, $y \oplus 0_I \neq y$ and $y \odot 1_I = y$. Thus, $1^{'}_I \neq 0_I$ and $0_I^{/} \neq 1_I$.

5. For any $x \in$ NB,

$$x \oplus 1_I = (x \oplus 1_I) \odot 1_I \text{ or } \neq (x \oplus 1_I) \odot 1_I$$
$$= (x \oplus x') \odot (x \oplus 1_I) \text{ or } \neq (x \oplus x') \odot (x \oplus 1_I)$$
$$= x \oplus (x' \odot 1_I) \text{ or } \neq x \oplus (x' \odot 1_I)$$
$$= x \oplus x' \text{ or } \neq x \oplus x'$$
$$= 1_I \text{ or } \neq 1_I$$

Similarly,

$$x \odot 0_I = 0_I \text{ or } x \odot 0_I \neq 0_I$$

6. For any $x \in \mathrm{NB}$,

$$x \oplus x = (x \oplus x) \otimes 1_I \text{ or } \neq (x \oplus x) \otimes 1_I$$
$$= (x \oplus x) \otimes (x \oplus x') \text{ or } \neq (x \oplus x) \otimes (x \oplus x')$$
$$= x \oplus (x \otimes x') \text{ or } \neq x \oplus (x \otimes x')$$
$$= x \oplus 0_I \text{ or } \neq x \oplus 0_I$$
$$= x \text{ or } \neq x$$

Similarly, $x \odot x = x$ or $x \odot x \neq x$

7. For any $x, y \in \mathrm{NB}$,

$$x \odot (x \oplus y) = (x \oplus 0_I) \odot (x \oplus y) \text{ or } \neq (x \oplus 0_I) \odot (x \oplus y)$$
$$= x \oplus (0_I \odot y) \text{ or } \neq x \oplus (0_I \odot y)$$
$$= x \oplus 0_I \text{ or } \neq x \oplus 0_I$$
$$= x \text{ or } \neq x$$

Similarly, $x \oplus (x \odot y) = x$ or $x \oplus (x \odot y) \neq x$.

8. For any $x, y, z \in \mathrm{NB}$,

$$\left(x \oplus \left(y \oplus z\right)\right) \odot x' = x' \odot \left(x \oplus \left(y \oplus z\right)\right) \text{or} \neq x' \odot \left(x \oplus \left(y \oplus z\right)\right)$$
$$= \left(x' \odot x\right) \oplus x' \odot \left(y \oplus z\right) \text{or} \neq \left(x' \odot x\right) \oplus x' \odot \left(y \oplus z\right)$$
$$= 0 \oplus x' \odot \left(y \oplus z\right) \text{or} \neq 0 \oplus x' \odot \left(y \oplus z\right)$$
$$= x' \odot \left(y \oplus z\right) \text{or} \neq x' \odot \left(y \oplus z\right)$$

Again,

$$\left(\left(x \oplus y\right) \oplus z\right) \odot x' = x' \odot \left(x \oplus y\right) \oplus x' \odot z \text{ or} \neq x' \odot \left(x \oplus y\right) \oplus x' \odot z$$
$$= \left(x' \odot x \oplus x' \odot y\right) \oplus x' \odot z \text{ or} \neq \left(x' \odot x \oplus x' \odot y\right) \oplus x' \odot z$$
$$= x' \odot \left(y \oplus z\right) \text{or} \neq x' \odot \left(y \oplus z\right)$$

Thus,

$$x \oplus \left(y \oplus z\right) = \left(x \oplus y\right) \oplus z \text{ or } x \oplus \left(y \oplus z\right) \neq \left(x \oplus y\right) \oplus z$$

Similarly,

$$x \odot \left(y \odot z\right) = \left(x \odot y\right) \odot z \text{ or } x \odot \left(y \odot z\right) \neq \left(x \odot y\right) \odot z$$

9. For any $x, y \in \text{NB}$,

$$\left(x \oplus y\right) \oplus \left(x' \odot y'\right) = \left(\left(x \oplus y\right) \oplus x'\right) \odot \left(\left(x \oplus y\right) \oplus y'\right) \text{or} \neq \left(\left(x \oplus y\right) \oplus x'\right) \odot \left(\left(x \oplus y\right) \oplus y'\right)$$
$$= \left(\left(x \oplus x'\right) \oplus y\right) \odot \left(\left(y \oplus y'\right) \oplus x\right) \text{or} \neq \left(\left(x \oplus x'\right) \oplus y\right) \odot \left(\left(y \oplus y'\right) \oplus x\right)$$
$$= \left(1_I \oplus y\right) \odot \left(1_I \oplus x\right) \text{or} \neq \left(1_I \oplus y\right) \odot \left(1_I \oplus x\right)$$
$$= 1_I \odot 1_I \text{ or} \neq 1_I \odot 1_I$$
$$= 1_I \text{ or} \neq 1_I \text{..........}(A)$$

Again,

$$\left(x \oplus y\right) \odot \left(x^{/} \odot y^{/}\right) = \left(\left(x \oplus y\right) \odot x^{/}\right) \odot y^{/} \text{ or } \neq \left(\left(x \oplus y\right) \odot x^{/}\right) \odot y^{/}$$
$$= \left(\left(x^{/} \odot x\right) \oplus \left(x^{/} \odot y\right)\right) \odot y^{/} \text{ or } \neq \left(\left(x^{/} \odot x\right) \oplus \left(x^{/} \odot y\right)\right)$$
$$= \left(x^{/} \odot y\right) \odot y^{/} \text{ or } \neq \left(x^{/} \odot y\right) \odot y^{/}$$
$$= x^{/} \odot \left(y \odot y^{/}\right) \text{or } \neq x^{/} \odot \left(y \odot y^{/}\right)$$
$$= 0_{I} \text{ or } \neq 0_{I} \ldots\ldots\ldots (B)$$

From (A) and (B), $\left(x \oplus y\right)^{'} = x^{'} \odot y^{'}$ or $\left(x \oplus y\right)^{'} \neq x^{'} \odot y^{'}$

Similarly, it can be proven that

$$\left(x \odot y\right)^{'} = x^{'} \oplus y^{'} \text{or } \left(x \odot y\right)^{'} \neq x^{'} \oplus y^{'}$$

CONCLUSION

The main purpose to present this chapter is to disclose a new algebraic structure known as NBA. An NBA is introduced to take care of those issues that cannot be handled by the classical Boolean algebra (CBA). Unlike CBA, in NBA, state of a variable may be express in more than two ways. Due to the involvement of uncertainty present in various disciplined, researchers and Mathematicians are realized the fact that the two-valued logic is no more an authentic tool to address an uncertain issue. So, NBA is the extension of CBA. Some important properties and results based on NBA are proposed in this chapter.

FUTURE SCOPE

After the rigorous study of NBA, in the future, there is a huge opportunity for the researchers to develop another structure known as Anti-Boolean algebra and thus to establish the triplet $\left\langle Boolean\, algebra, neutro\, boolean\, algebra, anti\; Boolean\, algebra \right\rangle$ that helps us to introduce neutro Boolean group, anti-Boolean group, neutro Boolean ring, anti-Boolean ring, neutro Boolean ideal, neutro Boolean BCK-algebra, neutro Boolean CI-algebra, neutro Boolean homomorphism, neutro Boolean isomorphism, etc. Also, for practical implications, the NBA is useful to design the indeterministic(default) switching circuits.

ACKNOWLEDGMENT

The author is grateful to the anonymous referees for their valuable suggestions.

REFERENCES

Agboola, A. A. A. (2020a). Introduction to neutrogroups. *International Journal of Neutrosophic Science*, *6*, 41–47.

Agboola, A. A. A. (2020b). Introduction to neutrorings. *International Journal of Neutrosophic Science*, *7*, 62–73.

Agboola, A. A. A. (2021). Introduction to antigroups. *International Journal of Neutrosophic Science*, *12*, 71–80.

Agboola, A. A. A., Ibrahim, M. A., & Adeleke, E. O. (2020). Elementary examination of neutroalgebras and antialgebras viz-a-viz the classical number systems. *International Journal of Neutrosophic Science*, *4*, 16–19. doi:10.54216/IJNS.040102

Atanassov, K. T. (1986). Intuitionistic fuzzy sets. *Fuzzy Sets and Systems*, *20*(1), 87–96. doi:10.1016/S0165-0114(86)80034-3

Barnett, J. H. (2011). *Applications of Boolean algebra: Claude Shannon and circuit design*. Available from the webpage http://www. cs. nmsu. edu/historical-projects

Hailperin, T. (1981). Boole's algebra isn't Boolean algebra. *Mathematics Magazine*, *54*, 173–184.

Hanf, W. (1975). The Boolean algebra of logic. *Bulletin of the American Mathematical Society*, *81*(3), 587–589. doi:10.1090/S0002-9904-1975-13747-5

Huntington, E. V. (1904). Sets of independent postulates for the algebra of logic. *Transactions of the American Mathematical Society*, *5*(3), 288–309. doi:10.1090/S0002-9947-1904-1500675-4

Ibrahim, M. A., & Agboola, A. A. A. (2020). Introduction to neutrohyperGroups. *Neutrosophic Sets and Systems*, *38*, 15–32.

Jiménez, D. S., Mayorga, J. A. V., Ubilla, M. E. R., & Hernández, N. B. (2021). Neutroalgebra for the evaluation of barriers to migrants' access in Primary Health Care in Chile based on PROSPECTOR function. *Neutrosophic Sets and Systems*, *39*, 1–9.

Mohammadzadeh, E., & Rezaei, A. (2020). On neutronilpotent groups. *Neutrosophic Sets and Systems*, *38*, 33–40.

Muller, D. E. (1954). Application of Boolean algebra to switching circuit design and to error detection. *Transactions of the IRE Professional Group on Electronic Computers*, *3*, 6-12. 10.1109/IREPGELC.1954.6499441

Shannon, C. E. (1949). The synthesis of two-terminal switching circuits. *The Bell System Technical Journal*, *28*(1), 59–98. doi:10.1002/j.1538-7305.1949.tb03624.x

Smarandache, F. (1998). *A unifying field in logics. Neutrosophy: neutrosophic probability, set and logic*. American Research Press.

Smarandache, F. (2002). Neutrosophy, a new branch of philosophy, multiple valued logic. *International Journal*, *8*, 297–384.

Smarandache, F. (2005a). Neutrosophic set-a generalization of the intuitionistic fuzzy set. *International Journal of Pure and Applied Mathematics*, *24*, 287–297.

Smarandache, F. (2015b). Neutrosophic quadruple numbers, refined neutrosophic quadruple numbers, absorbance law, and the multiplication of neutrosophic quadruple numbers. *Neutrosophic Sets and Systems*, *10*, 96–98.

Smarandache, F. (2019a). Introduction to neutroalgebraic structures and antialgebraic structures (revisited). *Neutrosophic Sets and Systems*, *31*, 1–16.

Smarandache, F. (2019b). *Introduction to neutroalgebraic structures, in advances of standard and nonstandard neutrosophic theories*. Pons Publishing House.

Smarandache, F. (2020). Neutroalgebra is a generalization of partial algebra. *International Journal of Neutrosophic Science*, *2*, 8–17. doi:10.54216/IJNS.020103

Smarandache, F., & Hamidi, M. (2020). Neutro-BCK-Algebra. *International Journal of Neutrosophic Science*, *8*, 110–117.

Smarandache, F., Rezaei, A., Agboola, A. A. A., Jun, Y. B., Borzooei, R. A., Davvaz, Borumand Saeid, A., Akram, M., Hamidi, M., & Mirvakili, S. (2021). *On Neutro Quadruple Groups*. Infinite Study.

Smarandache, F., Rezaei, A., & Kim, H. S. (2020). *A New Trend to Extensions of CI-algebras*. Academic Press.

Zadeh, L. A. (1965). Fuzzy sets. *Information and Control*, *8*(3), 338–353. doi:10.1016/S0019-9958(65)90241-X

Chapter 6
NeutroAlgebra and NeutroGeometry for Dealing the Heteroclinic Patterns

Prem Kumar Singh
https://orcid.org/0000-0003-1465-6572
Gandhi Institute of Technology and Management, Visakhapatnam, India

ABSTRACT

Recently, dealing the heteroclinic data and its pattern finding has been considered as one of major concern for the researchers. It becomes more complex in the case of dynamic or periodic uncertainty in data. These types of data form heteroclinic orbit, which changes the space from equilibrium to non-equilibrium and vice versa. In this case, several parallel lines can be drawn from the given point to measure the indeterminacy, which is used to be find flight paths. It means these types of data do not follow the fifth law. To deal with them, following the NeutroAlgebra, a new geometry called as NeutroGeometry is studied in this chapter to characterize them based on Euclidean, non-Euclidean, and indeterminate zones.

INTRODUCTION

The geometrical visualization of any data sets based on Euclidean postulates from past centuries (Birkhoff, G., D.1932). However Russell (Russell, B.1897) opposed the fifth Euclidean postulates as failure of spherical representation. Lobachevsky (Lobachevsky, N. 2010) provided hyper space where more than two parallel lines can be drawn from the given point with reference to the given line as shown in Figure 1.

DOI: 10.4018/978-1-6684-3495-6.ch006

Figure 1. There are many lines that can be drawn parallel to BC from A

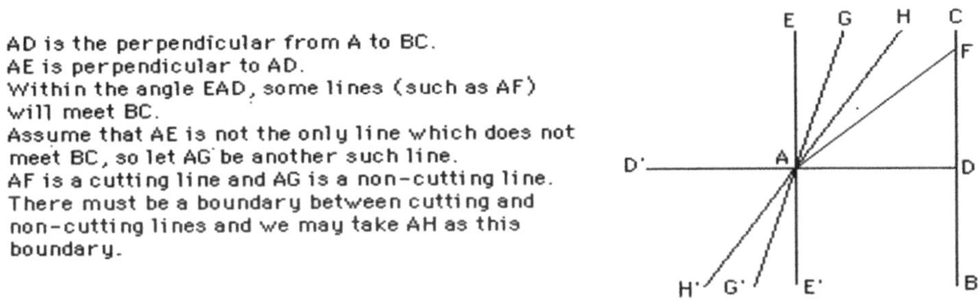

To understand this concept, let us consider an example of flights which crosses same point parallel way given to a reference line (Sala et al., 2019). Second example can be taken as any balloon played by any child. The balloon used to be flat whenever given to child for the play. It means the triangle drawn on uninflated make 180^0 angle. The inflated balloon leaves the flat surface and become spherical where 180angle concept of Euclidean fails as shown in Figure 2 (Calvo, n.d.). It means the Euclidean space fails in dealing the data structures that having hierarchical ordering or tree representation (Popov, M., R. 1996; Wikipedia, 2022a).There are many examples exists which cannot be deal with Euclidean geometry like Periodic strings, hetero clinic orbit (Wikipedia, 2020). To deal with these types of data sets recently NeutroGeometry was introduced by Smarandache (Smarandache, F. 2021). The current paper tries to put forward effort for dealing the NeutroGeometry data and processing.

Figure 2. The Non-Euclidean geometry to draw triangle in spherical space.

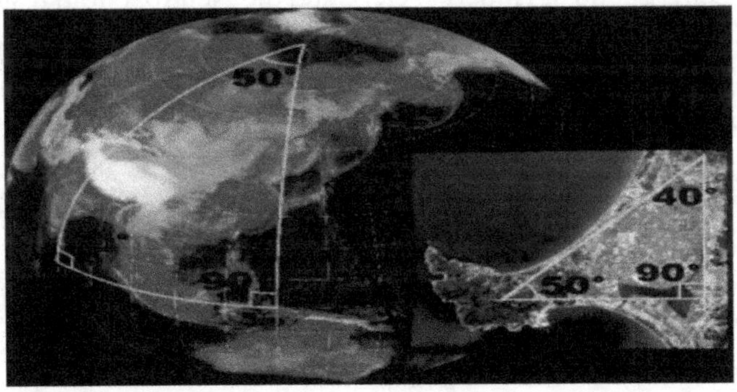

The Smarandache Geometries (SG) are mixed types of Non-Euclidean geometries(Coxeter, H.S.M.1942) and sometimes partially Non-Euclidean and partially Euclidean ones, as in our real geometric spaces that are hybrid and heterogeneous (Bhattacharya, S. 2004).An SG is a geometry which has at least one Smarandache denied axiom (1969), i.e., an axiom that behaves in at least two different ways within the same space, i.e., validated and invalided, or only invalided but in multiple distinct ways (Linfan, M. 2006; Kuciuk, L. &Antholy, M. 2005). The reason is dealing equilibrium in hetero clinic data sets is computationally expensive tasks because of its bipolar conditionality as discussed by Singh (Singh, P., K. 2021e). However many real life examples exist for Non-Euclidean geometry, like surface growth, epitaxial growth of nanostructures, 3D printing, growth of plants, and cell mobility requirement(James, A., W. 2005; Tong, 2019, Cetica, 2021; Wikipedia, 2022b)). Even two or more human can thought in same quanta information as parallel discussed by Turiyam set (Singh, P., K. 2021a,b).In 2019 Smarandache(Smarandache, F. 2019) has generalized the classical Algebraic Structures to NeutroAlgebraic Structures (or NeutroAlgebras) {whose operations and axioms are partially true, partially indeterminate, and partially false}as extensions of Partial Algebras, and to AntiAlgebraic Structures (or AntiAlgebras) {whose operations and axioms are totally false} and on 2020 he continued to develop them for precise analysis of these data sets(Smarandache, F. 2020).One knows that in the classical algebraic structures, all operations are totally (100%) well-defined, and all axioms are totally (100%) true, but in our world, in many cases these restrictions are too strict, since there are many more things that only partially verify some operations or some laws. He then founded the NeutroGeometry and AntiGeometry(Smarandache, F. 2021) that are analogous, on a geometric space, to the NeutroAlgebra and AntiAlgebra, in the following way: "While the Non-Euclidean Geometries resulted from the total negation of only one specific axiom (Euclid's Fifth Postulate), the AntiGeometry results from the total negation of any axiom and even of more axioms from any geometric axiomatic system (Euclid's, Hilbert's, etc.), and the NeutroAxiom results from the partial negation of one or more axioms [and no total negation of no axiom] from any geometric axiomatic system. Therefore, the NeutroGeometry and AntiGeometry are respectively alternatives and generalizations of the Non-Euclidean Geometries."It plays major role while dealing understanding theory of Relativity proposed by Einstein[2]. Same time the hyperbolic geometry can be used while dealing with the field of Topology and ternary relations. Let us consider a set X whose elements are represented as circular format. One of suitable examples is the clock watch $X=\{1, 2, 3, 4, 5, 6, 7, 8, 9, 10, 11, 12\}$. In this case, the ternary relation can be defined as $X^3=X\times X\times X$ where the relation $R(x_1,x_2,x_3)$exists iff traversal of $x_1$to$x_3$in a clockwise must passes throughx_2. Let us consider, the ternary relation $R(8, 12, 4)$ and $R(12, 8, 4)$. The first one holds whereas the second ternary relation does not hold. In this precise characterization of time (1, 0, 0), anti-time (0, 0, 1) and neutro-time (t, i, f) is major issues till now. Some of the researchers tried to represent them via fourth dimension using Turiyam context (Singh, P., K. 2021a, b). This paper focused on dealing these types of uncertainty and its representation via NeutroGeometry with an illustrative example.

Remaining part of the paper is structured as follows: Section 2 provides background about Non-Euclidean geometry and its examples. Section 3 provides examples of NeutroGeometry and its Illustration in Section 4. Section 5 provides conclusions followed by Acknowledgements and references.

BACKGROUND

There are many examples of non-Euclidean geometry like Surface growth, fractal and brain image analysis etc. Even the earth is spherical which is non-Euclidean. In this section, some examples will be given to understand the Non-Euclidean geometry:

Example 1: Most of the time researchers have confusion about geometrical representation of Euclidean and Non-Euclidean. Same time their difference while measuring the triangle. To resolve this issue Figure 3 represents the elliptical, Euclidean and Hyperbolic space for data representation. It can be observed that the Euclidean space contains sum of triangle as 180^0 whereas in other space this condition fails.

Figure 3. The difference between Hyperbolic, Euclidean and Elliptic geometry

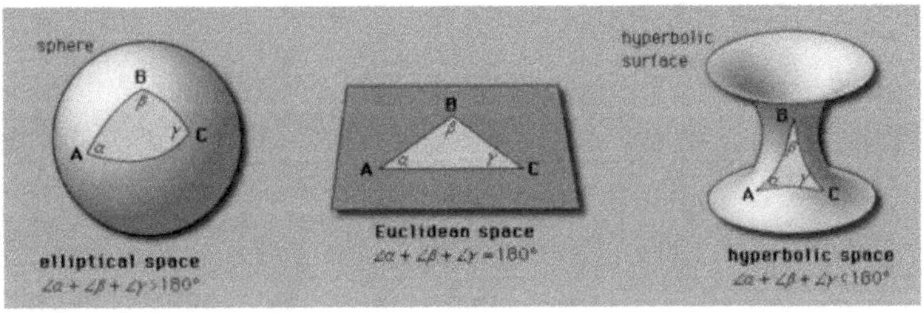

Example 2: The problem arises with the expert while understanding the elliptical, Euclidean, Hyperbolic space that which type of data is Non-Euclidean. Is there any mathematical graph exists which is non-Euclidean. To understand this issue the Figure 4 represents the Non-Euclidean geometrical space[4].

Figure 4. An example of Non-Euclidean geometry

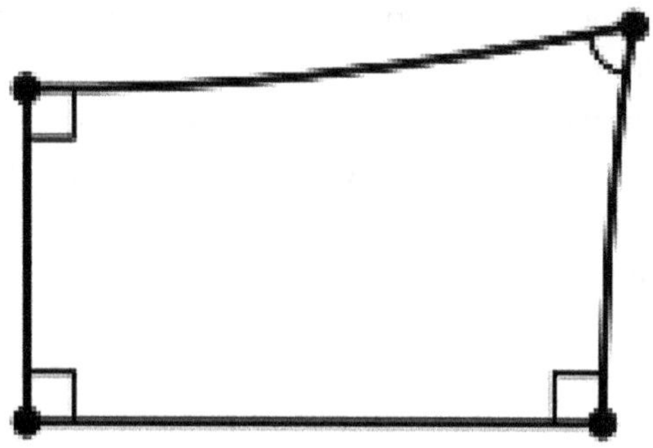

Example 3: The problem arises how to characterize this geometry is Euclidean or Non-Euclidean. To solve this issue nature of the failure of fifth postulates of Euclidean geometry need to be focused i.e. nature of parallel lines from the given point as shown in Figure 5(Bhattacharya, S. 2004).

Figure 5. An example of Non- Euclidean geometry

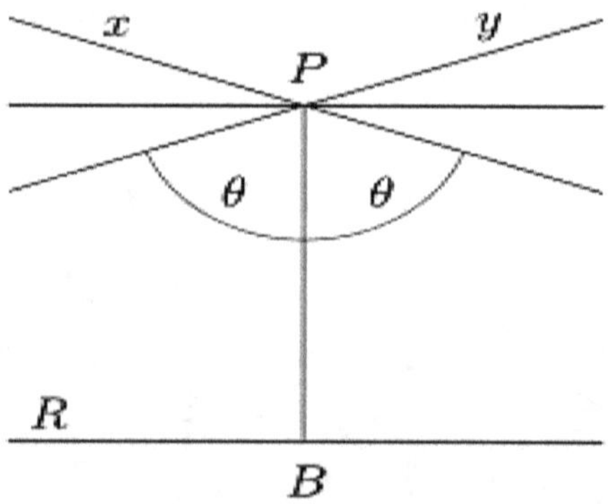

Example 4: The next problem arises when two or more experts characterize the parallel lines and its uncertainty before infinity as parallel, not-parallel, or uncertain shown in Figure 5. It creates conflict among them about Euclidean and Non-Euclidean space. It can be written in form of context as shown in Table 1. The characterization of this cognition based on Euclidean, non-Euclidean or uncertain is major issues for graphical visualization (Singh, P., K. 2021c; Singh, P. K., 2022a).To deal with it NeutroGeometry and its concept is indeed required as discussed by Smarandache recently (Smarandache, F.

Table 1. The expert (y_1) and (y_2) opinion about lines x, y and their parallel condition shown in Figure 5

Contradiction degree	0	0.5	0.8		0.5
Cognition about Lines	Parallel	Not parallel	Uncertain		*Euclidean*
Expert (y_1)	0.6	0.4	0.8		0.7
Expert (y_2)	0.7	0.5	0.9		0.6

2021). Hence this paper focused on studying the concept of NeutroGeometry in the next section.

NEUTROGEOMETRY

This section provides the characterization of Non-Euclidean geometry and its data representation using NeutroAlgebra and its properties. Let us suppose any data sets where the fifth postulates fail means consider as Non-Euclidean geometry. The fifth postulate fails means there are many lines can be drawn from the given point as parallel to the reference line shown in Figure 6.

It means the NeutroGeometry exists while dealing the Flight path as shown in Figure 7. The problem arises while characterization of flight path based on direct path, Navaid path and NeutroPath as follows:

1. Direct-Path (1, 0, 0): The flight covers same path define for one or both route endpoints for the given latitude and longitude.
2. Navaid-Path (0, 0, 1): There is no path that can exist between given two ends.

Figure 6. The non-Euclidean geometry and its condition

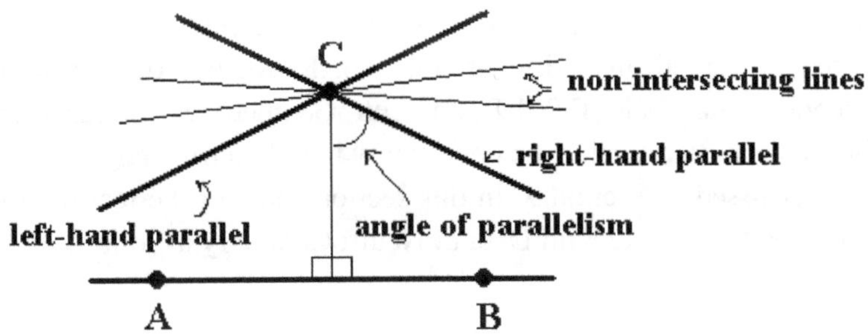

Figure 7. The NeutroGeometry representation of flight path from A to B

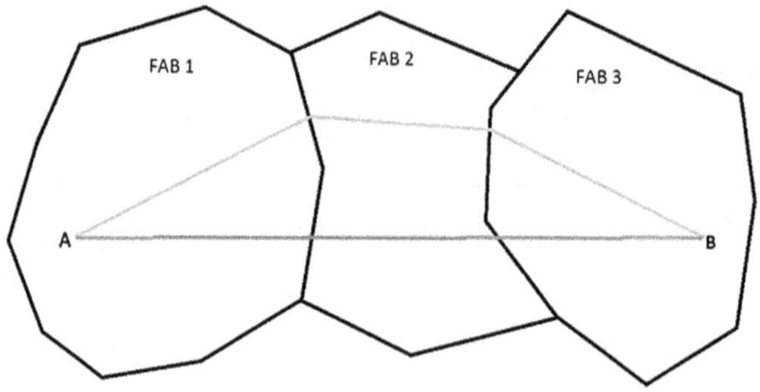

3. Neutro-Path (t, i, f): The flight covers neutropath and stay at some place in case of fuel filling, security checking or weather conditions which can be decided by critical point and angle.

In this case the NeutroGeometry will help while adequate characterization of uncertainty in given data sets and its analysis using NeutroAlgebra. It defined a function $f: X \rightarrow Y$ where three possibilities exist (Smarandache, F.):

1. A well-defined mapping exists among X and Y i.e.t.
2. A undefined or outer-defined mapping exists among X and Y i.e. f.
3. It is unknown that the mapping exists or not and what is the value for the mapping among X and Y i.e.i.

Hence the NeutroGeometry is useful while dealing the Non-Euclidean data sets and its characterization.

ILLUSTRATIONS OF NEUTROGEOMETRY

Recently, dynamic uncertainty and its representation is considered as one of the major issues (Singh, P., K. 2021a, b, c, d, e, f). The problem arises when the data sets are Non-Euclidean as discussed by Singh (2022a). To achieve this goal NeutroAlgebra(Smarandache, F. 2019) and NeutroGeometry(Smarandache, F. 2021) was established. To characterize the uncertainty based on true, false and uncertain concepts as discussed in Section 3. In this section some of the real life examples will be discussed for dealing with Data of NeutroGeometry as follows:

1. **Reviewer Cognition analysis**(Singh, P., K. 2021d)): Recently, a problem was addressed that each reviewer thinks in different way about the given paper and its quality. Same time the reviewer opinion used to change after the revision. It might be possible that two or more reviewer comments exist parallel for the given paper and its methodology which is Non-Euclidean. The problem arises with editors while acceptance (1, 0, 0), rejection (0, 0, 1) and indeterminacy condition of paper. To deal with these types of data sets NeutroGeometry is required.
2. **Domain based expert Analysis**(Singh, P. K. 2020): One of the most prominent problem is finding some of the potential researchers in the given field for brain drain analysis. This problem arises when many monkeys and ghost researchers publish several papers in the given areas. In this case, the precise analysis of researchers of the given domain as expert, non-expert, or Neutro-Expert (partially expert and partially not)the concept of NeutroGeometry is helpful. The researchers having defined papers in the given area can be considered as expert (1, 0, 0). The researchers who do not have any papers in the given area can be considered as an anti-Expert (0, 0, 1). The researchers who published papers in undefined areas, multiple areas with many collaborations then they can be defined as Neutro-experts (t, i, f).
3. **NeutroLaws**(Smarandache, F. 2021): It is well known that every citizen contradicts about the law and order of their country. In democratic country like India the laws differ from people to people, gender to gender, and politician to politician. The Law applied fully for the person who opposes the government i.e. Law(1, 0, 0), the law is not applied with the person who supports the given political party i.e. Anti-Law (0, 0, 1) and for some people it is undefined i.e. NeutroLaw (t, i, f). It applied in job, recruitment and promotions also where people get job easily in case he/she supports the given government, given caste, or given regions. In this case they useor bring many laws or incorporate any things to give the job to particular person i.e. Concept(1, 0, 0). The given person does not support the government, does not belongs to same case or region they try to manipulate the system for not recruiting the particular person i.e. Anti-Concept(0, 0, 1). The third case is undefined when they do not know anything about the given person or might happen that due to political mileage they can give job to anyone. These types of case is undefined can be considered as Neutro-Concept(t, i, f).
4. Molecule Structure (Cetica, 2021): There are several fields like chemistry, biology, botany and Anisotropy[8]data contains lots of Non-Euclidean geometry as shown in Figure 8. One of the reasons is non-Euclidean data and its interpretation contains inductive biasness and depends on intuition as well as type of data

and its changing structure as shown in Figure 8 (Singh. P. K. 2021f; Singh, P. K., 2022a).

Figure 8. The Non-Euclidean data in several fields.

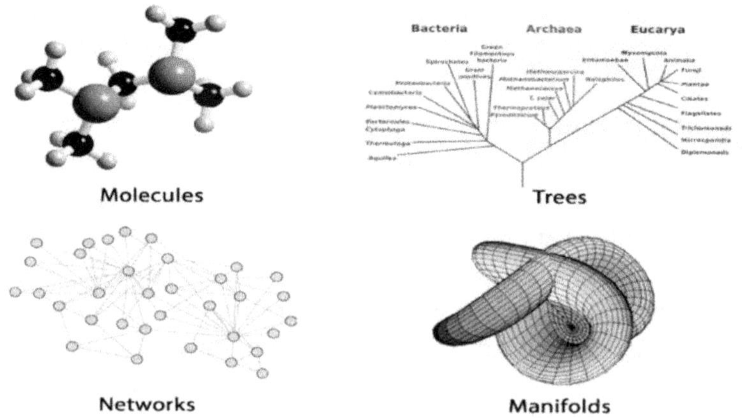

5. **Fractals:** The fractal analysis is another example of non-Euclidean geometry. It contains self repeating patterns which allow infinite number of zoom in and zoom out the given data for investigating the given pattern. In this case, there are many heteroclinic or homoclinic orbit exists. One of the reasons is the fractal contains internal homothety that any given object can be dilated or shrink but its original shape does not change. These types of geometry used to found in leaf, snowflakes, broccoli, brain image as well as algae data base also (Pandey, L. K., Ojha, K. K., Singh, P.K., Singh, C. S., Dwivedi, S., &Bergey, E.A, 2016). It can be represented using the Mandelbrot set (Wikipedia, 2022c) having orbit for the critical z=0 for values $x \in X$. It can be represented using the quadratic map as follows: $Z_{n+1} = Z_n^2 + X$.
6. **Citation Analysis (Wikipedia, 2022d)**: The citation analysis is another issue for the research communities. The problem arises when the technical paper, review paper, Scopus paper, Google scholar paper, Journal paper or conference papers citation used to consider as equal(Singh, P. K. 2020). In this case, the characterization of citation based on true, false and uncertain regions and analyzes them in NeutroAlgebra. A paper cited by the domain expert can be considered as citation (1, 0, 0). A paper cited by non-expert, irrelevant to given topic, retracted paper from the given journal, posthumous author publications citation can be considered as Anti-Citation(0, 0, 1). In case the paper is cited with interdisciplinary, self-citations, influenced citations, honors citations

etc. can be considered as Neutro-citation(t, i, f). The citation analysis and its visualization can be accomplished using the NeutroGeometry.

7. **NeutroLaw:** The human cognition also not like Euclidean geometry which discussed by Turiyam set (Singh, P., K. 2021a,b). The human cognition and its interpretation changes from Masculine, Feminine, and Transgender. It used to observe in case of any act done by men, woman or transgender as there is vague IT Act defined in country like India (Kapoor P, Prem Kumar Singh, Ch. Aswani Kumar 2020). The interpretation used to changes from gender to gender. In case men will complaint about sexual harassment none of the people consider his situation. The women will complaint about sexual assault everyone will believe immediately without proof also. In case the transgender will complaint about the same issue many will not listen. The characterization of human cognition and its judgement is totally uncertain which can be represented precisely by NeutroLaw (t, i, f). It can be applied in similar manner for IT Act also. However to accomplish this goal the extensive properties of NeutroAlgebra(Smarandache, F. 2020; 2021) and its graphical analytics (Singh, P., K.2021c, d) is required for the given phase of time (Singh, P.,K., 2022b).

FUTURE RESEARCH DIRECTIONS

In near future, author will focus on graphical representation of Non-Euclidean data and its characterization based on NeutroGeometry will be discussed for various applications.

CONCLUSION

This paper focused on studying Non-Euclidean Geometry data sets and its uncertainty measurement (which are particular cases of the AntiGeometry, while the NeutroGeometries are alternatives of the Non-Euclidean Geometry) with a real-life examples. To characterize the data sets and its uncertainty in true, false and uncertain regions, independently for adequate analysis of pattern exists beyond Euclidean geometry.

A method is proposed in Section 3 using the NeutroAlgebra and NeutroGeometry. The concept, anti-concept and neutro-concept are also discussed for data analysis in AntiGeometry and NeutroGeometry in Section 4.

FUNDING

Author declares that, there is no funding for this paper.

CONFLICT OF INTERESTS

The author declares there is no conflict of interest.

ACKNOWLEDGMENT

Author thanks Dr. M. Al-Tahan for the invitation on this topic.

REFERENCES

Bhattacharya, S. (2004). *A model to a Smarandache Geometry*. http://fs.unm.edu/ModelToSmarandacheGeometry.pdf

Birkhoff, G. D. (1932). A Set of Postulates for Plane Geometry (Based on Scale and Protractors*). Annals of Mathematics*, 33.

Calvo, S. C. (n.d.). *The Shape of Space: The Beginning of Non-Education Geometry*. Retrieved from https://www.encyclopedia.com/science/encyclopedias-almanacs-transcripts-and-maps/shape-space-beginning-non-euclidean-geometry

Cetica, M. (2021). Fractals - A Non-Euclidean Geometry. *Marco Cetica*. Retrieved from https://marcocetica.com/posts/non_euclidean_geometry/

Coxeter, H.S.M. (1942). *Non-Euclidean Geometry*. University of Toronto Press.

James, A. W. (2005). *Hyperbolic Geometry* (2nd ed.). Springer.

Kapoor, P., & Prem Kumar Singh, C. (2020). IT act Crime Pattern Analysis Using Regression and Correlation Matrix. *Proceedings of 8th International Conference on Reliability, Infocom Technologies and Optimization*, 1102-1106. doi: 10.1109/ICRITO48877.2020.9197835

Kuciuk, L., & Antholy, M. (2005). An introduction to the Smarandache geometries. *JP Journal of Geometry & Topology*, *5*(1), 77–81. http://fs.unm.edu/IntrodSmGeom.pdf

Linfan, M. (2006). *Pseudo-Manifold geometries with applications*. Cornell University. https://arxiv.org/pdf/math/0610307

Lobachevsky, N. (2010). Pangeometry (A. Papadopoulos, Ed. & Trans.). European Mathematical Society.

Pandey, L. K., Ojha, K. K., Singh, P. K., Singh, C. S., Dwivedi, S., & Bergey, E. A. (2016). Diatoms image database of India (DIDI): a research tool. *Environmental Technology & Innovation, 5*, 148-160. doi:10.1016/j.eti.2017.02.005

Popov, M. R. (1996). The Smarandache Non-Geometry. *Abstracts of Papers Presented to the American Mathematical Society Meetings, 17*(3), 595.

Russell, B. (1897). *Introduction: An essay on the foundations of geometry*. Cambridge University Press.

Sala, F., Chami, I., Wolf, A., Gu, A., & Ré, C. (2019). *Into the Wild: Machine Learning In Non-Euclidean Spaces*. Stanford University. Retrieved from https://dawn.cs.stanford.edu/2019/10/10/noneuclidean/

Singh, P. K. (2020). Multi–granular based n–valued neutrosophic contexts analysis. *Granular Computing, 5*(3), 287-301. doi:10.1007/s41066-019-00160-y

Singh, P. K. (2021a). Turiyam set a fourth dimension data representation. *Zeitschrift für Angewandte Mathematik und Physik, 9*(7), 1821–1828. doi:10.4236/jamp.2021.97116

Singh, P. K. (2021b). Fourth dimension data representation and its analysis using Turiyam Context. *Journal of Computer and Communications, 9*(6), 222–229. doi:10.4236/jcc.2021.96014

Singh, P. K. (2021c). Single-valued Plithogenic graph for handling multi-valued attribute data its context. *International Journal of Neutrosophic Sciences, 15*(2), 98–112. doi:10.5281/zenodo.5482171

Singh, P. K. (2021d). Complex multi-–fuzzy context analysis at different granulation. *Granular Computing, 6*(1), 191–206. doi:10.100741066-019-00180-8

Singh, P. K. (2021e). Bipolarity in multi–way fuzzy context and its analysis using m–way granulation. *Granular Computing, 2021*. Advance online publication. doi:10.100741066-021-00277-z

Singh, P. K. (2021f). AntiGeometry and NeutroGeometry Characterization of Non-Euclidean Data Sets. *Journal of Neutrosophic and Fuzzy Systems, 1*(1), 24–33.

Singh, P. K. (2022a). Data with Non-Euclidean Geometry and its Characterization. *Journal of Artificial Intelligence and Technology, 2*(1), 3–8. doi:10.37965/jait.2021.12001

Singh, P. K. (2022b). Complex Plithogenic Set. *International Journal of Neutrosophic Sciences, 18*(1), 57–72.

Smarandache, F. (2019). Introduction to NeutroAlgebraic Structures and AntiAlgebraic Structures. In *Advances of Standard and Nonstandard Neutrosophic Theories*. Pons Publishing House. http://fs.unm.edu/AdvancesOfStandardAndNonstandard.pdf

Smarandache, F. (2020). NeutroAlgebra is a Generalization of Partial Algebra. *International Journal of Neutrosophic Science, 2*, 8–17.

Smarandache, F. (2021). NeutroGeometry&AntiGeometryare alternatives and generalizations of the Non-Euclidean Geometries. *Neutrosophic Sets and Systems, 46*, 456-476. http://fs.unm.edu/NSS/NeutroGeometryAntiGeometry31.pdf

Tong, F. (2019). What is Geometric Deep Learning? *Medium*. https://flawnsontong.medium.com/what-is-geometric-deep-learning-b2adb662d91d

Wikipedia. (2022a). Non-Euclidean Geometry. In *Wikipedia*. Retrieved from https://en.wikipedia.org/wiki/Non-Euclidean_geometry

Wikipedia. (2022b). Anisotrophy. In *Wikipedia*. https://en.wikipedia.org/wiki/Anisotropy

Wikipedia. (2022c). Mandelbrot Set. In *Wikipedia*. https://en.wikipedia.org/wiki/Mandelbrot_set

Wikipedia. (2022d). Citation Analysis. In *Wikipedia*. https://en.wikipedia.org/wiki/Citation_analysis

Chapter 7
AntiHyperstructures:
Applications to AntiSemihypergroups and AntiHypergroups

Mariam Hariri
https://orcid.org/0000-0002-8995-5457
Zayed University, UAE

Osman M. Anis
Lebanese International University, Lebanon

Madeleine Ali Al-Tahan
Lebanese International University, Lebanon

Darjan Karabašević
University Business Academy in Novi Sad, Serbia

ABSTRACT

AntiSophication is the process through which AntiAlgebraic structures can be generated from any given classical algebraic structures. The concept of AntiHyperStructures has been studied in previous work. The aim of this chapter is to study AntiSemihypergroups, AntiQuasihypergroups, AntiSubSemihypergroups, AntiHypergroups, and their properties. Some new definitions and theorems will be presented and supported by illustrative examples. In addition, the authors prove the existence of an infinite number of AntiSemihypergroups and AntiHypergroups.

DOI: 10.4018/978-1-6684-3495-6.ch007

INTRODUCTION

Smarandache (1998) initiated the field of Neutrosophy, which is a new branch of Philosophy that studies the origin, nature, and scope of neutralities and has applications in many different fields of science. Inspired by this new field, Smarandache (2020) constructed for the first time the neutrosophic triplet corresponding to the Algebraic Structures, which is: (Algebraic Structure, NeutroAlgebraic Structure, AntiAlgebraic Structure).

NeutroAlgebraicStructures and AntiAlgebraicStructures can be generated from classical algebraic structures through processes called NeutroSophication and AntiSophication respectively. In the classical algebraic structures that are constructed on a given set, the axioms are totally true for all elements in the set. However, in NeutroAlgebraicStructures (or NeutroAlgebras), axioms are partially true, partially indeterminate, and partially false. While in AntiAlgebraicStructures (or AntiAlgebras), axioms are totally false for all elements of the given set.

This motivated several researchers to study NeutroAlgebraicStructures and AntiAlgebraicStructures and present new findings and results. Agboola et al. (2020) examined the concepts of NeutroAlgebras and AntiAlgebras viz-a-viz the Classical Number Systems, and the concepts of AntiRings, Rezaei and Smarandache (2020) introduced the concept of Neutro-BE-algebras and Anti-BE-algebras, and Al-Tahan et al. (2021) studied NeutroOrderedAlgebra considering its applications to semigroups and NeutroHyperStructures.

Hyperstructures are a generalization of classical algebraic structures. The concept of hyperstructures was introduced in 1934 by F. Marty at the eighth Congress of Scandinavian Mathematicians (Marty, 1934). Hyperstructures can be constructed from a non-empty set together with a hyperoperation. The class of algebraic hyperstructures is larger than that of algebraic structures where the operation on two elements in the latter is again an element, whereas the hyperoperation of two elements in the first class is a non-void set. Ibrahim and Agboola (2020) extended the concept of NeutroAlgebraicStructures to define NeutroHyperstructure and NeutroHypergroups. This chapter specifically focuses on AntiSemihypergroups, AntiQuasihypergroups, AntiSubSemihypergroups, and AntiHypergroups. It investigates some of their properties and provides non-trivial illustrative examples. The chapter is concluded by suggesting ideas for future research.

BACKGROUND

In this section, some definitions and examples about Algebraic Hyperstructures and NeutroHyperstructures are presented. For more details about hyperstructure theory, it is recommended to refer to (Corsini, 1993), (Davvaz, 2013), (Davvaz & Leoreanu-Fotea, 2008), (Vougiouklis, 1991), and (Vougiouklis, 1994).

Definition 1 (Corsini, 1993): Let H be a non-empty set and $\circ : H \times H \to P^*(H)$ be a hyperoperation. The couple (H, $\circ$) is called a hypergroupoid. For any two non-empty subsets A and B of H and $x \in H$; we define

$$A \circ B = \bigcup_{a \in A, b \in B} a \circ b, A \circ x = A \circ \{x\} and\, x \circ B = \{x\} \circ B$$

Definition 2 (Corsini, 1993): A hypergroupoid (H, $\circ$) is called a semihypergroup if the associative axiom is satisfied. i.e, for all a, b, c of H we have $(a \circ b) \circ c = a \circ (b \circ c)$ which means

$$\bigcup_{u \in a \circ b} u \circ c = \bigcup_{v \in b \circ c} a \circ v$$

Definition 3 (Corsini, 1993: A hypergroupoid (H, $\circ$) is called a quasihypergroup if for all a of H we have $a \circ H = H \circ a = H$. This condition is also called the reproduction axiom.

Definition 4 (Corsini, 1993): A hypergroupoid (H, $\circ$) which is both a semihypergroup and a quasihypergroup is called a hypergroup.

Definition 5 (Vougiouklis, 1994): A hypergroupoid $(H, \circ)$ is called an H_v -semigroup if the weak associative axiom is satisfied. i.e.,

$$(x \circ (y \circ z)) \cap ((x \circ y) \circ z) \neq \varnothing$$

for all $x, y, z \in H$.

Definition 6 (Al-Tahan et al., 2021): Let A be any non-empty set and " $\bigcirc$ " be a map. Then " $\bigcirc$ " is called a NeutroHyperoperation on A if some (or all) of the following conditions hold in a way that $(T, I, F) \notin \{(1,0,0), (0,0,1)\}$.

1. There exist $x, y \in A$ with $x \bigcirc y \subseteq A$. (This condition is called degree of truth, "T").
2. There exist $x, y \in A$ with $x \bigcirc y \subsetneq A$. (This condition is called degree of falsity, "F").
3. There exist $x, y \in A$ with $x \bigcirc y$ is indeterminate in A. (This condition is called degree of indeterminacy, "I").

The couple $(A,\bigcirc)$ is called a NeutroHypergroupoid.

Definition 7 (Al-Tahan et al., 2021): Let A be any non-empty set and "$\bigcirc$" be a map. Then "$\bigcirc$" is called a NeutroAssociative on A if some (or all) of the following conditions hold in a way that

$$(T,I,F) \notin \{(1,0,0),(0,0,1)\}.$$

1. There exist $x,y,z \in A$ with $x\bigcirc(y\bigcirc z) = (x\bigcirc y)\bigcirc z$. (This condition is called degree of truth, "T").
2. There exist $x,y,z \in A$ with $x\bigcirc(y\bigcirc z) \neq (x\bigcirc y)\bigcirc z$. (This condition is called degree of falsity, "F").
3. There exist $x,y,z \in A$ with $x\bigcirc(y\bigcirc z)$ is indeterminate or $(x\bigcirc y)\bigcirc z$ is indeterminate or we cannot find if $x\bigcirc(y\bigcirc z)$ and $(x\bigcirc y)\bigcirc z$ are equal. (This condition is called degree of indeterminacy, "I").

The couple $(A,\bigcirc)$ is called a NeutroSemiHypergroup.

Definition 8 (Al-Tahan et al., 2021): Let A be any non-empty set and $\bigcirc: A\times A \to P^*(A)$ be a map. Then $(A,\bigcirc)$ is called NeutroSemihypergroup if "$\bigcirc$" is a NeutroAssociative but not an AntiHyperoperation.

ANTIHYPERSTRUCTURES

Definitions and Examples

Definition 9 (Al-Tahan et al., 2021): Let A be any non-empty set and $\bigcirc: A\times A \to P^*(A)$ be a map. Then "$\bigcirc$" is called an AntiHyperoperation on A if $x\bigcirc y \subsetneq A$ for all $x,y \in A$. The couple $(A,\bigcirc)$ is called an AntiHypergroupoid.

Example 1: Let Z be the set of integers and $\bigcirc$, $*: Z\times Z \to P^*(Z)$ defined as

$$m\bigcirc n = \left\{n, \frac{1}{m^2+3}\right\} and\, m*n = \begin{cases} Z & if\ m=n, \\ m-n & otherwise. \end{cases}$$

for all $m,n \in Z$.

Then $(Z,\bigcirc)$ is AntiHypergroupoid and $(Z,*)$ is hypergroupoid.

Definition 10 (Al-Tahan et al., 2021): Let A be any non-empty set and $\bigcirc: A \times A \to P^*(A)$ be a map. Then "$\bigcirc$" is called an AntiAssociative on A if $x\bigcirc(y\bigcirc z) \neq (x\bigcirc y)\bigcirc z$ for all $x,y,z \in A$.

Definition 11: Let A be any non-empty set and $\bigcirc: A \times A \to P^*(A)$ be a map. Then $(A,\bigcirc)$ is called AntiSemihypergroup if "$\bigcirc$" is an AntiHyperoperation or is AntiAssociative.

Example 2: Let Z be the set of integers and define $\square: Z \times Z \to P^*(Z)$ as follows for all $x,y \in Z$:

$$x\square y = \begin{cases} [x,y] & if\, x < y, \\ [y,x] & if\, y < x, \\ \sqrt{2} & if\, x = y = 0, \\ \frac{1}{5} & otherwise. \end{cases}$$

It is clear that $(Z,\square)$ is an AntiHypergroupoid. Therefore, $(Z,\square)$ is an AntiSemihypergroup.

Example 3: Let R be the set of real numbers, and $A = R - \{-1,0,1\}$. Define $*: A \times A \to P^*(A)$ as follows for all $x,y \in A$:

$$x * y = \frac{x}{y}$$

$(A,*)$ is a hypergroupoid and "$*$" is AntiAssociative on A. This is clear as $x*(y*z) = \frac{xz}{y}$ and $(x*y)*z = \frac{x}{yz}$ are not equal for all $x \in A$. Therefore, $(A,*)$ is an AntiSemihypergroup.

Example 4: Let $O = \{m\}$ and define $\cdot: O \times O \to P^*(O)$ as follows for all $m \in O$.

$$m \cdot m = \{a, m\}.$$

Since $(O,\cdot)$ is an Antihypergroupoid, then $(O,\cdot)$ is an AntiSemihypergroup.

Definition 12: Let A be any non-empty set and $\bigcirc$: $A \times A \to P^*(A)$ be a map. Then "$\bigcirc$" is called an AntiWeakAssociative on A if $x\bigcirc(y\bigcirc z) \cap (x\bigcirc y)\bigcirc z = \varnothing$ for all $x, y, z \in A$.

Definition 13: Let A be any non-empty set and $\bigcirc$: $A \times A \to P^*(A)$ be a map. Then $(A,\bigcirc)$ is called Anti H_v Semihypergroup if "$\bigcirc$" is an AntiHyperoperation or is AntiWeakAssociative.

Remark 1: Every Anti H_v Semihypergroup is an AntiSemihypergroup.

Theorem 1: Let $(A_1,\bigcirc_1),(A_2,\bigcirc_2)$ be AntiSemihypergroups,

$$A_1 \times A_2 = \{(a,m) : a \in A_1 \text{ and } m \in A_2\}.$$

Then $(A_1 \times A_2,\bigcirc)$ is an AntiSemihypergroup.

Proof: For $a = (x_1, y_1) \in A_1, m = (x_2, y_2) \in A_2$ define:

$$a\bigcirc m = ((x_1\bigcirc_1 x_2),(y_1\bigcirc_2 y_2)).$$

Two cases are considered here.

- Case $\bigcirc_1$ or $\bigcirc_2$ is an AntiHyperoperation. It is clear that $\bigcirc$ is also an AntiHyperoperation.
- Case $\bigcirc_1$ or $\bigcirc_2$ is an AntiAssociative. Without loss of generality, suppose without loss of generality that $\bigcirc_1$ is AntiAssociative. Then for all

$$x_1, x_2, x_3 \in A_1,\ (x_1\bigcirc_1 x_2)\bigcirc_1 x_3 \neq x_1\bigcirc_1(x_2\bigcirc_1 x_3).$$

The latter implies that for any triplets

$$(x_1, y_1),(x_2, y_2),(x_3, y_3) \in A_1 \times A_2,$$

$$((x_1, y_1)\bigcirc(x_2, y_2))\bigcirc(x_3, y_3) \neq (x_1\bigcirc_1(x_2\bigcirc_1 x_3), y_1\bigcirc_2(y_2\bigcirc_2 y_3)).$$

Thus, $\bigcirc$ is AntiAssociative. Therefore, $(A_1 \times A_2,\bigcirc)$ is an AntiSemihypergroup.

Example 5. Let $(A,*)$ be the AntiSemihypergroup in Example 3. Theorem 1 asserts that $(A\times A,\bigcirc)$ is an AntiSemihypergroup. Here, $(x,y)\bigcirc(z,w)=\left(\frac{x}{z},\frac{y}{w}\right)$ for all $(x,y),(z,w)\in A\times A$.

Definition 14: Let $(A,\bigcirc)$ be an AntiSemihypergroup (Anti H_v Semihypergroup) and $M\neq\varnothing\subseteq A$. Then M is an AntiSubsemihypergroup (Anti H_v Subsemihypergroup) if $(M,\bigcirc)$ is an AntiSemihypergroup (Anti H_v Semihypergroup).

Lemma 1: Every non-empty subset of an AntiSemihypergroup (Anti H_v Semihypergroup) is an AntiSubsemihypergroup (Anti H_v Subsemihypergroup).

Proof: Let $(A,\bigcirc)$ be an AntiSemihypergroup and $M\neq\varnothing\subseteq A$. If "$\bigcirc$" is AntiAssociative on A then "$\bigcirc$" is AntiAssociative on M, and hence, $(M,\bigcirc)$ is an AntiSemihypergroup. If $(A,\bigcirc)$ is an AntiHypergroupoid then $(M,\bigcirc)$ is an AntiHypergroupoid, and hence, $(M,\bigcirc)$ is an AntiSemihypergroup.

Lemma 2: Let $(H,\bigcirc)$ be an AntiSemihypergroup and $(J,*)$ be a semihypergroup. The extension $H[J]$ of H by J is given as $H[J]=H\cup J$ and the map "$\circledast$" is defined as follows:

$$x\circledast y=\begin{cases}x\bigcirc y & \text{if } x,y\in H,\\ x*y & \text{if } x,y\in J,\\ x & \text{if } x\in H, y\in J,\\ y & \text{if } x\in J, y\in H.\end{cases}$$

Then $(H[J],\circledast)$ is a NeutroSemihypergroup.

Proof: Let $x,y,z\in J$. Then

$$x\circledast(y\circledast z)=x\circledast(y*z)=x*(y*z)$$

and

$$(x\circledast y)\circledast z=(x*y)\circledast z=(x*y)*z.$$

Therefore, $x \circledast (y \circledast z) = (x \circledast y) \circledast z$ since $(J, *)$ is a semihypergroup. The latter implies that $\circledast$ is not AntiAssociative. Moreover, one can easily see that $\circledast$ is not an AntiHyperoperation.

If $\bigcirc$ is AntiHyperoperation then it is clear that $(H[J], \circledast)$ is a NeutroSemihypergroup. If $\bigcirc$ is AntiHyperoperation then it is AntiAssociative. The latter implies that there exist $x, y, z \in H$ such that

$$x \bigcirc (y \bigcirc z) \neq (x \bigcirc y) \bigcirc z.$$

Having

$$x \circledast (y \circledast z) = x \circledast (y \bigcirc z) = x \bigcirc (y \bigcirc z)$$

and

$$(x \circledast y) \circledast z = (x \bigcirc y) \circledast z = (x \bigcirc y) \bigcirc z$$

implies that

$$x \circledast (y \circledast z) \neq (x \circledast y) \circledast z.$$

Therefore, $(H[J], \circledast)$ is a NeutroSemihypergroup.

Definition 15: Let H be any non-empty set and "$\bigcirc$" be a map. Then $(H, \bigcirc)$ is called AntiQuasihypergroup if $a \bigcirc H \neq H$ or $H \bigcirc a \neq H$ or $a \bigcirc H \neq H \bigcirc a$ for all $a \in H$.

Example 6: Let $M = \{o, m, a\}$, and define $\heartsuit$: $M \times M \to P^*(M)$ as presented in Table 1.

Since $M \heartsuit o = \{o, m\} \neq M$, $M \heartsuit m = \{a, m\} \neq M$, and $M \heartsuit a = \{a, m\} \neq M$, it follows that $(M, \heartsuit)$ is an AntiQuasihypergroup.

Definition 16: Let A be any non-empty set and "$\bigcirc$" be a map. Then $(A, \bigcirc)$ is called AntiHypergroup if $(A, \bigcirc)$ is an AntiHypergroupoid or an AntiSemihypergroup or an AntiQuasihypergroup.

Table 1. The AntiQuasihypergroup $(M,♥)$

♥	o	m	a
o	m	{a,m}	m
M	{o,m}	a	{a,m}
A	m	{a,m}	m

Remark 2: Every AntiHypergroupoid is an AntiSemihypergroup and AntiQuasihypergroup, and every AntiSemihypergroup (AntiQuasihypergroup) is an AntiHypergroup.

Remark 3: Not every AntiSemihypergroup (AntiQuasihypergroup or AntiHypergroup) is an AntiHypergroupoid.

Remark 2 is illustrated in the following example.

Example 7: Let $(M,♥)$ be the AntiQuasihypergroup presented in Example 6. $(M,♥)$ is a hypergroupoid but is an AntiQuasihypergroup.

Existence of an Infinite Number of AntiSemihypergroups (AntiHypergroups)

Let $(H,\bigcirc)$ be an AntiHypergroupoid, and J be a non-empty set such that $H \cap J = \varnothing$, $(H \bigcirc H) \cap J = \varnothing$, and m be an element that is neither in H nor in J. The extension $H[J]$ of H by J is given as $H[J] = H \cup J$ and the map "$\circledcirc$" is defined as follows:

$$x \circledcirc y = \begin{cases} x \bigcirc y & \text{if } x, y \in H, \\ m & \text{otherwise.} \end{cases}$$

Theorem 2: Let $(H,\bigcirc)$ be an AntiHypergroupoid, and J a non-empty set such that $H \cap J = \varnothing$, $(H \bigcirc H) \cap J = \varnothing$, and m be an element that is neither in H nor in J. Then $(H[J],\circledcirc)$ is an AntiHypergroupoid.

Proof: Let $x, y \in H[J]$. If $x, y \in H$ then $x \bigcirc y \notin H$. Having $H \cap J = \varnothing$ and $(H \bigcirc H) \cap J = \varnothing$ implies that $x \circledcirc y = x \bigcirc y \notin H[J]$. If $x \in J$ or $y \in J$

then $x \circledcirc y = m \notin H\left[J\right]$ as m is an element that is neither in H nor in J. Therefore, $\left(H\left[J\right], \circledcirc\right)$ is an AntiHypergroupoid.

Theorem 3: Let $n \geq 1$ be an integer. Then there exists an AntiHypergroupoid of cardinality n.

Proof: Example 4 asserts that there exists an AntiHypergroupoid of cardinality 1. Let $k = n - 1$, $m_i, m, a \neq o$, $m \neq m_i$, and $a \neq m_i$, for $i = 1, \cdots, k$, and $J = \{m_1, \cdots, m_k\}$. Theorem 2 asserts that $\left(O\left[J\right], \circledcirc\right)$ is an AntiHypergroupoid of cardinality equal to n.

Corollary 1: There are infinitely many AntiHypergroupoids.

Proof: The proof follows from Theorem 3.

Theorem 4: Let $n \geq 1$ be an integer. Then there exists an AntiSemihypergroup of cardinality n.

Proof: The proof follows from having every AntiHypergroupoid is an AntiSemihypergroup and from Theorem 3.

Corollary 2: There are infinitely many AntiSemihypergroups.

Proof: The proof follows from Theorem 4.

Theorem 5: Let $n \geq 1$ be an integer. Then there exists an AntiHypergroup of cardinality n.

Proof: The proof follows from having every AntiHypergroupoid is an AntiHypergroup and from Theorem 3.

Corollary 3: There are infinitely many AntiHypergroups.

Proof: The proof follows from Theorem 5.

FUTURE RESEARCH DIRECTIONS

The results in this chapter may be considered as a base for any possible future study in the field of AntiHyperstructures. The authors suggest the following research ideas in this direction:

1. Find all AntiSemihypergroups of small order (up to a certain type of equivalence).
2. Define other AntiHyperstructures.

CONCLUSION

In this chapter, the notions of Antihypergroupoids, AntiSemihypergroups, AntiQuasihypergroups, and Antihypergroups were introduced. New definitions, properties, and theorems were presented and were illustrated by several non-trivial illustrative examples. Finally, the existence of an infinite number of AntiSemihypergroups and AntiHypergroups was shown.

FUNDING

This research received no specific grant from any funding agency in the public, commercial, or not-for-profit sectors.

ACKNOWLEDGMENT

The authors are grateful to the referees for their valuable comments and suggestions.

REFERENCES

Agboola, A. A. A., & Ibrahim, M. A. (2020). Introduction to AntiRings. *Neutrosophic Science and System*, *36*, 293–307.

Agboola, A. A. A., Ibrahim, M. A., & Adeleke, E. (2020). Elementary Examination of NeutroAlgebras and AntiAlgebras viz-a-viz the Classical Number Systems. *International Journal of Neutrosophic Science*, *4*, 16–19. doi:10.54216/IJNS.040102

Al-Tahan, M., Davvaz, B., Smarandache, F., & Anis, O. (2021). On some NeutroHyperstructures. *Symmetry*, *13*(4), 535. doi:10.3390ym13040535

Al-Tahan, M., Smarandache, F., & Davvaz, B. (2021). NeutroOrderedAlgebra: Applications to semigroups. *Neutrosophic Sets and System*, *39*, 133–147.

Corsini, P. (1993). Prolegomena of Hypergroup Theory. Udine Aviani Editore.

Davvaz, B., & Leoreanu-Fotea, V. (2008). *Hyperring Theory and Applications*. International Academic Press.

Heidari, D., & Davvaz, B. (2011). On ordered hyperstructures. *Politehn. Univ. Bucharest Sci. Bull. Ser. A Appl. Mathematics and Physics*, *73*(2), 85–96.

Ibrahim, M. A., & Agboola, A. A. A. (2020). Introduction to NeutroHyperGroups. *Neutrosophic Sets and System*, *38*, 15–32.

Marty, F. (1934). Sur une generalization de la notion de groupe. *Proceedings of the 8th Congress on Mathmatics, Scandinaves*, 45–49.

Rezaei, A., & Smarandache, F. (2020). On Neutro-BE-algebras and Anti-BE-algebras. *International Journal of Neutrosophic Science*, *4*, 8–15.

Rezaei, A., Smarandache, F., & Mirvakili, S. (2021). Applications of (Neutro/Anti) sophications to Semihypergroups. *Journal of Mathematics*, *2021*, 1–7. https://doi.org/10.1155/2021/6649349

Smarandache, F. (1998). Neutrosophy. Neutrosophic Probability, Set, and Logic, 105.

Smarandache, F. (2020). Introduction to NeutroAlgebraic structures and AntiAlgebraic structures (revisited). *Neutrosophic Sets and System*, *31*, 1–16.

Smarandache, F. (2020). Introduction to NeutroAlgebraic structures and AntiAlgebraic structures (revisited). *Neutrosophic Sets and System*, *31*, 1–16.

Smarandache, F. (2020). NeutroAlgebra is a generalization of partial algebra. *International Journal of Neutrosophic Science*, *2*, 8–17.

Smarandache, F. (2020). Introduction to NeutroAlgebraic structures and AntiAlgebraic structures (revisited). *Neutrosophic Sets and System*, *31*, 1–16.

Vougiouklis, T. (1991). The fundamental relation in hyperrings. The general hyperfield. *Proc. of the 4th Int. in Congress on Algebraic Hyperstructures and Appl.,* 203-211.

Vougiouklis, T. (1994). *Hyperstructures and Their Representations*. Hadronic Press, Inc.

ADDITIONAL READING

Al-Tahan, M., & Davvaz, B. (2017). Algebraic Hyperstructures Associated to Biological Inheritance. *Mathematical Biosciences*, *285*, 112–118. doi:10.1016/j.mbs.2017.01.002 PMID:28108295

Al-Tahan, M., Davvaz, B., Smarandache, F., & Anis, O. (2021). Properties of Productional NeutroOrderedSemigroups. *Neutrosophic Sets and System*, *42*, 178–190.

Davvaz, B., & Omidi, S. (2016). Basic notions and properties of ordered semihyperrings. *Categ. Gen. Algebr. Struct. Appl.*, *4*(1), 43–62.

Hamidi, M., & Smarandache, F. (2020). Neutro-BCK-Algebra. *International Journal of Neutrosophic Science*, *8*, 110–117. doi:10.54216/IJNS.080204

Smarandache, F. (2020). Extension of HyperGraph to n-SuperHyperGraph and to Plithogenic n-SuperHyperGraph, and Extension of HyperAlgebra to n-ary (Classical-/Neutro-/Anti-) HyperAlgebra. *Neutrosophic Sets and Systems*, *33*, 290–296.

Chapter 8 Theoretical Study of the NeutroAlgebra Generated by the Combining Function in Prospector and Some Pedagogical Notes

Noel Batista Hernández
Asociación Latinoamericana de Ciencias Neutrosóficas, Ecuador

Erick González Caballero
Technological University of Havana, Cuba

Lilia Esther Valencia Cruzaty
Universidad Estatal Península de Santa Elena, Ecuador

Wilmer Ortega Chávez
Universidad Nacional Intercultural de la Amazonia, Peru

Carmen Flor Padilla Huarac
Universidad Nacional José Faustino Sánchez Carrión, Peru

Selene Luz Chijchiapaza Chamorro
Universidad Nacional Daniel Alcides Carrión, Peru

ABSTRACT

NeutroAlgebras consist of algebras containing at least one NeutroAxiom. In the literature, there is a method for decision-making in a migrational situation that uses a NeutroAlgebra generated by the combining function in Prospector. This method is an important hit in showing both the utility and theoretical significance

DOI: 10.4018/978-1-6684-3495-6.ch008

INTRODUCTION

The theory of NeutroAlgebras, which has recently been introduced by F. Smarandache, generalizes the classical theory of Algebras and Partial Algebras within the framework of Neutrosophy (Smarandache, 2020a, 2020b). NeutroAlgebras preserve the study of algebraic structures defined as ordered pairs formed by a set of elements and algebraic operations. The main difference between the NeutroAlgebras concerning algebras and partial algebras is that the former ones contain at least one NeutroAxiom which is an axiom where there are two types of elements, those that comply with the axiom and those that do not comply with it.

Following the main idea of Neutrosophy (Smarandache, 2005), given an Algebra (Axiom) <A>, there is a triad (<A>, <NeutA>, <AntiA>) where the Algebra (Axiom) <A> is 100% true or true for all the elements, NeutroAlgebras (NeutroAxioms) <NeutA> are satisfied only by a part of the elements, while AntiAlgebras (AntiAxioms) <AntiA> are not satisfied by any of the elements of the set, (Smarandache, 2020a).

This new approach to one of the most classical branches of mathematics constitutes a challenge for the understanding of these novel ideas. Classical algebra is based on mathematical logic where only axioms 100% true are permitted. From a theoretical point of view, there may be doubts about the usefulness of modeling indeterminacy and uncertainty in a well-established and solid mathematical branch such as algebra.

The other controversial aspect is about the usefulness of studying NeutroAlgebras due to a large number of applications of classical algebras in other sciences, such as Physics, Chemistry, Engineering, among others. All these sciences need knowledge on the theory of groups, Galois theory, theory of linear spaces, among other sub-branches of the classical algebras. Thus, some scientists could argue that these classical theories are sufficient to model scientific and real-life problems, and NeutroAlgebras can be considered unnecessary, (Isaacs, 2008).

In the literature review on NeutroAlgebras carried out by the authors, some examples of NeutroGroups, NeutroRings, among others, were found, which essentially only constitute theoretical approaches to these subjects, (Abobala, 2021; Agboola, 2020a; Gayen et al., 2020; Ibrahim & Agboola, 2020; Sankari & Abobala, 2020). The only exception of one NeutroAlgebra with practical application is the one based on the uninorm defined from the combining function in the Prospector expert system. This consists in a decision-making method that was designed for modeling the study of migration, (Silva-Jiménez, Valenzuela-Mayorga, Roja-Ubilla, & Batista-Hernández, 2021).

Uninorms are operators defined for the first time in the framework of fuzzy logic, which generalizes the axiomatic of t-norm, representing the logical AND, and t-conorm, representing the logical OR (Fodor, Yager, & Rybalov, 1997). In addition to the axioms of commutativity, associativity, non-decreasing, in uninorms, the axiom of the existence of a neutral element is extended to a value other than 0 and 1. This type of operator has been applied in countless problems of many kinds, (Bordignon & Gomide, 2014; González-Hidalgo, Massanet, Mir, & Ruiz-Aguilera, 2014; Yager, 2001; Yager & Kreinovich, 2003; Yager & Rybalov, 1996). It has also been generalized in other frameworks such as the theories of intuitionistic fuzzy sets (Deschrijver & Kerre, 2004), neutrosophic sets (González-Caballero, Leyva-Vázquez, & Smarandache, 2021), and neutrosophic offsets (González-Caballero, Smarandache, & Leyva-Vázquez, 2019).

On the other hand, the combining function in Prospector has transcended the framework of this expert system and is itself a subject of study (De-Baets & Fodor, 1999). Prospector expert system is dedicated to mineral search prediction (Hart, Duda, & Einaudi, 1978; Marík & Kouba, 1991; Zhang & Luo, 1999).

The application of a NeutroAlgebra to solve a real-life problem in (Silva-Jiménez et al., 2021) served to ratify the idea presented by Smarandache that NeutroAlgebras are important for modeling in the so-called Soft Sciences, where the existence of 100% true axioms is not the generality. That is why from the pedagogical point of view these NeutroAlgebras are a good starting point to teach the general theory of NeutroAlgebras to students of careers where mathematics are essentials, and which there is an important social component, for example, in economics and political sciences.

The method in (Silva-Jiménez et al., 2021) is an example of defining an abstract structure, which is generated in a simple way and which has a tangible application in a real-life problem. It is an opportunity to motivate students from different careers to learn about NeutroAlgebras, both in elective undergraduate and postgraduate courses. In addition, the method created by those authors can be used for decision-making in pedagogical problems. Moreover, this method interconnects several theories such as uninorms, NeutroAlgebras, and decision-making.

The contribution of this chapter is to deepen the theoretical aspects of NeutroAlgebra defined by Silva et al. (Silva-Jiménez et al., 2021), where an interesting practical application is performed, but which is not sufficiently characterized from the theoretical point of view. Specifically, the authors of this chapter identify this NeutroAlgebra as a NeutroGroup, and therefore NeutroGroups are the algebraic structures we focus on in the following, (Agboola, 2020a, 2020b).

The theoretical study carried out by the authors in this chapter will improve the teaching of this subject in students where modeling with NeutroAlgebras can be useful. The characterization that will be made of the NeutroAlgebra based on the combining function in Prospector will allow us discovering the potential of this method to solve decision-making problems in other fields of knowledge. Moreover, the theoretical study of this NeutroAlgebra is the contribution that the authors make to the general theory of NeutroAlgebras.

This chapter has the following structure; the next section contains the background of the fundamental concepts necessary to understand this chapter, such as those of NeutroAlgebra, Group, NeutroGroup, uninorm, the combining function in Prospector, and other important and necessary aspects. The following section contains the theoretical development proposed in the chapter as it was previously explained. The last section is the conclusion of this chapter.

BACKGROUND

This section contains a background of the most important concepts used in this chapter. We begin with the theory of groups.

Theory of Groups

Within classical algebra, the group theory has been developed over centuries and has been widely applied in fields such as Physics, Chemistry, Cryptography, Geometry, among others.

Below is the formal definition of this algebraic structure and others.

Definition 1: (Isaacs, 2008) Given the ordered pair $\left(\mathrm{G}, \oplus\right)$ of the nonempty set G with the operation $\oplus : \mathrm{G} \times \mathrm{G} \rightarrow \mathrm{G}$ satisfying the following axioms:

1. $\forall x, y \in \mathrm{G}$ it is fulfilled $x \oplus y \in \mathrm{G}$ (it is an *internal operation*).
2. $\forall x, y, z \in \mathrm{G}$ it is fulfilled $x \oplus \left(y \oplus z\right) = \left(x \oplus y\right) \oplus z$ (*associativity*).
3. $\exists e \in \mathrm{G}$ which is called the *neutral element* such that $\forall x \in \mathrm{G}$ fulfills $e \oplus x = x \oplus e = x$ (*Existence of a neutral element*).
4. $\forall x \in \mathrm{G}$, $\exists -x \in \mathrm{G}$ such that $x \oplus \left(-x\right) = \left(-x\right) \oplus x = e$ (Existence *of an opposite or inverse element*).

If $\left(\mathrm{G}, \oplus\right)$ satisfies conditions 1 and 2, we say it is a *semigroup*. A semigroup that fulfills axiom 3 is called is a *monoid*. A monoid satisfying condition 4 is a *group*.□

A group that fulfills the property $\forall x, y \in \mathrm{G} \quad x \oplus y = y \oplus x$, is said to be a *commutative group* or an *Abelian group*.

If $\text{card}(\text{G}) < \infty$ the order of the group is said to be finite and equal to n, it is denoted by $o(\text{G}) = n$. Another definition of the order of a group is the maximum $m \in \mathbb{N}$ in the set of $x \in \text{G}$, such that

$$x^{m} = \underbrace{x \oplus x \oplus \cdots \oplus x}_{m\,times} = e$$

is a minimum for this x. If such m exists it is said that the order of the group is m, otherwise it is said that the group has infinite order.

Elementary examples of groups are the usual sum in the domain of integer, rational, real, and complex numbers, $(\mathbb{Z},+)$, $(\mathbb{Q},+)$, $(\mathbb{R},+)$, and $(\mathbb{C},+)$, respectively. In these cases the pairs $(-x,x)$ are the opposites.

Interesting examples are the groups $(\mathbb{Z}_{n},+)$ and $(\mathbb{Z}_{n},\cdot)$ defined like the sum and the product modulo n, respectively, of the set $\mathbb{Z}_{n} = \{0,1,2,\cdots,n-1\}$. They are finite, Abelian, and cyclic groups of order n. Cyclic means that $\exists a \in \mathbb{Z}_{n}$, such that $\mathbb{Z}_{n} = \{a^{n} : n \in \mathbb{Z}\}$, where $a^{-k} = (a^{-1})^{k}$.

Within the teaching of higher mathematics, few careers study the theory of groups, however, this subject can be introduced illustrating with examples of groups defined from terms with which the student feels familiar.

For example, if A is the set of alphanumeric strings that contains the empty word that is denoted by $\langle\square\rangle$. Then the operation $conc : \text{A} \times \text{A} \to \text{A}$, such that if p, q are two alphanumeric strings then $conc(\text{p},\text{q})$ is the result of concatenating them, where

$$conc(\text{p},\langle\square\rangle) = conc(\langle\square\rangle,\text{p}) = \text{p}.$$

This is a monoid because it is a mapping that fulfills conditions 1-3.

More specifically, if $\text{p} = \langle\text{abxm45}\rangle$ and $\text{q} = \langle\text{9rtgds}\rangle$, then,

$$conc(\text{p},\text{q}) = \langle\text{abxm459rtgds}\rangle,$$

moreover

$conc\left(\mathrm{p},\left\langle\square\right\rangle\right)=conc\left(\left\langle\square\right\rangle,\mathrm{p}\right)=\left\langle\mathrm{abxm45}\right\rangle.$

However, it is not a group because an inverse or opposite was not defined for each string. It is not commutative either because

$$conc\left(\mathrm{p},\mathrm{q}\right)=\left\langle\mathrm{abxm459rtgds}\right\rangle\neq conc\left(\mathrm{q},\mathrm{p}\right)=\left\langle\mathrm{9rtgdsabxm45}\right\rangle.$$

This example can be didactic for Computer Science students; in view that computational programming depends on artificial languages.

Other didactical examples of groups are those formed by the rotations of figures in the plane or that of the crystal classes in physics. By explaining these groups, teachers can motivate the students to understand why could be necessary to incorporate indeterminacy. For example, scientists which study a crystal whose structure is not exactly known.

Some additional important concepts within group theory are the following:

Definition 2: (Isaacs, 2008) A nonempty subset H of a group $\left(\mathrm{G},\oplus\right)$ is a subgroup of G if $\left(\mathrm{H},\oplus\right)$ is a group itself or equivalently, $e\in\mathrm{H}$, and $x,y\in\mathrm{H}$ implies $x\oplus\left(-y\right)\in\mathrm{H}$.

Two trivial subgroups of a given group $\left(\mathrm{G},\oplus\right)$ are G itself and $\left\{e\right\}$.

If $\left(\mathrm{H},\oplus\right)$ is a subgroup of the group $\left(\mathrm{G},\oplus\right)$, then the relation $\mathrm{R}\left(x,y\right):\left(-x\right)\oplus y\in\mathrm{H}$ is an equivalence relation on a set G.

Definition 3: (Isaacs, 2008) Let $\left(\mathrm{H},\oplus\right)$ be a subgroup of the group $\left(\mathrm{G},\oplus\right)$ and $x\in\mathrm{G}$. The set $x\mathrm{H}=\left\{x\oplus y\mid y\in\mathrm{H}\right\}$ is called the *left coset of H in G*, whereas $\mathrm{H}x=\left\{y\oplus x\mid y\in\mathrm{H}\right\}$ is called the *right coset of H in G.*

Definition 4: (Isaacs, 2008) Let $\left(\mathrm{H},\oplus\right)$ be a subgroup of the group $\left(\mathrm{G},\oplus\right)$, we say that H is a *normal subgroup* of G when $\forall x\in\mathrm{G},x\mathrm{H}=\mathrm{H}x$. It is denoted by $\mathrm{H}\triangleleft\mathrm{G}$.

See that if H is normal we have $\mathrm{H}x=y\mathrm{H}$ if and only if $\left(-x\right)\oplus y\in\mathrm{H}$ and the equivalence class of *x* is the right coset of H in G. Similarly, the left coset of H in G is also an equivalence class.

Trivially, every subgroup of an Abelian group is normal.

Definition 5: (Isaacs, 2008) Let $\left(\mathrm{H},\oplus\right)$ be a normal subgroup of the group $\left(\mathrm{G},\oplus\right)$, the *quotient of the set G by the equivalence relation* $\left(-x\right)\oplus y\in H$ is defined as $\left(x\mathrm{H}\right)\oplus\left(y\mathrm{H}\right)=\left(x\oplus y\right)\mathrm{H}$ and it is denoted by G/H.

It is easy to prove that G / H is a group.

Definition 6: (Isaacs, 2008) Let $(G, \oplus_1)$ and $(H, \oplus_2)$ be two groups. A mapping $f : G \to H$ such that

$$\forall x, y \in G, f(x \oplus_1 y) = f(x) \oplus_2 f(y)$$

is called a *homomorphism of G in H*.

A bijective homomorphism is called an *isomorphism*, and G and H are said they are *isomorphic* and it is denoted by $G \cong H$.

Definition 7: (Isaacs, 2008) Let $f : G \to H$ a homomorphism of the group G in the group H.

$$f(G) = \{y \in H \mid y = f(x) \wedge x \in G\} \subseteq H$$

is called the *image of f* and is denoted by $Im(f)$, whereas, the set of $\{x \in G \mid f(x) = e\}$ is called the *Kernel of f* and it is denoted by $Ker(f)$.

$Im(f)$ is a subgroup of H and $Ker(f)$ is a subgroup of G.

Theorem 1: (Isaacs, 2008) Suppose $N \triangleleft G$, if *f* is a homomorphism of G into H whose kernel contains N, and π is a natural map from G to G/H, i.e., $\pi(x) = xH$, then there is a unique homomorphism $\bar{f} : G / N \to H$ such that $\bar{f} \circ \pi = f$.

Theorem 2: (Isaacs, 2008) If $f : G \to H$ is a homomorphism with kernel K, then $Im(f) \cong G / K$.

NeutroAlgebra

A NeutroAlgebra is an algebraic structure that contains at least one NeutroAxiom and no AntiAxiom, (Smarandache, 2020a, 2020b). Especially if any of the NeutroAxioms corresponds to one of Definition 1, the structure is said to be a NeutroGroup (Agboola, 2020a, 2020b; Kandasamy, Kandasamy, & Smarandache, 2020).

In (Silva-Jiménez et al., 2021) it is defined $\bar{P}(x, y) = P(x, y)$ for all

$$(x, y) \in [-1, 1]^2 \setminus \{(-1, 1), (1, -1)\},$$

$$\bar{P}(-1, 1) = \bar{P}(1, -1) = \text{undefined},$$

$$\bar{P}\left(\text{undefined}, \text{undefined}\right) = \text{undefined},$$

and

$$\bar{P}\left(\text{undefined}, x\right) = \bar{P}\left(x, \text{undefined}\right) = \begin{cases} \text{undefined}, \text{if } x > 0 \\ x, \text{if } x \le 0 \end{cases}.$$

In this function, the aggregation of the extreme values is explicitly considered as undefined because it is an indicator of contradiction. When undefined is aggregated with any $x > 0$ it is considered undefined, while it performs as a neutral element for any $x \le 0$.

Definition 8: (Silva-Jiménez et al., 2021) Let S be a finite set defined as

$$S = \left\{\left(x, y\right) : x, y \in \left\{k, \text{undefined}\right\}, k \in \mathbb{Z} \cap \left[-10, 10\right]\right\}.$$

The operator $\odot$ is defined for every $\left(x, y\right) \in S$, such that:

If $\bar{P}\left(x, y\right)$ is not undefined, then

$$x \odot y = \frac{round\left(\bar{P}\left(\frac{x}{10}, \frac{y}{10}\right) * 10\right)}{10},$$

where *round* is the function that outputs the integer nearest to the argument.□

It is shown that this definition corresponds to a finite NeutroAlgebra since the NeutroAssociativity is fulfilled.

These concepts of NeutroAlgebras are difficult to teach in an environment where pure mathematics prevails, however, analyzing that in the previous method it is justified not to comply with the associativity for some cases, those that include the value undefined, then it is understandable to consider NeutroAlgebras within the learning of mathematics.

Uninorms

The uninorms are defined as follows:

Definition 9: (Fodor et al., 1997) The mapping $\mathcal{U}:\left[a,b\right]^2 \to \left[a,b\right]$, where $\left[a,b\right] \subset \mathbb{R}$ is a closed interval, is a *uninorm* if it satisfies the following axioms, $\forall x,y,z \in \left[a,b\right]$:

1. $\mathcal{U}\left(x,y\right)=\mathcal{U}\left(y,x\right)$ (Commutativity),
2. $\mathcal{U}\left(x,\mathcal{U}\left(y,z\right)\right)=\mathcal{U}\left(\mathcal{U}\left(x,y\right),z\right)$ (Associativity),
3. $y \leq z$ implies $\mathcal{U}\left(x,y\right) \leq \mathcal{U}\left(x,z\right)$ (Non-decreasing),
4. There is $e \in \left[a,b\right]$, called the *neutral element*, such that $\forall x \in \left[a,b\right], \mathcal{U}\left(e,x\right)=x$.

It is generally assumed $\left[a,b\right]=\left[0,1\right]$.

Uninorms can also be generated from some functions.

Let $h:\left[a,b\right] \to \bar{\mathbb{R}}$ be a strictly increasing and continuous mapping and $e \in \left(a,b\right)$ such that $h\left(a\right)=-\infty$, $h\left(e\right)=0$, and $h\left(b\right)=\infty$. Then $h\left(x\right)$ is called an *additive generator* that generates a uninorm called a *representable uninorm* and has the neutral element *e* as follows:

$$\mathcal{U}\left(x,y\right)=h^{-1}\left(h\left(x\right)+h\left(y\right)\right)\forall\left(x,y\right)\in\left[a,b\right]^2 \setminus \left\{\left(a,b\right),\left(b,a\right)\right\} \tag{1}$$

And also either $\mathcal{U}\left(a,b\right)=\mathcal{U}\left(b,a\right)=a$ or $\mathcal{U}\left(a,b\right)=\mathcal{U}\left(b,a\right)=b$.

Uninorms have been extended to the neutrosophic framework as is shown below:

Definition 10: (González-Caballero et al., 2021) A *neutrosophic uninorm* $\mathbf{U}_{\mathrm{N}}$ is commutative, increasing, and associative mapping,

$$\boldsymbol{U}_{\mathrm{N}}:\left(\left]{}^{-}_{\square}0,1^{+}\right[\times\left]{}^{-}_{\square}0,1^{+}\right[\times\left]{}^{-}_{\square}0,1^{+}\right[\right)^2 \to \left]{}^{-}_{\square}0,1^{+}\right[\times\left]{}^{-}_{\square}0,1^{+}\right[\times\left]{}^{-}_{\square}0,1^{+}\right[,$$

such that:

$$\boldsymbol{U}_{\mathrm{N}}\left(x\left\langle T_x,I_x,F_x\right\rangle,y\left\langle T_y,I_y,F_y\right\rangle\right)=\left\langle \boldsymbol{U}_{\mathrm{N}}\mathrm{T}\left(x,y\right),\boldsymbol{U}_{\mathrm{N}}\mathrm{I}\left(x,y\right),\boldsymbol{U}_{\mathrm{N}}\mathrm{F}\left(x,y\right)\right\rangle,$$

where $\mathbf{U}_{\mathrm{N}}\mathrm{T}$ means the degree of membership, $\mathbf{U}_{\mathrm{N}}\mathrm{I}$ the degree of indeterminacy and $\mathbf{U}_{\mathrm{N}}\mathrm{F}$ the degree of non-membership of both, *x* and *y*. Additionally, there exists a neutral element

$$e \in \left]_{\square}^{-}0,1^{+}\right[\times \left]_{\square}^{-}0,1^{+}\right[\times \left]_{\square}^{-}0,1^{+}\right[,$$

where

$$\forall x \in \left]_{\square}^{-}0,1^{+}\right[\times \left]_{\square}^{-}0,1^{+}\right[\times \left]_{\square}^{-}0,1^{+}\right[, \ \boldsymbol{U}_{\mathrm{N}}\left(e,x\right) = x.$$

Definition 11: (González-Caballero et al., 2019) Let c be a neutrosophic component (T_O, I_O or F_O). $c : M_o \rightarrow [\Psi, \Omega]$, where $\Psi \leq 0$ and $\Omega \geq 1$. The neutrosophic component *N-offuninorm* $N_O^u : [\Psi, \Omega]^2 \rightarrow [\Psi, \Omega]$ satisfies the following conditions for any elements x, y, and $z \in M_O$.

1. There exists $c(e)M_o$, such that $N_O^u\left(c(x), c(e)\right) = c(x)$ (Identity),
2. $N_O^u\left(c(x), c(y)\right) = N_O^u\left(c(y), c(x)\right)$ (Commutativity),
3. If $c(x) \leq c(y)$ then $N_O^u\left(c(x), c(z)\right) \leq N_O^u\left(c(y), c(z)\right)$ (Monotonicity),
4. $N_O^u\left(N_O^u\left(c(x), c(y)\right), c(z)\right) = N_O^u\left(c(x), N_O^u\left(c(y), c(z)\right)\right)$ (Associativity).

This definition contains the same elements of the neutrosophic uninorms, with the particularity that the intervals where these aggregators are defined are generalized to $[\Psi, \Omega] \subset \mathbb{R}$, beyond the classical interval $[0,1]$.

Let (T_O, I_O or F_O), $c_o : M_o \rightarrow [\Psi, \Omega]$ and (T, I or F), $c_N : M_N \rightarrow [0,1]$ be a neutrosophic component N-offset and a neutrosophic component, respectively. There exists a bijective mapping that transforms every neutrosophic component N-offuninorm to a neutrosophic component uninorm and vice versa (González-Caballero et al., 2019).

The first equations to define are the following:

$$\varphi_3\left(c(x)\right) = \frac{c(x) - \Psi}{\Omega - \Psi} \tag{2}$$

$$\varphi_3^{-1}\left(c(x)\right) = \left(\Omega - \Psi\right)c(x) + \Psi \tag{3}$$

They are bijective mappings.

If $\hat{\mathrm{U}}_{\mathrm{N}}(\cdot,\cdot)$ is a neutrosophic uninorm, then we can define the neutrosophic component N-offuninorm $\widehat{\mathrm{No}}^{\mathrm{u}}(\cdot,\cdot)$ as follows:

$$\hat{N}_O^u\left(c_O(x),c_O(y)\right)=\varphi_3^{-1}\left(\hat{U}_N\left(\varphi_3\left(c_O(x)\right),\varphi_3\left(c_O(y)\right)\right)\right) \tag{4}$$

Conversely, given $\widehat{\mathrm{No}}^{\mathrm{u}}(\cdot,\cdot)$ then $\hat{\mathrm{U}}_{\mathrm{N}}(\cdot,\cdot)$ can be defined as follows:

$$\hat{U}_N\left(c_N(x),c_N(y)\right)=\varphi_3\left(\widehat{No}^u\left(\varphi_3^{-1}\left(c_N(x)\right),\varphi_3^{-1}\left(c_N(y)\right)\right)\right) \tag{5}$$

Note that Equation 4 allows us to obtain a neutrosophic offuninorm in $[\Psi,\Omega]$ from a neutrosophic uninorm in $[0,1]$.

Combining Function in Prospector

The uninorms have been applied in the resolution of a multiplicity of problems.

A special type of uninorm defined on the interval $[-1,1]$ is the combining function in Prospector $P:[-1,1]^2\to[-1,1]$ with neutral element 0, as it is shown below (De-Baets & Fodor, 1999; Hart et al., 1978):

$$P(x,y)=\frac{x+y}{1+xy} \tag{6}$$

The interpretation of $x<0$ is that the expert believes that he/she has evidence against the hypothesis that is evaluated; whereas $x>0$ means that the expert has evidence in favor of the hypothesis. $x=0$ is used when the expert has no evidence either for or against the hypothesis. The absolute value of x means the degree of belief in the evidence, where -1, 1 are the maximum degrees, the former one against the evidence and the second one in favor. $P(x,y)$ is a function that aggregates two evaluations. This is an Archimedean uninorm, that is $P(x,y)<\min(x,y)$ if $x,y<0$ and $P(x,y)>\max(x,y)$ if $x,y>0$. -1 and 1 are the two veto values, which are called *extremal elements* in (Hajék, 1985).

Two separate cases are $P(-1,1)$ and $P(1,-1)$, where it is evident that there is a contradiction because this means there is absolute evidence for and against the hypothesis at the same time. In some texts, it is defined by $P(-1,1) = P(1,-1) = 0$, since formula 6 is undefined for this pair of values.

So, the ordered pair $([-1,1],\oplus)$ which is defined for

$$x,y \in [-1,1] \setminus \{(-1,1),(1,-1)\},$$

such that $x \oplus y = P(x,y)$ and additionally $(-1)\oplus 1 = 1 \oplus (-1) = 0$, is an ordered Abelian group.

STUDY OF THE NEUTROALGEBRA GENERATED BY THE COMBINING FUNCTION IN PROSPECTOR

From now on, we will study the algebraic properties of the NeutroAlgebra obtained by Silva et al. Specifically, we begin analyzing the following algebraic structure: $([-1,1]\cup\{\mathrm{I}\},\bar{\oplus})$ such that if it is denoted by $\mathrm{NG} = [-1,1]\cup\{\mathrm{I}\}$, we have:

$$x \bar{\oplus} y = \begin{cases} P(x,y), \text{if } (x,y) \in [-1,1] \setminus \{(-1,1),(1,-1)\} \text{ and } x,y \neq \mathrm{I} \\ \mathrm{I}, \text{if } (x = y = \mathrm{I}) \text{ or } (x = \mathrm{I} \text{ and } y \geq 0) \text{ or } (x \geq 0 \text{ and } y = \mathrm{I}) \text{ or } (x,y) \in \{(-1,1),(1,-1)\} \\ x, \text{if } x < 0 \text{ and } y = \mathrm{I} \\ y, \text{if } y < 0 \text{ and } x = \mathrm{I} \end{cases}$$

Where $P(x,y)$ is obtained from Equation 6. Here I means indeterminate.

Proposition 3: $([-1,1]\cup\{\mathrm{I}\},\bar{\oplus})$ is an Abelian NeutroGroup.

Proof:

Evidently $\bar{\oplus}$ is well defined.

$\bar{\oplus}$ is associative for some values, e.g., if $x,y,z \in (-1,1)$ they satisfy associativity because $P(x,y)$ is associative.

In addition, for $x < 0 < y$ and $x \bar{\oplus} y \neq x$ (e.g., $y = -x$ satisfies this condition) then $x \bar{\oplus} (\mathrm{I} \bar{\oplus} y) = x \bar{\oplus} \mathrm{I} = x$ and $(x \bar{\oplus} \mathrm{I}) \bar{\oplus} y = x \bar{\oplus} y \neq x$. So, $\bar{\oplus}$ is not associative for the triad *x*, I, *y*.

There exists a neutral element *e* = 0, where $x \bar{\oplus} 0 = 0 \bar{\oplus} x = x$, $\forall x \in \mathrm{NG}$.

If $x \in (-1,1)$, then $-x$ is its opposite. Whereas, $-1, \mathrm{I}, 1$ have not opposite.

It is Abelian, from the definition of $\bar{\oplus}$ and the commutativity of $P(x,y)$. See that it is also a NeutroSemigroup and a NeutroMonoid.

Proposition 4: The order of $([-1,1]\cup\{\mathrm{I}\},\bar{\oplus})$ is ∞.

Proof:

For example, there is $m\in\mathbb{N}$, such that if x>0, $\underbrace{x\,\bar{\oplus}\,x\,\bar{\oplus}\cdots\bar{\oplus}\,x}_{m\,times}=1$ and if x<0, $\underbrace{x\,\bar{\oplus}\,x\,\bar{\oplus}\cdots\bar{\oplus}\,x}_{m\,times}=-1$, because $P(x,y)$ is Archimedean and ±1 are the extremal elements. Therefore, for some elements, there is not a natural number where $\underbrace{x\,\bar{\oplus}\,x\,\bar{\oplus}\cdots\bar{\oplus}\,x}_{m\,times}=0$. Indeed $x=0$ is the only element in NG that fulfills this condition.□

Proposition 5: The only subgroup of $([-1,1]\cup\{\mathrm{I}\},\bar{\oplus})$ is $\{0\}$.

Proof.

Given $\mathrm{H}\subset[-1,1]\cup\{\mathrm{I}\}$ and $\mathrm{H}\neq\varnothing,\{0\}$. If H is a subgroup then it satisfies condition 4 of Definition 1, therefore, $\mathrm{H}=[-\mathrm{L},\mathrm{L}]$. $\mathrm{I}\notin\mathrm{H}$ because associativity must be satisfied.

Also, L<1 because if $\mathrm{H}=[-1,1]$ it is a NeutroGroup. See that if -1 and 1 belong to H so also $\mathrm{I}=(-1)\bar{\oplus}1\in\mathrm{H}$.

However, given any $x\in\mathrm{H}$ and $x>0$ there is $m\in\mathbb{N}$, such that $\underbrace{x\,\bar{\oplus}\,x\,\bar{\oplus}\cdots\bar{\oplus}\,x}_{m\,times}=1$, therefore condition 1 of Definition 1 is not fulfilled, so there is not any subgroup of $([-1,1]\cup\{\mathrm{I}\},\bar{\oplus})$ except for the trivial one.□

It is easy to prove that $\mathrm{H}_1=[-1,0]$ and $\mathrm{H}_2=[0,1]$ are submonoids.

For decision-making, Silva et al. explicitly considered indeterminate the aggregation of $(-1)\bar{\oplus}1=\mathrm{I}$, trivially $\mathrm{I}\,\bar{\oplus}\,\mathrm{I}=\mathrm{I}$. They considered x<0 is less preferred than I, and I is less preferred than x>0, and $\bar{\oplus}$ aggregates obtaining the less preferred option. This is slightly different when $(-1)\bar{\oplus}1=-1$, however, Silva et al.'s approach is the most accurate.

It can be inferred from Proposition 5 how to define NeutroSubgroups from the NeutroGroup $([-1,1]\cup\{\mathrm{I}\},\bar{\oplus})$.

For example, $\mathrm{H}=[-\mathrm{L},\mathrm{L}]$ with L<1 does not satisfy Axiom 1 of groups for some values and it does for others.

If $\mathrm{H}=[\mathrm{s},\mathrm{t}]\subseteq[-1,1]$ where $\mathrm{s}\neq-\mathrm{t}$, it holds not Axioms 1 and 4.

If the symbol I is included in the domain of definition, the associativity of some of the elements is not fulfilled.

Proposition 6: The union of NeutroSubgroups is a NeutroSubgroup and the direct product of NeutroSubgroups is a NeutroSubgroup.

Proof:

Let H_1 and H_2 be two NeutroSubgroups, since $H_1 \subset H_1 \cup H_2$ and H_1 is a NeutroGroup we have there are elements of $H_1 \cup H_2$ (from H_1), satisfying the axiomatic of groups and others do not, so $H_1 \cup H_2$ is a NeutroGroup.

Given the NeutroSubgroups $\left(H_i, \oplus_i\right)$ (i=1, 2,…, n), the direct product of them is

$$H_1 \times H_2 \times \cdots \times H_n = \left\{\left(x_1, x_2, \cdots, x_n\right) \mid x_i \in H_i\right\},$$

such that for

$$\boldsymbol{x}, \boldsymbol{y} \in H_1 \times H_2 \times \cdots \times H_n,\ \boldsymbol{x} = \left(x_1, x_2, \cdots, x_n\right),\ \boldsymbol{y} = \left(y_1, y_2, \cdots, y_n\right)$$

we have

$$\boldsymbol{x} \oplus \boldsymbol{y} = \left(x_1 \oplus_1 y_1, x_2 \oplus_2 y_2, \cdots, x_n \oplus_n y_n\right).$$

Let $\boldsymbol{x}, \boldsymbol{y}, \boldsymbol{z} \in H_1 \times H_2 \times \cdots \times H_n$ such that every triple x_i, y_i, z_i satisfies the axiomatic of groups. They exist from the definition of NeutroGroups. Then, the triad $\boldsymbol{x}, \boldsymbol{y}, \boldsymbol{z}$ fulfills the axiomatic of groups.

On the other hand, we can select a triple

$$\overline{\boldsymbol{x}}, \overline{\boldsymbol{y}}, \overline{\boldsymbol{z}} \in H_1 \times H_2 \times \cdots \times H_n,$$

where at least one component $\overline{x}_i, \overline{y}_i, \overline{z}_i$ does not satisfy one axiom of groups. Therefore, $\overline{x}$, $\overline{y}$ or $\overline{z}$ do not satisfy the axiomatic of groups. Thus, $H_1 \times H_2 \times \cdots \times H_n$ is a NeutroGroup.□

In the following, we define the cosets of $\left(\left[-1,1\right] \cup \left\{I\right\}, \overline{\oplus}\right)$.

Remark 1:

1. Equivalently to Definition 3 for groups, let H be a NeutroSubgroup of $\left(\left[-1,1\right]\cup\left\{\mathrm{I}\right\},\bar{\oplus}\right)$, a left coset of H in NG is defined as $x\mathrm{H}=\left\{x\,\bar{\oplus}\,y \mid \mathrm{y}\in\mathrm{H}\right\}$ for $x\in\left[-1,1\right]\cup\left\{\mathrm{I}\right\}$.
2. They are Abelian NeutroGroups, so this indeed coincides with the right coset of H in NG $\mathrm{H}x=\left\{y\,\bar{\oplus}\,x \mid \mathrm{y}\in\mathrm{H}\right\}$.
3. Therefore, every NeutroSubgroup of NG is a normal NeutroSubgroup, following the equivalent for Definition 4 of Normal subgroups.

For example, for $\mathrm{H}_1=\left[-1,0\right]$ and $\mathrm{H}_2=\left[0,1\right]$, if it is fixed $x\in\left(-1,0\right]$, it is obtained $x\mathrm{H}_1=\mathrm{H}_1x=\mathrm{H}_1$ that is a submonoid; if $x=-1$ then $x\mathrm{H}_1=\mathrm{H}_1x=\left\{-1\right\}$ that is a subsemigroup.

In the same way for $\mathrm{H}_2=\left[0,1\right]$ and $x\in\left[0,1\right)$, it is obtained $x\mathrm{H}_2=\mathrm{H}_2x=\mathrm{H}_2$ that also is a submonoid; if $x=1$ then $x\mathrm{H}_2=\mathrm{H}_2x=\left\{1\right\}$ that is a subsemigroup.

Proposition 7: The set $C=\left\{x\mathrm{H} \mid x\in\mathrm{NG}\right\}$ form an Abelian NeutroGroup endowed with the operation $\bar{\oplus}$ such that if $x,y\in\mathrm{NG}$, then $\left(x\mathrm{H}\right)\bar{\oplus}\left(y\mathrm{H}\right)=\left(x\,\bar{\oplus}\,y\right)\mathrm{H}$. *C* satisfies the same axioms and NeutroAxioms of NG.

Proof:

This is trivial from Proposition 3. Here the neutral element is H and the opposite of $x\mathrm{H}$ is $\left(-x\right)\mathrm{H}$ if $x\in\left(-1,1\right)$. Similarly to NG, $\left(-1\right)\mathrm{H}$, IH, and $1\mathrm{H}$ have no opposite.

Revisiting the paper of Silva et al. where a scale in [-10, 10] was assumed instead of the original scale in [-1, 1], this change of scale had one reason, the authors of that paper considered [-10, 10] more appropriate for the respondents to express reliably their preferences.

There is a well-known relationship between the combining function in Prospector shown in Equation 6 and the equation of a uninorm in [0, 1], as can be seen in the following Equation:

$$P'\left(x,y\right)=\frac{xy}{\left(1-x\right)\left(1-y\right)+xy} \tag{7}$$

This mapping is a uninorm if this is defined as $P'\left(0,1\right)=P'\left(1,0\right)=\frac{1}{2}$. The neutral element of this uninorm is $e=\frac{1}{2}$. However, although formula 7 is defined for the values of the classical interval [0, 1] it appears less suitable than using [-1, 1] to aggregate the beliefs for or against the hypothesis.

In fact, in González et al. it is shown the formulas 2, 3, and 4, which allow us to obtain uninorms in any real closed interval $\left[\Psi,\Omega\right]$. These equations allow experts to use any scale, specifically the function for the uninorm in Equation 7 converts as follows:

$$P^{"}_{[\Psi,\Omega]}\left(x,y\right)=\frac{\left(\Omega-\Psi\right)\left(x-\Psi\right)\left(y-\Psi\right)}{\left(\Omega-x\right)\left(\Omega-y\right)+\left(x-\Psi\right)\left(y-\Psi\right)}+\Psi \tag{8}$$

Equation 8 can also be used as a generator for NeutroGroups if they are defined as: $\left(\left[\Psi,\Omega\right]\cup\left\{I\right\},\bar{\oplus}_{[\Psi,\Omega]}\right)$ such that if it is denoted by $NG_{[\Psi,\Omega]}=\left[\Psi,\Omega\right]\cup\left\{I\right\}$, we have:

$$x\,\bar{\oplus}_{[\Psi,\Omega]}\,y=\begin{cases}P''\left(x,y\right),\text{if}\left(x,y\right)\in\left[\Psi,\Omega\right]\setminus\left\{\left(\Psi,\Omega\right),\left(\Omega,\Psi\right)\right\}\text{and}\,x,y\neq\mathrm{I}\\ \mathrm{I},\text{if}\left(x=y=\mathrm{I}\right)\text{or}\left(x=\mathrm{I}\,\text{and}\,y\geq e\right)\text{or}\left(x\geq e\,\text{and}\,y=\mathrm{I}\right)\text{or}\left(x,y\right)\in\left\{\left(\Psi,\Omega\right),\left(\Psi,\Omega\right)\right\}\\ x,\text{if}\left(x<e\,\text{and}\,y=\mathrm{I}\right)\\ y,\text{if}\left(y<e\,\text{and}\,x=\mathrm{I}\right)\end{cases}$$

Where $e=\dfrac{\Psi+\Omega}{2}$ is the neutral element of $P^{"}_{[\Psi,\Omega]}\left(x,y\right)$ and therefore it is also neutral for the NeutroGroup $\mathrm{NG}_{[\Psi,\Omega]}$. Where the pairs of opposite elements when they exist are $x=e-\delta$, $-x=e+\delta$ or $x=e+\delta$, $-x=e-\delta$, and $\delta\in\left[0,\dfrac{\Omega-\Psi}{2}\right]$.

Homomorphisms can be defined between any pair of NeutroGroups $\left(\mathrm{NG}_{[\Psi_1,\Omega_1]},\bar{\oplus}_{[\Psi_1,\Omega_1]}\right)$ and $\left(\mathrm{NG}_{[\Psi_2,\Omega_2]},\bar{\oplus}_{[\Psi_2,\Omega_2]}\right)$.

Proposition 8: Let $\left(\mathrm{NG}_{[\Psi_1,\Omega_1]},\bar{\oplus}_{[\Psi_1,\Omega_1]}\right)$ and $\left(\mathrm{NG}_{[\Psi_2,\Omega_2]},\bar{\oplus}_{[\Psi_2,\Omega_2]}\right)$ be two NeutroGroups, let us define the following linear mapping:

$$f\left(z\right)=\begin{cases}\dfrac{\Omega_2-\Psi_2}{\Omega_1-\Psi_1}\left(z-\Psi_1\right)+\Psi_2,\text{if}\,z\in\left[\Psi_1,\Omega_1\right]\\ \mathrm{I},\text{if}\,z=\mathrm{I}\end{cases} \tag{9}$$

$f(z)$ is a homomorphism from $\mathrm{NG}_{[\Psi_1,\Omega_1]}$ into $\mathrm{NG}_{[\Psi_2,\Omega_2]}$, moreover, it is an isomorphism.

Proof:

It is easy to check that

$$f\left(P^{"}_{[\Psi_1,\Omega_1]}(x,y)\right) = P^{"}_{[\Psi_2,\Omega_2]}\left(f(x),f(y)\right),$$

moreover

$$f\left(\frac{\Omega_1+\Psi_1}{2}\right) = \frac{\Omega_2+\Psi_2}{2},$$

so $f(z)$ maps the neutral element in one NeutroGroup into the neutral element of the other one and also using the arithmetic defined for I we have

$$f\left(x\,\bar{\oplus}_{[\Psi_1,\Omega_1]}\,y\right) = f(x)\,\bar{\oplus}_{[\Psi_2,\Omega_2]}\,f(y).$$

$f(z)$ is a linear map, then it is bijective, and it is an isomorphism of NeutroGroups, where

$$f\left(x\,\bar{\oplus}_{[\Psi_1,\Omega_1]}\,y\right) = f(x)\,\bar{\oplus}_{[\Psi_2,\Omega_2]}\,f(y).$$

The result of this proposition, which is that all $\bar{\oplus}_{[\Psi_1,\Omega_1]}$ are isomorphic each other, guarantees that whatever scale is taken to generate the NeutroGroups the results will be equivalent. Therefore, each designer of the study for decision-making can choose the scale of measurement that is most favorable for the interviewees to evaluate the hypothesis.

As $f(z)$ is an isomorphism of NeutroGroups, we have $Im(f) = \mathrm{NG}_{[\Psi_2,\Omega_2]}$ and $Ker(f) = \{e\}, e = \frac{\Omega_1+\Psi_1}{2}$

On the other hand, it is not difficult to prove that theorems 1 and 2 of groups are also true for these NeutroGroups.

In Silva et al. a finite NeutroGroup is defined from a continuous NeutroGroup in [-10, 10]. This discrete modeling allows decision makers to aggregate discrete values, which labels could be associated with, e.g., the value 0.1 that makes more sense than 0.101 semantically speaking. The latter one is only a more precise approximation of the former. That was modeled by the auxiliary function *round*(x), defined for any real value, which returns the nearest integer number, and also if the distance from x to two integers is the same, the largest one is selected.

Specifically, it can be defined

$$x \oplus_1 y = \frac{round\left(10 \oplus \left(x \,\bar{\oplus}_{[-1,1]}\, y\right)\right)}{10}$$

for

$$x, y \in \{-1, -0.9, \cdots, -0.1, 0, \mathrm{I}, 0.1, 0.2, \cdots, 1\}.$$

Additionally, we can define:

$$x \oplus_n y = round\left(x \,\bar{\oplus}_{[-n,n]}\, y\right) \tag{10}$$

For

$$x, y \in \{-n, -n+1, \cdots, -1, 0, \mathrm{I}, 1, 2, \cdots, n\},\ n \in \mathbb{N}\text{ , n>1.}$$

We can show they are Abelian NeutroGroups that satisfy the same axioms and NeutroAxioms of $NG_{[-n,n]}$. They have finite orders because they are finite, although in the other sense the order is infinite for the reasons explained in Proposition 4, e.g. see that

$$\underbrace{n \,\bar{\oplus}_{[-n,n]}\, n \cdots n \,\bar{\oplus}_{[-n,n]}\, n}_{m\,times} = n \neq 0.$$

For example, if n = 5, we have the Cayley table shown in Table 1.

Table 1. Cayley table corresponding to $\oplus_5$

$\oplus_5$	**-5**	**-4**	**-3**	**-2**	**-1**	**0**	**I**	**1**	**2**	**3**	**4**	**5**
-5	-5	-5	-5	-5	-5	-5	-5	-5	-5	-5	-5	I
-4	-5	-5	-5	-5	-4	-4	-4	-4	-3	-2	0	5
-3	-5	-5	-4	-4	-4	-3	-3	-2	-1	0	2	5
-2	-5	-5	-4	-3	-3	-2	-2	-1	0	1	3	5
-1	-5	-4	-4	-3	-2	-1	-1	0	1	2	4	5
0	-5	-4	-3	-2	-1	0	I	1	2	3	4	5
I	-5	-4	-3	-2	-1	I	I	I	I	I	I	I
1	-5	-4	-2	-1	0	1	I	2	3	4	4	5
2	-5	-3	-1	0	1	2	I	3	3	4	5	5
3	-5	-2	0	1	2	3	I	4	4	4	5	5
4	-5	0	2	3	4	4	I	4	5	5	5	5
5	I	5	5	5	5	5	I	5	5	5	5	5

Source: Authors.

These NeutroGroups satisfy Propositions 3-6 demonstrated for their equivalent NeutroGroups in the continuous interval. Now, let us obtain the Cayley table for n = 3, see Table 2.

Table 2. Cayley table corresponding to $\oplus_3$

$\oplus_3$	**-3**	**-2**	**-1**	**0**	**I**	**1**	**2**	**3**
-3	-3	-3	-3	-3	-3	-3	-3	I
-2	-3	-3	-2	-2	-2	-1	0	3
-1	-3	-2	-2	-1	-1	0	1	3
0	-3	-2	-1	0	I	1	2	3
I	-3	-2	-1	I	I	I	I	I
1	-3	-1	0	1	I	2	2	3
2	-3	0	1	2	I	2	3	3
3	I	3	3	3	I	3	3	3

Source: Authors.

Evidently, there is not any isomorphism of NeutroGroups between these two NeutroGroups above. Therefore there exist a homomorphism $f(z)$ which is not surjective if

$$f:\{-3,-2,-1,0,\mathrm{I},1,2,3\}\rightarrow\{-5,-4,-3,-2,-1,0,\mathrm{I},1,2,3,4,5\}$$

and is not injective if

$$f:\{-5,-4,-3,-2,-1,0,\mathrm{I},1,2,3,4,5\}\rightarrow\{-3,-2,-1,0,\mathrm{I},1,2,3\},$$

therefore these finite NeutroGroups do not satisfy Proposition 8.

DISCUSSION AND CONCLUSION

According to the authors' knowledge, the approach to NeutroAlgebras to solve practical problems is available for the first time in Silva et al. (Silva-Jiménez et al., 2021), where a finite NeutroAlgebra generated by the uninorm associated with the combining function in Prospector is used. This function is the aggregator in the mining expert system of this name. In Silva et al. it is applied to solve a decision-making problem on the immigration issue. The other examples of NeutroAlgebras that can be consulted in the literature are derived from variations of classical algebraic structures, with an exclusively theoretical intention, (Agboola, 2020a, 2020b; Kandasamy et al., 2020).

The authors of this chapter take advantage of the example in Silva et al. to delve further into its theoretical properties. This will allow us the dissemination of this recent theory that could be controversial among the scientific community of Neutrosophy and beyond since mathematics is dependent on bivalent logic, where the axioms must be fulfilled for 100% of the elements of the algebraic structure. However, this NeutroAlgebra establishes certain decision-making criteria, where the symbol I is explicitly used to denote the contradiction that exists when there is absolute evidence for and against a hypothesis. Obviously, it is considered that when two criteria evaluated as I are aggregated, its output is I.

Less obvious is when the indeterminate value is aggregated with a negative number, and then the negative value is obtained; whereas when it is aggregated with a positive number, we obtain indeterminate. We can interpret these criteria like an implicitly defined order where the positive evaluations are the most preferred, followed by 0 and I, and finally, there are the negative numbers. The rule consists of output the less preferred of two values when one of them is indeterminate. This criterion was successfully used in the decision-making proposed by Silva et al., even though this means sacrificing some classical axioms of algebra.

The authors of this chapter take the proposed method as an example to illustrate that NeutroAlgebras can be based on an interesting theoretical basis and also can have a practical application that would not be obtained from classical algebraic structures since it was necessary to give up an axiom such as associativity which may not be fulfilled for all elements. In this case, the criterion was to prioritize the decision-maker's common sense over mathematical elegance, that was what they did when explicitly they assumed and symbolized the contradiction.

Didactically speaking, this is the opportunity to motivate undergraduate and postgraduate students in sciences such as mathematics, management, economics, among others, to delve into these new theories, because it simultaneously has the elegance of mathematical theories and the applicability to solve real-life problems.

Specifically, we have shown that what was called NeutroAlgebra in Silva et al. can be defined as an Abelian NeutroGroup in a continuous interval, which has infinite order, and although it is redefined on any scale, the results always will be equivalent because there is a linear isomorphism between them.

As in the original method, finite NeutroGroups can be defined from those defined in continuous intervals. This chapter explains how to generalize both cases to any scale. The finite NeutroGroups preserve most of the properties of their equivalent continuous cases, however, in the finite case, there are no isomorphisms between NeutroGroups of different scales.

In future works, the authors will intend to find other applications of this theory in other problems of real-life or in the foundations of mathematics and physics. This is due to the ample applicability of classical groups that also we expect to find in this new theory.

REFERENCES

Abobala, M. (2021). Semi Homomorphisms and Algebraic Relations between Strong Refined Neutrosophic Modules and Strong Neutrosophic Modules. *Neutrosophic Sets and Systems*, *39*, 107–119.

Agboola, A. A. A. (2020a). Introduction to NeutroGroups. *International Journal of Neutrosophic Science*, *6*(1), 41–47.

Agboola, A. A. A. (2020b). On Finite NeutroGroups of Type-NG [1,2,4]. *International Journal of Neutrosophic Science*, *10*(2), 84–95.

Bordignon, F., & Gomide, F. (2014). Uninorm based evolving neural networks and approximation capabilities. *Neurocomputing*, *127*, 13–20. doi:10.1016/j.neucom.2013.04.047

De-Baets, B., & Fodor, J. (1999). Van Melle's combining function in MYCIN is a representable uninorm: An alternative proof. *Fuzzy Sets and Systems*, *104*(1), 133–136. doi:10.1016/S0165-0114(98)00265-6

Deschrijver, G., & Kerre, E. E. (2004). Uninorms in L*-fuzzy set theory. *Fuzzy Sets and Systems*, *148*, 243–262. doi:10.1016/j.fss.2003.12.006

Fodor, R., Yager, R., & Rybalov, A. (1997). Structure of uninorms. *International Journal of Uncertainty, Fuzziness and Knowledge-based Systems*, *5*(04), 411–427. doi:10.1142/S0218488597000312

Gayen, S., Smarandache, F., Jha, S., Singh, M. K., Broumi, S., & Kumar, R. (2020). Soft Subring Theory Under Interval-valued Neutrosophic Environment. *Neutrosophic Sets and Systems*, *36*, 193–219.

González-Caballero, E., Leyva-Vázquez, M., & Smarandache, F. (2021). On Neutrosophic Uninorms. *Neutrosophic Sets and Systems*, *45*, 340–348.

González-Caballero, E., Smarandache, F., & Leyva-Vázquez, M. (2019). On Neutrosophic Offuninorms. *Symmetry*, *11*(9), 1136. doi:10.3390ym11091136

González-Hidalgo, M., Massanet, S., Mir, A., & Ruiz-Aguilera, D. (2014). *A new edge detector based on uninorms*. Paper presented at the 15th International Conference on Information Processing and Management of Uncertainty in Knowledge-Based Systems. 10.1007/978-3-319-08855-6_19

Hajék, P. (1985). Combining functions for certainty degrees in consulting systems. *International Journal of Man-Machine Studies*, *22*(1), 59–76. doi:10.1016/S0020-7373(85)80077-8

Hart, P. E., Duda, L. O., & Einaudi, M. T. (1978). PROSPECTOR-A Computer-Based Consultation System for Mineral Exploration. *Mathematical Geology*, *10*(5), 589–610. doi:10.1007/BF02461988

Ibrahim, M. A., & Agboola, A. A. A. (2020). NeutroVector Spaces I. *Neutrosophic Sets and Systems*, *36*, 328–351.

Isaacs, I. M. (2008). *Finite Group Theory: Graduate Studies in Mathematics* (Vol. 92). American Mathematical Society.

Kandasamy, V., Kandasamy, I., & Smarandache, F. (2020). NeutroAlgebra of Neutrosophic Triplets using {Zn, ×} NeutroAlgebra of Neutrosophic Triplets using {Zn, ×}. *Neutrosophic Sets and Systems*, *38*, 509–523.

Marík, V., & Kouba, Z. (1991). Some knowledge-acquisition methods for Prospector-like systems. *Knowledge-Based Systems*, *4*(4), 225–230. doi:10.1016/0950-7051(91)90052-4

Sankari, H., & Abobala, M. (2020). AH-Homomorphisms in Neutrosophic Rings and Refined Neutrosophic Rings. *Neutrosophic Sets and Systems*, *38*, 524–536.

Silva-Jiménez, D., Valenzuela-Mayorga, J. A., Roja-Ubilla, M. E., & Batista-Hernández, N. (2021). NeutroAlgebra for the evaluation of barriers to migrants' access in Primary Health Care in Chile based on PROSPECTOR function. *Neutrosophic Sets and Systems*, *39*, 1–9.

Smarandache, F. (2005). *Unifying Field in Logics: Neutrosophic Logic. Neutrosophy Set, Neutrosophic Probability and Statistics*. American Research Press.

Smarandache, F. (2020a). Introduction to NeutroAlgebraic Structures and AntiAlgebraic Structures (revisited). *Neutrosophic Sets and Systems*, *31*, 1–16.

Smarandache, F. (2020b). NeutroAlgebra is a Generalization of Partial Algebra. *International Journal of Neutrosophic Science, 2*(1), 8-17.

Yager, R. R. (2001). Uninorms in fuzzy system modeling. *Fuzzy Sets and Systems*, *122*(1), 167–175. doi:10.1016/S0165-0114(00)00027-0

Yager, R. R., & Kreinovich, V. (2003). Universal approximation theorem for uninorm-based fuzzy systems modeling. *Fuzzy Sets and Systems*, *140*(2), 331–339. doi:10.1016/S0165-0114(02)00521-3

Yager, R. R., & Rybalov, A. (1996). Uninorm aggregation operators. *Fuzzy Sets and Systems*, *80*(1), 111–120. doi:10.1016/0165-0114(95)00133-6

Zhang, C., & Luo, X. (1999). Isomorphic Transformations of Uncertainties for Incorporating EMYCIN-Style and PROSPECTOR-Style Systems into a Distributed Expert System. *Journal of Computer Science and Technology*, *14*(4), 386–392. doi:10.1007/BF02948741

ADDITIONAL READING

Agboola, A. (2020). Introduction to AntiGroups. *International Journal of Neutrosophic Science*, *12*(2), 71–80.

Agboola, A., Ibrahim, M., & Adeleke, E. (2020). Elementary Examination of NeutroAlgebras and AntiAlgebras viz-a-viz the Classical Number Systems. *International Journal of Neutrosophic Science*, *4*, 16–19. doi:10.54216/IJNS.040102

Al-Tahan, M., Smarandache, F., & Davvaz, B. (2021). NeutroOrderedAlgebra: Applications to Semigroups. *Neutrosophic Sets and Systems*, *39*, 133–147.

Ibrahim, M. A., & Agboola, A. A. A. (2020). *Introduction to NeutroHyperGroups* (Vol. 38). Infinite Study.

Shafarevich, I. R. (1987). *Algebra I: Basic Notions of Algebra*. Springer-Verlag.

Smarandache, F. (2020). *Neutroalgebra is a generalization of partial algebra*. Infinite Study.

Smarandache, F. (2020). Generalizations and Alternatives of Classical Algebraic Structures to NeutroAlgebraic Structures and AntiAlgebraic Structures. *Journal of Fuzzy Extension & Applications*, *1*(2), 85–87.

Smarandache, F., Єahin, M., Bakbak, D., Uluçay, V., & Kargn, A. (2021). *NeutroAlgebra Theory Volume I*. Infinite Study.

KEY TERMS AND DEFINITIONS

Abelian Group: Commutative group.

AntiAxiom: Axiom of an algebraic structure that is not satisfied by any of the elements.

Combining Function: Function defined to aggregate assessments in expert systems like Prospector.

Group: Algebraic structure consisting of a set of elements and an operation satisfying a law of composition, associativity, the existence of a neutral element, and every element has an opposite or an inverse.

NeutroAlgebra: Algebraic structure containing at least one NeutroAxiom and none AntiAxiom.

NeutroAxiom: Axiom of an algebraic structure that is satisfied by only one part of the elements.

NeutroGroup: NeutroAlgebra based on the structure of a group.

Prospector: Expert system designed to predict in the search of minerals.

Uninorm: An operator of fuzzy logic satisfying the axioms of t-norm and t-conorms like commutativity, associativity, non-decreasing, and the existence of a neutral element *e*. The main difference among uninorms with t-norms and t-conorms is that *e* can vary in (0, 1).

Chapter 9
Neutro Algebra and Neutro Group

Suman Das
Tripura Univerity, India

Rakhal Das
Tripura Univerity, India

Surapati Pramanik
https://orcid.org/0000-0002-8167-7026
Nandalal Ghosh B. T. College, India

ABSTRACT

This chapter improves and extends several definitions, properties, and applications of neutro-algebras. In classical algebraic structures, the laws of compositions on a given set are well-defined. In fuzzy algebraic structures, the laws of compositions on a given set are not well-defined exactly but with a membership value. This chapter examines those structures which have huge applications or strong potential for applications outside of mathematics, namely computer science or engineering. The study emphasizes that the nature of neutro-algebra and anti-neutro algebra differs from that of the classical algebra and fuzzy algebra. This chapter presents Venn diagram for neutro-function and establishes some relations between the partial algebra and neutro-algebra. The main contribution of the chapter is the introduction of the notion of neutro-group.

DOI: 10.4018/978-1-6684-3495-6.ch009

INTRODUCTION

The early years of the 20th century brought forth uncertainty in scientific study. In 1965, Zadeh (1965) defined the Fuzzy Set (FS) to deal with uncertainty in non-stochastic sense. Then, FS has been extended and generalized to incorporate indeterminacy and inconsistency in many directions (Atanassov, 1986; Smarandache, 1998). Single-Valued Neutrosophic Set (Wang, Smarandache, Zhang, & Sunderraman, 2010) was proposed as a subclass of Neutrosophic Set (NS) (Smarandache, 1998). Smarandache showed extreme interest in neutrosophy (Smarandache, 1998) and published a plethora of new theories dealing with neutrosophics (Smarandache 2013; 2017a; 2017b; 2019; 2020a; 2020b; 2021). The theory of NSs got enormous popularity after the publication of international journal, namely, "Neutrosophic Sets and Systems". The main thrusts of NSs in research areas include decision making (Peng & Dai, 2020), conflict resolution (Pramanik & Roy, 2014), image processing (Koundal, Gupta & Singh, 2016), video tracking (Hu, Ye, Fan, Shen, Huang & Pi, 2017), fault diagnosis (Ye, 2017), air surveillance (Fan, Hu & Li, 2019), cluster analysis (Ye, 2014), and so on.

Theories and applications of NSs can be found in the studies (Smarandache & Pramanik, 2016; 2018; El-Hefenawy et al., 2016; Broumi et al., 2018; Khan et al., 2018; Pramanik, Mallick & Dasgupta, 2018; Nguyen, Son, Ashour & Dey, 2019; Peng & Dai, 2020; Pramanik, 2020; Muzaffar, Nafis, & Sohail, 2020).

Smarandache (2019) introduced the neutro-defined and anti-defined laws, as well as the neutro-axiom and anti-axiom, inspired by neutrosophy (Smarandache, 1998), that helped in creating the new research areas called neutro-structures and anti-structures. Classical algebraic structure is an algebraic structure dealing only with classical axioms which are totally true.

Then, an neutro-algebraic structure (Smarandache, 2020a) is an algebraic structure that has at least one neutro-axiom and one anti-axioms. Again, in all classical algebraic structures, the axioms (associativity, commutativity, etc.) defined on a set are totally true, but it is again a restrictive case, because there are numerous situations in science and in any domain of knowledge when an axiom defined on a set may be only partially-true (and partially-false), that are called neutro-axiom, or totally false that are called anti- axiom. Therefore, a neutro-algebra (Smarandache, 2020b) is an algebra that has at least one neutro-operation or one neutro-axiom (axiom that is true for some elements, Indeterminate for other elements, and false for the other elements).

In 2013, Agboola and Davvaz (2013) deduced the connections between NSs and algebraic hyper-structures. Several researchers presented the neutrosophic hyper-group (Agboola & Davvaz, 2013), neutrosophic hyper vector spaces (Agboola & Akinleye, 2015), neutro hyper-group (Ibrahim & Agboola, 2020a), neutrosophic canonical hyper-group (Agboola & Davvaz, 2014) and neutrosophic hyper-rings (Ibrahim & Agboola, 2020b).

Agboola (2020a, 2020b, 2020c) introduced formally the notions of neutro-groups, neutro-sub-groups, neutro-rings, neutro-subrings, neutro-ideal, neutro-quotient-rings, anti-groups and proved a few properties of these structures and their sub-structures. Rezaei and Smarandache (2020) presented the notion of neutro-BE-algebras and anti-BE-algebras. Agboola and Ibrahim (2020) introduced the concept of anti-rings and established some of their properties. Ibrahim and Agboola (2020c) introduced the ideas of neutro-vector spaces.

A partial algebra (Burmeister, 1982; 1993) is an algebra that has at least one partial operation, and all its axioms are classical axioms, which are true for all elements. Smarandache (2020b) established that neutro-algebra is a generalization of partial algebra, and provided a few examples of neutro-algebras that are not partial algebras. In the same study, Smarandache (2020b) introduced the neutro-function and neutro-operation.

This chapter establishes that neutro-algebra is a generalization of partial algebra, and provides a few examples of neutro-algebras that are not partial algebras. This chapter also analyzes the neutro-function and neutro-operation. The main contribution of the chapter is the introduction of the notion of neutro-group.

The rest of the chapter is organized as follows:

Next section presents the preliminaries of neutro-algebra, partial algebra, neutro-functions, etc. The following next section presents the main results of the chapter namely, Venn diagram of neutro-function, relation between neutro-algebra and partial algebra and neutro-group. The following section concludes the chapter by stating the contribution and future scope of the research.

PRELIMINARIES

This section presents some basic definitions, results to develop of the present study.

Definition 1: (Smarandache, 2019) A neutro algebra is an algebra which has at least one neutro- operation or one neutro- axiom (axiom that is true for some elements, indeterminate for other elements, and false for other elements), and no anti-axiom.

Note 1: The neutro-algebra is a generalization of partial algebra, which is an algebra that has at least one partial operation, while all its axioms are totally true (classical axioms).

Note 2: Comparison between the partial algebra and the neutro-algebra (Smarandache, 2020b)

1. When the neutro-algebra has no neutro-axiom and no outer-operation, it coincides with the partial algebra.
2. There are neutro-algebras that have no neutro-operations, but have neutro-axioms. These are different from partial algebras.
3. Neutro-algebras that have both neutro-operations and neutro-axioms.
4. These are different from partial algebras too.

Definition 2: (Reichel, 1984) A function f: X→Y is called a partial function if it is well-defined for some elements in X, and undefined for all the other elements in X. Therefore, there exist some elements $a \in$ X such that $f(a) \in$ Y (well-defined), and for all other elements b ∈ X, one has f(b) is undefined.

Definition 3: (Smarandache, 2019) A function f: X→Y is called a neutro-function if it has elements in X for which the function is well-defined {degree of truth (T)}, elements in X for which the function is indeterminate {degree of indeterminacy (I)}, and elements in X for which the function is outer-defined {degree of falsehood (F)}, where T, I, F ∈ [0, 1], with (T, I, F) ≠ (1, 0, 0) that represents the (Total) Function, and (T, I, F) ≠ (0, 0, 1) that represents the anti-function.

Definition 4: (Smarandache, 2020a) A function $f(x)$ is said to be classical function if $f(x)$ is well-defined for all the elements in its domain of definition.

Definition 5: (Smarandache, 2020a) A function $f(x)$ is said to be neutro-function if the function is partially well-defined (T), partially indeterminate (I), and partially outer-defined (F) on its domain of definition, where (T, I, F) are different from (1, 0, 0), and (0,0, 1).

Definition 6: (Smarandache, 2019) A function $f(x)$ is said to be anti-function if the function is outer-defined for all the elements in its domain of definition.

Definition 7: (Smarandache, 2020a) The (classical) universal algebra (or general algebra) is a branch of mathematics that studies classes of (classical) algebraic structures.

Definition 8: (Smarandache, 2020a) An algebraic structure (or algebra) is a nonempty set A endowed with some (totally well-defined) operations (functions) on A, and satisfying some (classical) axioms (totally true) - according to the universal algebra.

Definition 9: (Smarandache, 2020a) A partial algebra is an algebra defined on a nonempty set P_A that is endowed with some partial operations (or partial functions: partially well-defined, and partially undefined). While the axioms (laws) defined on a partial algebra are all totally (100%) true.

Classification of Algebras

Smarandache (2020b) classified algebras as follows:

1. A classical algebra is a nonempty set C_A that is endowed with total operations (or total functions, i.e., true for all set's elements) and (classical) axioms (also true for all set's elements).
2. A neutro-algebra (or neutro-algebraic structure) is a nonempty set N_A that is endowed with at least one neutro-operation (or neutro-function), or one neutro-axiom that is referred to the set's (partial-, neutro-, or total-) operations, and no anti-operation and no anti-axiom.
3. An anti-algebra (or anti-algebraic structure) is a nonempty set A_A that is endowed with at least one anti-operation (or anti-function) or at least one anti-axiom.

Therefore, the neutrosophic triplet is of the form:

<Algebra, neutro-algebra (which includes the partial algebra), anti-algebra>.

Definition 10: (Smarandache, 2020b) The universal neutro-algebra (or general neutro-algebra) is a branch of neutrosophic mathematics that studies classes of neutro- algebras and anti-algebras.

MAIN RESULTS

This section establishes the main results of this chapter.

Venn-Diagram of Neutro-Function

Let

$$\mathcal{U} = \{ a_1, a_2, a_3, a_4, a_5, a_6, a_7, a_8, a_9, a_{10}, a_{11}, a_{12}, a_{13}, a_{14}, a_{15} \}$$

be a universe of discourse, and two of its nonempty subsets

$$\mathcal{X} = \{ a_1, a_2, a_3, a_4, a_5, a_6, a_7, a_8 \},\ \mathcal{Y} = \{ a_8, a_9, a_{10}, a_{11}, a_{12}, a_{13} \},$$

and the function ψ constructed as follows:

$\psi : \mathcal{X} \to \mathcal{Y}$ such that

$\psi(a_1) = a_8 \in \mathcal{Y}$ (well-defined);

Figure 1. Venn diagram for neutro function

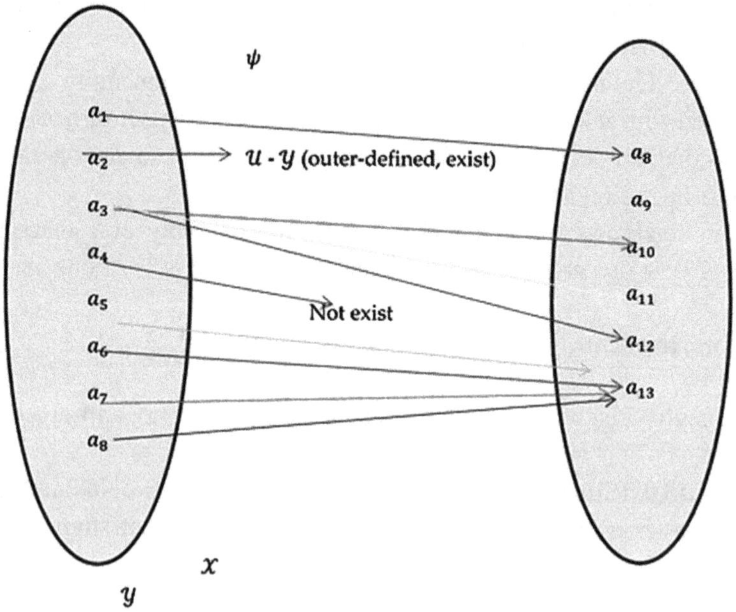

$\psi(a_2) = a_7 \in \mathcal{U} - \mathcal{Y}$ (outer-defined);

$\psi(a_3)$ = undefined (doesn't exist);

$\psi(a_4) = a_{10}$ or a_{11} or a_{12}

(It does exist, but it is not known exactly), therefore $\psi(a_4)$ = indeterminate;

ψ (Some element among a_5, a_6, a_7, a_8) = a_{13}, {i.e., it can be $\psi(a_5) = a_{13}$ or $\psi(a_6) = a_{13}$, it is not sure about}, therefore ψ (indeterminate) = a_{13}.The Venn-diagram of the above neutro-algebra function is

Shown in Figure 1.

Theorem 1: The neutro -algebra is a generalization of partial algebra.

Proof. Partial algebra is equipped with partially defined operations, in partial algebra that is endowed with some partial operations or partial functions: partially well-defined, and partially undefined. While the axioms (laws) defined on a partial algebra are all totally (almost 100%) true. Whereas, a neutro-algebra is a nonempty set that is endowed with at least one neutro-operation or neutro-function, or one neutro-axiom that is referred to the set's partial-, neutro-, or total-operations, in that

Table 1.

*	*x*	y
x	y	*x*
y	x	*x* or *y*

case undefined and neutro membership will be greater than 0%. Hence, they are neutro-algebras according with the above definition of neutro-algebras.

Theorem 2: The partial algebra is a neutro-algebra but the converse is not true in general.

Proof: The partial algebra is a neutro-algebra is proved in Theorem 3.1.

Remark 1: The converse of the above theorem may not be true in general, which follows from the following counter-example.

Example 1: Let $\mathcal{U} = \{x, y, z\}$ be a universe of discourse, and $\mathcal{X} = \{x, y\}$ be one of its nonempty subsets structure $\mathcal{X} = (\mathcal{X}, *)$, constructed as shown in Table 1 using Cayley Table.

Here, * is an indeterminate-operation. Since, y*y = x or y (indeterminate), but for any element other than y*y, x*y is well-defined. The same, because * is not partially defined (since y*y ≠ undefined), $\mathcal{X}$ cannot be a partial algebra.

Similarly, the axiom of commutatively is totally true, since x*y and y*x are defined, and they are equal: $x*y = x = y*x$.

Therefore, $\mathcal{X}$ equipped with the axiom of commutativity is a neutro-algebra).

Neutro-Group

Definition 11: Let $\mathcal{U}$ be a universe of discourse, and * be the neutro-operation on $\mathcal{U}$. Then the neutro algebra ($\mathcal{U}$, *) is called a neutro-group if the following conditions are satisfied:

1. A classical **axiom** defined on a nonempty set is an axiom that is totally true (i.e., true for all set's elements).
2. A **neutro-axiom** (or neutrosophic axiom) defined on a nonempty set is an axiom that is true for some set's elements [degree of truth (T)], indeterminate for other set's elements [degree of indeterminacy (I)], and false for the other set's elements [degree of falsehood (F)], where $T, I, F \in [0,1]$, with $(T, I, F) \neq (1,0,0)$ that represents the classical axiom, and $(T, I, F) \neq (0,0,1)$ that represents the anti-axiom.

3. An **anti-axiom** defined on a nonempty set is an axiom that is false for all set's elements.

Therefore, we form the neutrosophic triplet as follows:

neutrosophic triplet = **<axiom, neutro-axiom, anti-axiom>**

= **<axiom(1, 0, 0), neutro-axiom(t, i, f), anti-axiom(0, 0, 1)>,**

where $(T, I, F) \notin \{(1, 0, 0), (0, 0, 1)\}$.

Definition 12: An neutro-algebra structure is called an neutro-group if the following properties hold:

Neutro-Closure Property

For any two elements *x*, *y* under the neutro-algebraic operation * on the neutro-algebra set $\mathcal{U}$ is neutro closed, i.e., *x* * *y* has the set's (partial-, neutro-, or total) membership value with classical partial and neutro-algebraic membership values.

Neutro-Associative Property

1. there are some elements a, b, c such that a*(b*c) = (a*b)*c (degree of truth T),
2. there are some elements d, e, f such that d*(e*f) or (d*e)*f are indeterminate (degree of indeterminacy I),
3. there are some elements g, h, j such that g*(h*j) is different from (g*h)*j (degree of falsehood F), where (T, I, F) is different from (1,0,0) that represents the classical axiom, and (T, I, F) is different from (0, 0, 1) that represents the anti-axiom.

Existence of Neutro Unit Element

There exists an element in which the total membership value of x*y= y*x= x is more than 0.5.

Existence of Neutro Inverse Element

If there exists the same unit-element x for x and y, or neut(x)=neut(y)=x, since x*x=x and y*x=x*y=y.

Definition 13: If an neutro-group is cumulative under the neutro operation *, then it is called neutro-commutative group.

Example 2: Let $\mathcal{U}$ = {x, y, z, t} be a universe of discourse. The subset $\mathcal{X}$ ={x, y, z}is constructed using Cayley Table as shown in Table 2.

Table 2.

*	x	y	z
x	x	z	z
y	x	x	x
z	z	x	t

Neutro-Defined Law of Composition

For example: x*y = z ∈ $\mathcal{X}$, but z*z = t ∉ $\mathcal{X}$.

Neutro-Associativity

For example: x*(x*z) = x*z = z and (x*x)*z = x*z = z;

While for example: x*(y*z) = x*x = x and (x*y)*z = z*z = t ≠ x.

Neutro-Commutativity

Because, for example: x*z = z*x = z, but x*y = z while y*x = x ≠ z.

Neutro-Unit Element

There exists the same unit-element x for x and z, or neut(x) = neut(z) = x, since x*x = x and z*x = x*z = z. But there is no unit element for b, because y*x = x, not y, for any x ∈ $\mathcal{X}$ (see the above Cayley Table).

Neutro-Inverse Element

From the composition table, with respect to the same unit element x, there exists an inverse element for x, which is x, or inv(x) = x, because x*x = x, and an inverse element for z, which is y, or inv(z) = y, because z*y = y*z = x. But there is no inverse element for y, since y has no unit element. Therefore, ($\mathcal{X}$, *) is a neutro defined neutro commutative neutro- group.

CONCLUSION

This chapter presents Venn diagram for neutro-function and establishes some relations between partial algebra and neutro-algebra. The main contribution of the chapter is the introduction of the notion of neutro-group. In future, the developed concept of neutro-group can further be investigated. Its core properties can deeply be explored further. The main impact of the notion of neutro-group will be in the applications in neutro-algebra and real-life problems.

FUNDING

This research received no specific grant from any funding agency in the public, commercial, or not-for-profit sectors.

CONFLICTS OF INTEREST

The authors declare that the research has been conducted in the absence of any commercial or financial relationships that could be construed as a potential conflict of interest.

ACKNOWLEDGMENT

The authors thank Prof. Florentin Smarandache, Mathematics, Physical and Natural Sciences Division, University of New Mexico, Gallup Campus, NM 87301, USA, for his constructive comments and suggestions which help immensely in developing the chapter.

REFERENCES

Agboola, A. A. A. (2020a). Introduction to neutrorings. *International Journal of Neutrosophic Science*, *7*(2), 62–73.

Agboola, A. A. A. (2020b). On finite neutrogroups of type-NG [1,2,4]. *International Journal of Neutrosophic Science*, *10*(2), 84–95.

Agboola, A. A. A. (2020c). Introduction to anti groups. *International Journal of Neutrosophic Science*, *12*(2), 71–80.

Agboola, A. A. A., & Akinleye, S. A. (2015). Neutrosophic hypervector spaces. *ROMAI Journal, 11*, 1–16.

Agboola, A. A. A., & Davvaz, B. (2013). Introduction to neutrosophic hypergroups. *ROMAI Journal, 9*(2), 1–10.

Agboola, A. A. A., & Davvaz, B. (2014). On neutrosophic canonical hypergroups and neutrosophic hyperrings. *Neutrosophic Sets and Systems, 2*, 34–41.

Agboola, A. A. A., & Ibrahim, M. A. (2020). Introduction to antirings. *Neutrosophic Sets and Systems, 36*, 293–307.

Atanassov, K. T. (1986). Intuitionistic fuzzy sets. *Fuzzy Sets and Systems, 20*(1), 87–96. doi:10.1016/S0165-0114(86)80034-3

Broumi, S., Bakali, A., Talea, M., Smarandache, F., Uluçay, V., Sahin, S., & Pramanik, S. (2018). Neutrosophic sets: An overview. In F. Smarandache & S. Pramanik (Eds.), *New trends in neutrosophic theory and applications* (Vol. 2, pp. 403–434). Pons Editions.

Burmeister, P. (1982). Partial algebras-survey of a unifying approach towards a two-valued model theory for partial algebras. *Algebra Universalis, 15*(1), 306–358. doi:10.1007/BF02483730

Burmeister, P. (1993). Partial algebras-An introductory survey. In I. G. Rosenberg & G. Sabidussi (Eds.), *Algebras and orders*. Springer. doi:10.1007/978-94-017-0697-1_1

El-Hefenawy, N., Metwally, M. A., Ahmed, Z. M., & El-Henawy, I. M. (2016). A review on the applications of neutrosophic sets. *Journal of Computational and Theoretical Nanoscience, 13*(1), 936–944. doi:10.1166/jctn.2016.4896

Fan, E., Hu, K., & Li, X. (2019, March). Review of neutrosophic-set-theory-based multiple-target tracking methods in uncertain situations. In *2019 IEEE International Conference on Artificial Intelligence and Computer Applications* (pp. 19-27). IEEE. 10.1109/ICAICA.2019.8873497

Hu, K., Ye, J., Fan, E., Shen, S., Huang, L., & Pi, J. (2017). A novel object tracking algorithm by fusing color and depth information based on single valued neutrosophic cross-entropy. *Journal of Intelligent & Fuzzy Systems, 32*(3), 1775–1786. doi:10.3233/JIFS-152381

Ibrahim, M. A., & Agboola, A. A. A. (2020a). Introduction to neutro hypergroups. *Neutrosophic Sets and Systems, 38*, 15–32.

Ibrahim, M. A., & Agboola, A. A. A. (2020b). Introduction to Neutrosophic Hypernear-rings. *International Journal of Neutrosophic Science*, *10*(1), 9–22.

Ibrahim, M. A., & Agboola, A. A. A. (2020c). Introduction to neutrovector space I. *Neutrosophic Sets and Systems*, *36*, 329–350.

Khan, M., Son, L. H., Ali, M., Chau, H. T. M., Na, N. T. N., & Smarandache, F. (2018). Systematic review of decision making algorithms in extended neutrosophic sets. *Symmetry*, *10*(8), 314. doi:10.3390ym10080314

Koundal, D., Gupta, S., & Singh, S. (2016). Applications of neutrosophic sets in medical image denoising and segmentation. In F. Smarandache & S. Pramanik (Eds.), *New trends in neutrosophic theory and application* (pp. 257–275). Pons Editions.

Muzaffar, A., Nafis, N. T., & Sohail, S. S. (2020). Neutrosophy logic and its classification: An overview. *Neutrosophic Sets and Systems*, *35*, 239–251.

Nguyen, G. N., Son, L. H., Ashour, A. S., & Dey, N. (2019). A survey of the state-of-the-arts on neutrosophic sets in biomedical diagnoses. *International Journal of Machine Learning and Cybernetics*, *10*(1), 1–13. doi:10.100713042-017-0691-7

Peng, X., & Dai, J. (2020). A bibliometric analysis of neutrosophic set: Two decades review from 1998 to 2017. *Artificial Intelligence Review*, *53*(1), 199–255. doi:10.100710462-018-9652-0

Pramanik, S. (2020). Rough neutrosophic set: An overview. In F. Smarandache & S. Broumi (Eds.), *Neutrosophic theories in communication, management and information technology* (pp. 275–311). Nova Science Publishers.

Pramanik, S., Mallick, R., & Dasgupta, A. (2018). Contributions of selected Indian researchers to multi-attribute decision making in neutrosophic environment. *Neutrosophic Sets and Systems*, *20*, 108–131.

Pramanik, S., & Roy, T. K. (2014). Neutrosophic game theoretic approach to Indo-Pak conflict over Jammu-Kashmir. *Neutrosophic Sets and Systems*, *2*, 82–101.

Reichel, H. (1984). *Structural induction on partial algebras*. Akademie-Verlag.

Rezaei, A., & Smarandache, F. (2020). On neutro-BE-algebras and anti-BE-algebras (revisited). *International Journal of Neutrosophic Science*, *4*(1), 8–15.

Smarandache, F. (1998). *Neutrosophy: neutrosophic probability, set, and logic: analytic synthesis & synthetic analysis*. American Research Press.

Smarandache, F. (2013). n-Valued refined neutrosophic logic and its applications in Physics. *Progress in Physics*, *4*, 143–146.

Smarandache, F. (2017a). *Neutrosophic perspectives: triplets, duplets, multisets, hybrid operators, modal logic, hedge algebras, and applications* (2nd ed.). Pons Publishing House.

Smarandache, F. (2017b). *Plithogeny, plithogenic set, logic, probability, and statistics.* Brussels: Pons ASBL.

Smarandache, F. (2019). *Advances of standard and nonstandard neutrosophic theories*. Pons Publishing House.

Smarandache, F. (2020a). Introduction to neutro algebraic structures and anti algebraic structures (revisited). *Neutrosophic Sets and Systems*, *31*, 1–16.

Smarandache, F. (2020b). Neutro algebra is a generalization of partial algebra. *International Journal of Neutrosophic Science*, *2*(1), 8–17.

Smarandache, F. (2021). Plithogenic probability & statistics are generalizations of multivariate probability & statistics. *Neutrosophic Sets and Systems*, *43*, 280–289.

Smarandache, F., & Pramanik, S. (Eds.). (2016). *New trends in neutrosophic theory and applications*. Pons Editions.

Smarandache, F., & Pramanik, S. (Eds.). (2018). *New trends in neutrosophic theory and applications* (Vol. 2). Pons Editions.

Wang, H., Smarandache, F., Zhang, Y., & Sunderraman, R. (2010). Single valued neutrosophic sets. *Review of the Air Force Academy*, *1*, 10–14.

Ye, J. (2014). Single valued neutrosophic minimum spanning tree and its clustering method. *Journal of Intelligent Systems*, *23*(3), 311–324. doi:10.1515/jisys-2013-0075

Ye, J. (2017). Single-valued neutrosophic similarity measures based on cotangent function and their application in the fault diagnosis of steam turbine. *Soft Computing*, *21*(3), 817–825. doi:10.100700500-015-1818-y

Zadeh, L. A. (1965). Fuzzy sets. *Information and Control*, *8*(3), 338–353. doi:10.1016/S0019-9958(65)90241-X

ADDITIONAL READING

Burris, S. N., & Sankappanavar, H. P. (2013). The Horn's theory of Boole's partial algebras. *The Bulletin of Symbolic Logic*, *19*(1), 97–105. doi:10.2178/bsl.1901040

Cayley, P. (1854). On the theory of groups, as depending on the symbolic equation $\theta^n = 1$. *Philosophical Magazine*, *7*, 40–47.

Foulis, D., & Bennett, M. (1994). Effect algebras and unsharp quantum logics. *Foundations of Physics*, *24*(10), 1331–1352. doi:10.1007/BF02283036

Chapter 10
NeutroGroups Generated by Uninorms:
A Theoretical Approach

Erick González Caballero
Technological University of Havana, Cuba

Maikel Leyva
Regional Autonomous University of Los Andes, Ecuador

Jesús Estupiñán Ricardo
Regional Autonomous University of Los Andes, Ecuador

Noel Batista Hernández
Latin American Association of Neutrosophic Sciences, Ecuador

ABSTRACT

NeutroAlgebras and particularly NeutroGroups are novel algebraic structures where at least one of the classical axioms is satisfied by not all the elements of the set. In the theoretical literature consulted by the authors, most of the NeutroGroups examples are obtained from the well known Z_n groups endowed with either product or sum operations. This chapter aims to propose a method for the generation of NeutroGroups in a simple and automatic way from uninorms, with special interest in uninorms based on combining functions in both Mycin and Prospector expert systems. The use of theses uninorms suggests the possible application of NeutroGroups in fields such as decision making as can be found in the consulted literature, as well as for Expert Systems, among others. NeutroGroups of finite or infinite cardinality will be generated, where the axioms of NeutroAsociativity or NeutroInverse are fulfilled, and also indeterminacy is symbolically included within this theory. One family of non-commutative NeutroGroups is proposed.

DOI: 10.4018/978-1-6684-3495-6.ch010

INTRODUCTION

Neutrosophy is the branch of philosophy that treats neutralities. F. Smarandache introduced this theory, where if <A> denotes a concept, idea, theory and so on, a triad (<A>, < neutA >, < AntiA >) could be defined, such that < AntiA > represents what is opposed to <A>, while <NeutA > represents what is neither <A> or <AntiA>, (Smarandache, 2005). Recently, the same author defined the NeutroAlgebras as new algebraic structures arising in the theory of Neutrosophy, so classical structures formed by elements well defined based on always true axioms are generalized to others that may be false or undefined in some cases, (Smarandache, 2020a; Smarandache, 2020b; Smarandache, 2020c). Specifically, in NeutroAlgebras, it is essential that the axioms of algebraic structures are not fulfilled for all elements of the structure, although at least one element must fulfill it. A Partial or Classical Algebra fulfills all the axioms, while the AntiAlgebra is defined such that at least one axiom is not fulfilled at all.

Among the most studied classical algebraic structures are groups, defined as a nonempty set of elements with an operator such that it is internal, which satisfies the associativity axiom, the existence of a neutral element and that each element of the set has an inverse (Isaacs, 2008). In the case of NeutroGroups, these axioms are not fulfilled by all the elements of the set, (Agboola, 2020a).

Classical group theory constitutes an important pillar in mathematical modeling within Physics, Chemistry and cryptography, (Isaacs, 2008). The NeutroAlgebras have been applied to resolve Decision Making problems, (Silva-Jiménez et al., 2021).

One of the purposes argued by F. Smarandache to extend the classical algebraic theory is that in the real world not all axioms are always fulfilled, nor are the operations always well defined.

Apart from these arguments, some well-known results from mathematics and physics indicate the need to include indeterminacy in these sciences that are based on mathematical logic. Some examples are the Uncertainty Principle of Heisenberg and the Gödel's theorems of incompleteness. These ideas suggest the relevance of developing a group theory where indeterminacy is part of the algebraic structure.

The main objective of this chapter is to propose the generation of NeutroGroups from uninorms, (Fodor et al., 1997). The uninorms are defined as the generalization of t-norms and t-conorms where the axioms of commutativity, associativity, increasing and existence of a neutral element are preserved, although in uninorms the neutral element may be a value in the range (0, 1), not 1 as in t- norms, nor 0 as in t-conorms. This aggregation operator was originally defined in the theory of fuzzy sets to intervals [0, 1], then it was extended to other frameworks like the intuitionistic fuzzy sets (Deschrijver & Kerre, 2004), neutrosophic sets (González-Caballero et al.,), and neutrosophic offsets (González-Caballero et al., 2019).

The advantage of what is proposed in this chapter is the automatic generation in a simple way of NeutroGroups of finite or infinite cardinality. The examples that are usually proposed in the literature are based on the groups $\mathbb{Z}_n$ (Kandasamy et al., 2020) endowed with the sum or product operation, in this chapter this universe of NeuroGroups is expanded. In addition, some generating uninorms use the aggregation functions in Mycin medical expert and Prospector mining expert systems, (Kaufmann et al., 1981; Shortliffe, 1976). These last uninorms suggest the possible application of these NeutroAlgebras in expert systems and decision making.

These functions can be extended to a more general field of elements where indeterminacy represented symbolically with the letter "I" is included. This is an advantage for these functions based on experts' evaluations and evidence, e.g., for aggregating totally opposite values, if it is believed that there is both, evidence totally against and evidence totally in favor of a hypothesis, sometimes the combined value is defined as 0, however, it is better to assess it as a contradiction and therefore an indeterminacy in an explicit way.

This chapter is divided according to the following structure; the next section contains the background of the concepts necessary for this chapter to be self-contained, such as the concepts of classical groups, NeutroAlgebras, NeutroGroups, combining functions in Mycin and Prospector. Later, next section contains the definitions, results, and examples for generating NeutroGroups. Last section is devoted to the conclusions. The appendices show one example of NeutroGroups of interest and some computer programs coded in MATLAB language used by authors.

BACKGROUND

Basic Concepts of Group Theory

Group theory within algebra has been extensively studied and there is very little to research on this subject. The groups have been applied in the mathematical modeling of particle physics, in chemistry and cryptography. Formally a group is defined as follows:

Definition 1: (Isaacs, 2008) The ordered pair $(\mathrm{G}, \otimes)$ of the nonempty set G with the operation $\otimes : \mathrm{G} \times \mathrm{G} \to \mathrm{G}$ is said to be a *group* if the following axioms are satisfied:

1. $\forall x, y \in \mathrm{G}$ it is fulfilled $x \otimes y \in \mathrm{G}$ (⊗ is well-defined).
2. $\forall x, y, z \in \mathrm{G}$ it is fulfilled $x \otimes (y \otimes \mathrm{z}) = (x \otimes y) \otimes \mathrm{z}$ (⊗ is *associative*).

3. $\exists e \in \mathrm{G}$ which is called the *neutral element* such that $\forall x \in \mathrm{G}$ they fulfill $e \otimes x = x \otimes e = x$.
4. $\forall x \in \mathrm{G}$, $\exists x^{-1} \in \mathrm{G}$ such that $x \otimes x^{-1} = x^{-1} \otimes x = e$.

Not necessarily the group with the operation ⊗ must be commutative or $\forall x, y \in \mathrm{G}$, $x \otimes y = y \otimes x$. If it fulfills this property, the group is said to be *commutative* or *Abelian*.

The *order* of the group is interpreted in two senses when the cardinality of G is finite, the *order* of G is considered to be the number of its elements, which is denoted by $o(\mathrm{G}) = \mathrm{card}(\mathrm{G})$.

In another sense, the order of $x \in \mathrm{G}$ is considered to be the minimum $m \in \mathbb{N}$ such that

$$x^m = \underbrace{x \otimes x \otimes \cdots \otimes x}_{m\,times} = e$$

and in general, the order of G is the maximum of the orders of its elements. In case there is no such m, it is said that $o(\mathrm{G}) = \infty$.

Example 1:

Classic examples of groups are listed below:

1. $(\mathbb{Z},+)$, $(\mathbb{Q},+)$, $(\mathbb{R},+)$, and $(\mathbb{C},+)$ are the domains of integer, rational, real, and complex, respectively, endowed with the usual sum operation.
2. The product defines commutative group structures in nonzero rational numbers, positive real numbers, and complex numbers modulo 1.
3. Given $n \in \mathbb{N}$, the groups $(\mathbb{Z}_n,+)$ or $(\mathbb{Z}_n,\cdot)$ are defined as the sum or product respectively of $\{0,1,2,\cdots,n-1\}$ modulo n. These are finite groups, Abelian, and cyclic or that can be generated by a single element, that is $\exists a \in \mathrm{G}$, such that $\mathrm{G} = \{a^n : n \in \mathbb{Z}\}$, where $a^{-k} = (a^{-1})^k$.

Basic Concepts of Uninorms

On the other hand, uninorms arose within the framework of fuzzy logic to generalize the concepts of t- norms, which endow the binary definition of AND with an axiomatic in [0, 1] and the t-conorms that do the same with the OR, (Fodor et al., 1997). This axiomatic allows representing algebraically the logical ANDs and ORs within truth values in [0, 1] instead of the classical {0, 1}. Uninorms are defined as below:

Definition 2: (Fodor et al., 1997) The mapping $\mathcal{U}:[0,1]^2 \to [0,1]$ is said to be a *uninorm* if it satisfies the following axioms $\forall x, y, z \in [0,1]$:

1. $\mathcal{U}(x,y) = \mathcal{U}(y,x)$ (Commutativity),
2. $\mathcal{U}(x,\mathcal{U}(y,z)) = \mathcal{U}(\mathcal{U}(x,y),z)$ (Associativity),
3. $y \leq z$ implies $\mathcal{U}(x,y) \leq \mathcal{U}(x,z)$ (Non-decreasing),
4. Thereexists $e \in [0,1]$,calledthe*neutralelement*,suchthat $\forall x \in [0,1], \mathcal{U}(e,x) = x$.

The uninorms generalize the t-norms ($e = 1$) and the t-conorms ($e = 0$). De Baets and Fodor generalized this definition to any closed interval $[a,b] \subset \mathbb{R}$, (De-Baets & Fodor, 1999). Later the authors will explain why.

Another interesting result is the fuzzy uninorms generated by functions as follows (De-Baets & Fodor, 1999; Fodor et al., 1997):

Let $h:[a,b] \to \bar{\mathbb{R}}$ be a mapping, strictly increasing and continuous and $e \in (a,b)$ such that $h(a) = -\infty$, $h(e) = 0$, and $h(b) = \infty$. Then $h(x)$ that is called *additive generator*, generates a uninorm called *representable uninorm* and with neutral element *e* with the help of the following formula:

$$\mathcal{U}(x,y) = h^{-1}(h(x) + h(y)) \; \forall (x,y) \in [a,b]^2 \setminus \{(a,b),(b,a)\} \tag{1}$$

And also either

$\mathcal{U}(a,b) = \mathcal{U}(b,a) = a$ or $\mathcal{U}(a,b) = \mathcal{U}(b,a) = b$.

Uninorms were generalized to other frameworks beyond fuzzy logic theory. When this definition was extended, the uninorms in L^*-Fuzzy set theory was introduced, which satisfy the following definition:

Definition 3: (Deschrijver & Kerre, 2004) The mapping $\mathcal{U}:(L^*)^2 \to L^*$ is said to be a *uninorm in* L^* if it satisfies the axioms of commutativity, associativity, it is non-decreasing and there is an element $e \in L^*$ called the *neutral element*, such that $\forall x \in L^*$ it is fulfilled $\mathcal{U}(x,e) = x$. Where

$L^* = \{(x_1, x_2) \mid (x_1, x_2) \in [0,1]^2 \wedge x_1 + x_2 \leq 1\}$.

In this case there are ordered pairs, where the first element represents the degree of truthfulness and the second element represents the degree of falsehood.

González et al. extended the uninorms to the framework of both the neutrosophic sets, see Definition 4, and the neutrosophic offsets, see Definition 5.

Definition 4: (González-Caballero et al.,) A *neutrosophic uninorm* $\mathbf{U}_{\mathrm{N}}$, is a commutative, increasing and associative mapping,

$$\boldsymbol{U}_{\mathrm{N}}:\left(\left]{}^{-}0,1^{+}\right[\times\left]{}^{-}0,1^{+}\right[\times\left]{}^{-}0,1^{+}\right[\right)^{2}\to\left]{}^{-}0,1^{+}\right[\times\left]{}^{-}0,1^{+}\right[\times\left]{}^{-}0,1^{+}\right[,$$

such that:

$$\boldsymbol{U}_{\mathrm{N}}\left(x\left\langle T_{x},I_{x},F\right\rangle_{x},y\left\langle T_{y},I_{y},F_{y}\right\rangle\right)=\left\langle \boldsymbol{U}_{\mathrm{N}}\mathrm{T}\left(x,y\right),\boldsymbol{U}_{\mathrm{N}}\mathrm{I}\left(x,y\right),\boldsymbol{U}_{\mathrm{N}}\mathrm{F}\left(x,y\right)\right\rangle,$$

where $\mathbf{U}_{\mathrm{N}}\mathrm{T}$ means the degree of membership, $\mathbf{U}_{\mathrm{N}}\mathrm{I}$ the degree of indeterminacy and $\mathbf{U}_{\mathrm{N}}\mathrm{F}$ the degree of non-membership of both, x and y. Additionally, there exists a neutral element

$$e\in\left]{}^{-}0,1^{+}\right[\times\left]{}^{-}0,1^{+}\right[\times\left]{}^{-}0,1^{+}\right[,$$

where

$$\forall x\in\left]{}^{-}0,1^{+}\right[\times\left]{}^{-}0,1^{+}\right[\times\left]{}^{-}0,1^{+}\right[,\ \boldsymbol{U}_{N}\left(e,x\right)=x\,.$$

Note that in this case two elements are aggregated with three components, which are the degrees of truthfulness, indeterminacy and falsehood.

Definition 5: (González-Caballero et al., 2019) Let *c* be a neutrosophic component ($\mathrm{T_O}$, $\mathrm{I_O}$ or $\mathrm{F_O}$). $c:\mathrm{M_o}\to\left[\Psi,\Omega\right]$, where $\Psi\le 0$ and $\Omega\ge 1$. The neutrosophic component *N-offuninorm* $\mathrm{N_O^u}:\left[\Psi,\Omega\right]^{2}\to\left[\Psi,\Omega\right]$ satisfies the following conditions for any elements x, y and z $\in\mathrm{M_O}$:

1. There exists $c\left(e\right)\in\mathrm{M_o}$, such that $\mathrm{N_O^u}\left(c\left(x\right),c\left(e\right)\right)=c\left(x\right)$ (Identity),
2. $\mathrm{N_O^u}\left(c\left(x\right),c\left(y\right)\right)=\mathrm{N_O^u}\left(c\left(y\right),c\left(x\right)\right)$ (Commutativity),
3. If $c\left(x\right)\le c\left(y\right)$ then $\mathrm{N_O^u}\left(c\left(x\right),c\left(z\right)\right)\le\mathrm{N_O^u}\left(c\left(y\right),c\left(z\right)\right)$ (Monotonicity),

4. $N_O^u\left(N_O^u\left(c(x),c(y)\right),c(z)\right)=N_O^u\left(c(x),N_O^u\left(c(y),c(z)\right)\right)$ (Associativity).

This definition contains the same elements of the neutrosophic uninorms, with the particularity that the intervals where these aggregators are defined are generalized to $[\Psi,\Omega]\subset\mathbb{R}$, beyond the classical interval $[0,1]$.

Specifically in González et al. (González-Caballero et al., 2019) the following equation is defined to be used in this chapter:

Let (T_O, I_O or F_O), $c_o : M_o \rightarrow [\Psi,\Omega]$ and (T, I or F), $c_N : M_N \rightarrow [0,1]$ be a neutrosophic component N-offset and a neutrosophic component, respectively. Thus, a bijective mapping exists that transforms every neutrosophic component N-offuninorm to a neutrosophic component uninorm and vice versa.

The equations are as follows:

$$\varphi_3\left(c(x)\right)=\frac{c(x)-\Psi}{\Omega-\Psi} \tag{2}$$

$$\varphi_3^{-1}\left(c(x)\right)=(\Omega-\Psi)c(x)+\Psi \tag{3}$$

Evidently, they are bijective mappings.

If $\hat{U}_N(\cdot,\cdot)$ is a neutrosophic uninorm, then we can define the neutrosophic component N-offuninorm $\widehat{N_O^u}(\cdot,\cdot)$ as follows:

$$\hat{N}_O^u\left(c_O(x),c_O(y)\right)=\varphi_3^{-1}\left(\hat{U}_N\left(\varphi_3\left(c_O(x)\right),\varphi_3\left(c_O(y)\right)\right)\right) \tag{4}$$

Conversely, given $\widehat{N_O^u}(\cdot,\cdot)$ then $\hat{U}_N(\cdot,\cdot)$ can be defined as follows:

$$\hat{U}_N\left(c_N(x),c_N(y)\right)=\varphi_3\left(\widehat{N_O^u}\left(\varphi_3^1\left(c_N(x)\right),\varphi_3^1\left(c_N(y)\right)\right)\right) \tag{5}$$

Note that Equation 4 allows to convert a neutrosophic offuninorm in $[\Psi,\Omega]$ from a neutrosophic uninorm in $[0,1]$.

Uninorms have countless applications in mathematical modeling in different fields such as decision making, fuzzy model systems, in evolving neural networks, digital image processing where morphological operators based on uninorms are defined, as aggregation operators, multi-agent decision making, game theory, among others, see (Appel et al., 2017; Bordignon & Gomide, 2014; González-Hidalgo et al., 2014; Pís & Mesiar, 1989; Sussner & Schuster, 2018; Yager, 2001; Yager & Kreinovich, 2003; Yager & Rybalov, 1996). In González et al. (González-Caballero et al., 2019) the application of neutrosophic offuninorms is exemplified in decision-making problems and game theory with uncertainty.

Combining Functions in Mycin and Prospector

Two uninorms that will be the focus of this chapter are the combining functions in Mycin and Prospector, (Hajék, 1985). The first one is used to aggregate evaluations in the interval $\left[-1,1\right]$ for a medical expert system, while the second one has the same function for decision making in mining, (De-Baets & Fodor, 1999; Hart et al., 1978; Marík & Kouba, 1991; Shortliffe, 1976; Tsadiras & Margaritis, 1998; Zhang & Luo, 1999).

In De Baets and Fodor it is shown that they are uninorms in $\left[a,b\right]=\left[-1,1\right]$ and furthermore they are representable.

Specifically, their formulas are:

$C:\left[-1,1\right]^2\to\left[-1,1\right]$, where:

$$C\left(x,y\right)=\begin{cases} x+y\left(1-x\right)\text{if }\min\left(x,y\right)\geq 0\\ \dfrac{x+y}{1-\min\left(\left|x\right|,\left|y\right|\right)}\text{if }\min\left(x,y\right)<0<\max\left(x,y\right)\\ x+y\left(1+x\right)\text{if }\max\left(x,y\right)\leq 0\end{cases} \tag{6}$$

For the combining function in Mycin, while:

$$P\left(x,y\right)=\frac{x+y}{1+xy} \tag{7}$$

It is used as a combining function in Prospector.

The evaluation $x<0$ means that the expert has evidence or beliefs against the proposition or hypothesis. On the contrary, if $x>0$ means that the expert has evidence in favor of the proposition that he/she is measuring. $x = 0$ means that there is no evidence either for or against the proposition.

Some properties of these functions are that they are Archimedean, that is, when two negative values are aggregated, a value less than the smallest of the aggregated values is obtained; when two positive values are aggregated, a value greater than the largest of the aggregated values is obtained. When a positive value and a negative one are aggregated, it is obtained an intermediate value. The only exception is when $x = -1, 1$, which perform like veto values. The value $e = 0$ is the neutral value. Both functions are strictly increasing, associative, and commutative functions.

The aggregation of the pairs (-1, 1) or (1, -1) undefines these functions, which is why some authors define it by 0 and others by -1.

Both functions are associated with uninorms in the interval $[0,1]$, when they are transformed using the following formulas:

$$C'(x,y) = h'^{-1}(h'(x) + h'(y)) \tag{8}$$

Where,

$$h' : [0,1] \to \bar{\mathbb{R}},\ h'(x) = h_C(2x-1),\ h_C(x) = \begin{cases} \ln(1+x) \text{if } x < 0 \\ -\ln(1-x) \text{if } x \geq 0 \end{cases}$$

and

$$h_C^{-1}(y) = \begin{cases} e^y - 1 \text{if } y < 0 \\ -e^{-y} + 1 \text{if } y \geq 0 \end{cases}.$$

And

$$P'(x,y) = \frac{xy}{(1-x)(1-y) + xy} \tag{9}$$

For the combining function in Mycin and Prospector, respectively. Both have the neutral value $e = \frac{1}{2}$.

Define on all $[-1,1]$ the operations $x \odot_C y = C(x,y)$, $x \odot_P y = P(x,y)$, such that $C(-1,1) = P(-1,1) = 0$. So, both $([-1,1], \odot_C)$ and $([-1,1], \odot_P)$ are Abelian groups with $e = 0$ as the neutral element and $(-x, x)$ as inverse or opposite elements.

Equivalently, $([0,1], \odot_{C'})$ and $([0,1], \odot_{P'})$, with the operations defined as $x \odot_{C'} y = C'(x,y)$, $x \odot_{P'} y = P'(x,y)$ form Abelian groups with the neutral element $e = \frac{1}{2}$ and the pairs $(x, 1-x)$ as inverse elements, where $C'(-1,1) = P'(-1,1) = \frac{1}{2}$.

NeutroAlgebras

F. Samrandache introduced NeutroAlgebras, (Smarandache, 2020c), especially NeutroGroups are algebraic structures where at least one of the conditions of Definition 1 is satisfied by some elements and not by others all the axioms are not totally unsatisfied.

Among the theory presented in the literature, NeutroGroups examples derive from the well-known finite groups $(\mathbb{Z}_n, +)$ or $(\mathbb{Z}_n, \cdot)$ modulo n, (Agboola, 2020a; Agboola, 2020b; Mohammadzadeh & Rezaei, 2020). However, although they were ingeniously obtained, these NeutroGroups do not suggest the possible practical applications of this theory. Apart from the fact that due to the ingenuity in its construction, it is difficult to create NeutroGroups in a simple and automatic way.

The objective of this chapter is to show the generation of NeutroGroups from uninorms in a simple and automatic way. Additionally, the use of the combining functions in Mycin and Propector suggests the possible applicability of what the authors propose. A preview of the applicability of NeutroGroups obtained in this way can be found in decision-making cases (Silva-Jiménez et al., 2021).

NEUTROGROUPS GENERATED BY UNINORMS

The contribution proposed in this chapter will be mainly based on Equations 2, 3 and 4. With the help of these formulas, authors will arrive at new equations of uninorms in the interval $[\Psi, \Omega]$, for example, for the combining functions in Mycin and Propector there are the following formulas:

To calculate the combining function in Mycin define the mapping $C'':[\Psi,\Omega]^2 \to [\Psi,\Omega]$ as follows: input the pair $x,y \in [\Psi,\Omega]$, each of them becomes in a value $x',y' \in [0,1]$ by using formula 2. Later it is calculated

$$C'(x',y') = h'^{-1}(h'(x') + h'(y')) \in [0,1],$$

(De-Baets & Fodor, 1999). Finally it is calculated

$$C''(x',y') = \varphi_3^{-1}(C'(x',y')) \in [\Psi,\Omega].\ C''(\Psi,\Omega) = \frac{\Psi+\Omega}{2}.$$

Moreover, for uninorm in Prospector there is a formula as shown in Equation 10, (De-Baets & Fodor, 1999).

$$P''(x,y) = \begin{cases} \frac{\Psi+\Omega}{2}, \text{if } x=\Psi \text{ and } y=\Omega \text{ or } x=\Omega \text{ and } y=\Psi \\ \frac{(\Omega-\Psi)(x-\Psi)(y-\Psi)}{(\Omega-x)(\Omega-y)+(x-\Psi)(y-\Psi)} + \Psi, \text{otherwise} \end{cases} \quad (10)$$

Note that in each of these cases, the neutral element is $e = \frac{\Psi+\Omega}{2}$.

Proposition 1: Let be a closed interval $[\Psi,\Omega] \subset \mathbb{R}$. The ordered pairs $([\Psi,\Omega], \odot_{C''})$ and $([\Psi,\Omega], \odot_{P''})$ are Abelian groups having $e = \frac{\Psi+\Omega}{2}$ as the neutral element, and are defined by $x \odot_{C''} y = C''(x,y)$ and $x \odot_{P''} y = P''(x,y)$, $\forall x,y \in [\Psi,\Omega]$.

Proof:

It is evident using the axioms of uninorm. The elements and their inverses are $x = e-\delta$, $x^{-1} = e+\delta$ or $x = e+\delta$, $x^{-1} = e-\delta$, where $\delta \in \left[0, \frac{\Omega-\Psi}{2}\right]$. □

Corollary 1:

If $C''(\Psi,\Omega) = v \neq e$ and $P''(\Psi,\Omega) = v \neq e$ $([\Psi,\Omega], \odot_{C''})$, $([\Psi,\Omega], \odot_{P''})$ are NeutroGroups that do not satisfy the axiom of the existence of an inverse.

Proof:

It is obvious.

This type of case presented in the Corollary 1 is not alien to the definition provided in De Baes and Fodor, where it is defined $C\left(-1,1\right)=-1$, although $C\left(-x,x\right)=0$ for every $x\neq -1,1$. This means that the contradiction between having both evidence totally against and totally in favor of a proposition is a contradiction that is aggregated as totally against instead of a null value.

Generally NeutroGroups can be defined from the groups above as follows:

Definition 6: Let $\left[\Psi,\Omega\right]\subset\mathbb{R}$ be a closed interval and let $\mathcal{U}\left(\cdot,\cdot\right)$ be a uninorm over $\left[\Psi,\Omega\right]$. Then the operation $\otimes$ is constructed as follows:

- Interval $\left[\Psi,\Omega\right]$ is generalized to $NG=\left[\Psi,\Omega\right]\cup\left\{\mathrm{I}\right\}$, where "I" is the symbol of indeterminacy.
- Group operation is redefined in NG such that $x\otimes y=\mathcal{U}\left(x,y\right)$ for every $x,y\in\left(\Psi,\Omega\right)$. Also, $\otimes=\otimes=\mathrm{I}$. This means that the aggregation of extreme evaluations is considered indeterminate because it is considered a contradiction.
- In addition it is defined

$$x\otimes\mathrm{I}=\mathrm{I}\otimes x=\begin{cases}\mathrm{I},\text{if } x\geq e \text{ or } x=\mathrm{I}\\ x,\text{if } x<e\end{cases}.$$

Proposition 2: The algebraic structure defined in Definition 6 is an Abelian NeutroGroup.

Proof

This structure is commutative.

Let

$$x\in\left[\Psi,\Omega\right],\ x\otimes\left(x\otimes e\right)=x\otimes x \text{ and } \left(x\otimes x\right)\otimes e=x\otimes x,$$

therefore, for this triple of values the associativity is fulfilled. However, for $x,y\in\left(\Psi,\Omega\right)$ such that $x<e$, $y>e$, and $x\otimes y\neq x$, only take $y=x^{-1}$ for these properties are met. So,

$$x\otimes\left(\mathrm{I}\otimes y\right)=x\otimes\mathrm{I}=x$$

and

$$\left(x \otimes \mathrm{I}\right) \otimes y = x \otimes y \neq x .$$

So for this triple the associativity is not fulfilled.

Apart from this, we have the properties of uninorms that $\forall x \in \left[\Psi, \Omega\right]$ $x \otimes e = x$, but also if $x < e$ it is true that $x \otimes \mathrm{I} = x$, therefore I is a neutral element for all $x < e$.

Additionally, Ψ and Ω are opposites with respect to *e*, but they also are with respect to I.

The idea of this definition was used in implementing NeutroAlgebras for decision making with the combining function in Prospector, see Equation 7, where it is considered $\bar{P}\left(-1,1\right) = \mathrm{I}$, to explicitly represent indeterminacy when aggregating -1 and 1. Moreover, it is rational to define $\bar{P}\left(\mathrm{I},\mathrm{I}\right) = \mathrm{I}$; when I is aggregated with a value > 0, it is taken as I, while if the value is < 0 it is taken as the value itself. This criterion for decision making sacrifices property $P\left(x, y\right)$ of generating an algebraic group, but it provides of more accurate valuations.

Example 2: In this example the authors intend to illustrate how to apply what was explained above from a different uninorm of $C\left(x, y\right)$ and $P\left(x, y\right)$. This uninorm is as follows:

$$U\left(x, y\right) = \begin{cases} 0, \text{if} \left(x, y\right) \in \left\{\left(0,1\right), \left(1,0\right)\right\} \\ 1 - \exp\left(-\frac{1}{c}\ln\left(1-x\right)\ln\left(1-y\right)\right), \text{otherwise} \end{cases}$$

The neutral element is $e = 1 - \exp\left(-c\right)$. This uninorm is generated by the mapping

$$h_c\left(x\right) = \ln\left(-\frac{1}{c}\ln\left(1-x\right)\right),\ x \in \left(0,1\right)$$

and

$$c > 0;\ h_c^{-1}\left(x\right) = 1 - \exp\left(-c \cdot \exp\left(x\right)\right).$$

This uninorm is converted to one on the interval $\left[\Psi,\Omega\right]$ using formulas 2, 3, and 4. See the following equation:

$$\bar{U}\left(x,y\right)=\begin{cases}\Psi,\text{if}\left(x,y\right)\in\left\{\left(\Psi,\Omega\right),\left(\Omega,\Psi\right)\right\}\\ \Omega-\left(\Omega-\Psi\right)\exp\left(-\frac{1}{c}\ln\left(\frac{\Omega-x}{\Omega-\Psi}\right)\ln\left(\frac{\Omega-y}{\Omega-\Psi}\right)\right),\text{otherwise}\end{cases}$$

With the neutral element $\bar{e}=\Omega-\left(\Omega-\Psi\right)\exp\left(-c\right)$.

If it is specified even more and are fixed $c=1$ and $\left[\Psi,\Omega\right]=\left[-10,10\right]$, we have:

$$\bar{U}_{10}\left(x,y\right)=\begin{cases}-10,\text{if}\left(x,y\right)\in\left\{\left(-10,10\right),\left(10,-10\right)\right\}\\ 10-20\exp\left(-\ln\left(\frac{10-x}{20}\right)\ln\left(\frac{10-y}{20}\right)\right),\text{otherwise}\end{cases}$$

With the neutral element $\bar{e}_{10}=10-20\exp\left(-1\right)$.

Now the algebraic structure $\left(\left[-10,10\right],\otimes_{10}\right)$ is defined such that for all $x,y\in\left[-10,10\right]$ then $x\otimes_{10}y=\bar{U}_{10}\left(x,y\right)$.

According to the previous definition if $x\in\left(-10,10\right)$ its inverse can be calculated as

$$x^{-1}=10-20\exp\left(\frac{1}{\ln\left(\frac{10-x}{20}\right)}\right),$$

it is easy to check that with this definition $x\otimes_{10}x^{-1}=\bar{e}_{10}$.

On the other hand, if $x=-10$ or $x=10$ the previous formula is undefined. Therefore the algebraic structure $\left(\left[-10,10\right],\otimes_{10}\right)$ is not a group for this detail, however it is Abelian, there is the neutral element and it is also associative due to the associativity of the uninorm.

If a new structure is defined in $NG_{10} = [-10,10] \cup \{\mathrm{I}\}$, such that the same definition is maintained for $\otimes_{10}$ $\forall x, y \in [-10,10]$, and then for $x \in [-10,10] \cup \{\mathrm{I}\}$, the following is established:

$$x \otimes_{10} \mathrm{I} = \mathrm{I} \otimes_{10} x = \begin{cases} \mathrm{I}, \text{if } x = \mathrm{I} \text{ or } x \geq e \\ x, \text{if } x < e \end{cases}.$$

If taken $x = 2, y = 3$ and $z = \mathrm{I}$, considering $\bar{e}_{10} \approx 2.6424$; it is

$$2 \otimes_{10} (\mathrm{I} \otimes_{10} 3) = 2 \otimes_{10} \mathrm{I} = 2 \text{ and } (2 \otimes_{10} \mathrm{I}) \otimes_{10} 3 = 2 \otimes_{10} 3 = 2.3570 \neq 2.$$

Therefore, for this triple of values, associativity is not fulfilled.

On the other hand, it can be shown that $x = y = 2, z = e$ does comply with associativity.□

Until now NeutroGroups of infinite cardinality has been obtained. To obtain finite NeutroGroups from uninorms, the function $round(x)$ that calculates the nearest integer value to *x* will be used and when there is the same distance from *x* to its nearest integers, the largest integer is taken.

Another useful function for domains of *x* where every value satisfies $|x| \leq 1$ is the function

$$round_d(x) = \frac{round(10^d x)}{10^d},$$

with $d \in \mathbb{N}$. For example, the range of $round_1(x)$ in $[-1,1]$ is the finite set $\{-1, -0.9, \ldots, 0, 0.1, \ldots, 1\}$.

If $U(x, y)$ is a uninorm, it can be defined

$$x \bar{\otimes} y = \frac{round(10^d U(x, y))}{10^d},$$

where it is used for $x, y \in [-1,1]$ or $x, y \in [0,1]$ with *d* digits after the comma. Other elements such as "I" can be included in these new operations in the same way with non-finite NeutroGroups.

Example 3: Suppose that the uninorm of Example 2 is generalized to the interval $\left[\Psi,\Omega\right]=\left[-1,2\right]$, obtaining the formula:

$$\bar{U}_{-1,2}\left(x,y\right)=\begin{cases}-1,\text{if}\left(x,y\right)\in\left\{\left(-1,2\right),\left(2,-1\right)\right\}\\ 2-3\exp\left(-\ln\left(\frac{2-x}{3}\right)\ln\left(\frac{2-y}{3}\right)\right),\text{otherwise}\end{cases}$$

Define

$$x\,\bar{\otimes}_{-1,2}\,y=round\left(\bar{U}_{-1,2}\left(x,y\right)\right)$$

if $x,y\in A=\left\{-1,0,1,2,\bar{e}_{-1,2}\right\}$, while for $\bar{A}=A\cup\left\{\mathrm{I}\right\}$ it is

$$x\otimes_{10}\mathrm{I}=\mathrm{I}\otimes_{10}x=\begin{cases}\mathrm{I},\text{if } x=\mathrm{I}\text{ or } x\geq\bar{e}_{-1,2}\\ x,\text{if } x<\bar{e}_{-1,2}\end{cases},$$

in this case the neutral element

$$\bar{e}_{-1,2}=2-3\exp\left(-1\right)\approx 0.89636\,.$$

The Cayley Table of this algebraic structure is as follows:

Example 4: In this example we will apply the generation of a finite NeutroGroup by combining $round_d\left(x\right)$ with $C\left(x,y\right)$. To do this, the authors work on the set

$$\bar{\mathrm{B}}=\left\{-1,-0.9,\ldots,0,\mathrm{I},0.1,\ldots,1\right\},$$

if

$$\mathrm{B}=\left\{-1,-0.9,\ldots,0,0.1,\ldots,1\right\}\ \forall x,y\in\bar{\mathrm{B}}$$

the algebraic structure formed by the pair $\left(\bar{\mathrm{B}},\odot\right)$ is defined:

Table 1. Cayley table of the finite NeutroGroup generated by $\bar{U}_{-1,2}(\cdot,\cdot)$.

$\overline{\otimes}_{-1,2}$	**-1**	**0**	$\bar{e}_{-1,2}$	**I**	**1**	**2**
-1	-1	-1	-1	-1	-1	-1
0	-1	-1	0	0	0	2
$\bar{e}_{-1,2}$	-1	0	$\bar{e}_{-1,2}$	I	1	2
I	-1	0	$\bar{e}_{-1,2}$	I	I	I
1	-1	-1	1	I	1	2
2	-1	2	2	I	2	2
$\overline{\otimes}_{-1,2}$	-1	0	$\bar{e}_{-1,2}$	I	1	2
-1	-1	-1	-1	-1	-1	-1
0	-1	-1	0	0	0	2
$\bar{e}_{-1,2}$	-1	0	$\bar{e}_{-1,2}$	I	1	2
I	-1	0	$\bar{e}_{-1,2}$	I	I	I
1	-1	-1	1	I	1	2
2	-1	2	2	I	2	2

Source: Authors

$$x \odot_C y = \begin{cases} round_1\left(C\left(x,y\right)\right), \text{if } x,y \in \mathrm{B} \\ \mathrm{I}, \text{if } \left(x = \mathrm{I} \wedge y = \mathrm{I}\right) \vee \left(x = \mathrm{I} \wedge y > 0\right) \vee \left(x > 0 \wedge y = \mathrm{I}\right) \\ x, \text{if } x \leq 0 \wedge y = \mathrm{I} \\ y, \text{if } y \leq 0 \wedge x = \mathrm{I} \end{cases}$$

The Cayley's table of this structure can be found in the appendices.

It seems from what has been developed so far that uninorms generate only Abelian NeutroGroups, however, a way to generate not Abelians Neutrogroups will be explained below:

Take $NG_{-1,1} = \left[-1,1\right] \cup \left\{\mathrm{I}\right\}$ and the set

$$\mathcal{M}_n\left(NG_{-1,1}\right) = \left\{M \mid M \text{ is a square matrix of order n with elements in } NG_{-1,1}\right\}.$$

Define the algebraic structure $\left(\mathcal{M}_n\left(NG_{-1,1}\right),\otimes_{\mathcal{M}_n}\right)$ such that $\boldsymbol{F}=\boldsymbol{D}\otimes_{\mathcal{M}_n}\boldsymbol{E}$ with $\boldsymbol{D},\boldsymbol{E},\boldsymbol{F}\in\mathcal{M}_n\left(NG_{-1,1}\right)$ and

$$\boldsymbol{F}_{ij}=\bar{U}\left(\boldsymbol{D}_{i1}\cdot\boldsymbol{E}_{1j},\boldsymbol{D}_{i2}\cdot\boldsymbol{E}_{2j},\cdots,\boldsymbol{D}_{in}\cdot\boldsymbol{E}_{nj}\right),\ \text{i,j} = 1, 2,\ldots,\text{n}.$$

Where, the definition of the uninorm includes the symbol "I" for indeterminacy. Let us illustrate with an example.

Example 5: Revisiting the combining function in Prospector, see Equation 7, let $\left(\mathcal{M}_2\left(NG_{-1,1}\right),\otimes_{\mathcal{M}_2}\right)$ be the algebraic structure formed by the set of square matrices of order 2 with elements in $NG_{-1,1}$ endowed with the operation $\otimes_{\mathcal{M}_2}$.

For example, let

$$\boldsymbol{X}=\begin{pmatrix}-0.1 & 0.3\\ -0.5 & 0.4\end{pmatrix},\ \boldsymbol{Y}=\begin{pmatrix}-1 & 0.3\\ 0.4 & 0.2\end{pmatrix},\ \text{and } \boldsymbol{Z}=\begin{pmatrix}0.3 & -0.1\\ 0.1 & 0.2\end{pmatrix},$$

then,

$$\boldsymbol{X}\otimes_{\mathcal{M}_2}\boldsymbol{Y}=\begin{pmatrix}0.217391 & 0.030054\\ 0.611111 & -0.070850\end{pmatrix},\ \boldsymbol{Y}\otimes_{\mathcal{M}_2}\boldsymbol{X}=\begin{pmatrix}-0.050761 & -0.186722\\ -0.139442 & 0.198098\end{pmatrix},$$

so $\boldsymbol{X}\otimes_{\mathcal{M}_2}\boldsymbol{Y}\neq\boldsymbol{Y}\otimes_{\mathcal{M}_2}\boldsymbol{X}$, therefore there is no associativity for this triple of values.

On the other hand,

$$\boldsymbol{X}\otimes_{\mathcal{M}_2}\left(\boldsymbol{Y}\otimes_{\mathcal{M}_2}\boldsymbol{Z}\right)=\begin{pmatrix}0.069066 & -0.015905\\ 0.190641 & -0.079523\end{pmatrix},$$

whereas

$$\left(\boldsymbol{X}\otimes_{\mathcal{M}_2}\boldsymbol{Y}\right)\otimes_{\mathcal{M}_2}\boldsymbol{Z}=\begin{pmatrix}0.068209 & -0.015730\\ 0.176478 & -0.075216\end{pmatrix},$$

so it is not associative for this triad of values.

In contrast, $W = \begin{pmatrix} 1 & 0 \\ 0 & 1 \end{pmatrix}$ satisfies

$$W \otimes_{\mathcal{M}_2} \left(W \otimes_{\mathcal{M}_2} W\right) = \left(W \otimes_{\mathcal{M}_2} W\right) \otimes_{\mathcal{M}_2} W = W,$$

therefore $\otimes_{\mathcal{M}_2}$ is not an AntiGroup. An interesting exercise is to find the inverse operation of $\otimes_{\mathcal{M}_2}$ and to check the conditions for its existence.

CONCLUSION

This chapter was dedicated to delving into the theory of NeutroAlgebras, specifically NeutroGroups. The chapter contributes to this theory from two points of view. Firstly, it explains and demonstrates how to simply and automatically generate different NeutroGroups from uninorms defined in different domains. This contributes to the way of creating this type of structure different of the NeutroGroups obtained from the groups $\left(\mathbb{Z}_n,+\right)$ or $\left(\mathbb{Z}_n,\cdot\right)$ which are the majority of those found in the literature so far. Especially, it is shown how to generate NeutroGroups of finite and infinite order, where associativity is only partially fulfilled. An example is also given in the generation of non-Abelian NeutroGroups. Secondly, the possible practical application of this theory in fields such as decision-making is suggested, although there are papers in the literature where this is shown, however in this chapter it is addressed in a formal way, especially as a consequence of the applicability of the uninorms. In the appendices, there is an example of one Cayley's table of a finite NeutroGroup generated by the combining function in Mycin and the used programs coded with Matlab language to perform the calculations provided in this chapter. In future works the authors intend to find applications of these results in the foundations of mathematics and particle physics.

REFERENCES

Agboola, A. A. A. (2020a). Introduction to NeutroGroups. *International Journal of Neutrosophic Science*, *6*(1), 41–47.

Agboola, A. A. A. (2020b). On Finite NeutroGroups of Type-NG [1,2,4]. *International Journal of Neutrosophic Science*, *10*(2), 84–95.

Appel, O., Chiclana, F., Carter, J., & Fujita, H. (2017). Cross-ratio uninorms as an effective aggregation mechanism in Sentiment Analysis. *Knowledge-Based Systems, 124*, 16–22. doi:10.1016/j.knosys.2017.02.028

Bordignon, F., & Gomide, F. (2014). Uninorm based evolving neural networks and approximation capabilities. *Neurocomputing, 127*, 13–20. doi:10.1016/j.neucom.2013.04.047

De-Baets, B., & Fodor, J. (1999). Van Melle's combining function in MYCIN is a representable uninorm: An alternative proof. *Fuzzy Sets and Systems, 104*(1), 133–136. doi:10.1016/S0165-0114(98)00265-6

Deschrijver, G., & Kerre, E. E. (2004). Uninorms in L*-fuzzy set theory. *Fuzzy Sets and Systems, 148*, 243–262. doi:10.1016/j.fss.2003.12.006

Duda, R. O., Hart, P. E., & Nilsson, N. J. (1981). Subjective Bayesian methods for rule-based inference systems. In M. Kaufmann (Ed.), *Readings in artificial intelligence* (pp. 192–199). Elsevier. doi:10.1016/B978-0-934613-03-3.50017-9

Fodor, R., Yager, R., & Rybalov, A. (1997). Structure of uninorms. *International Journal of Uncertainty, Fuzziness and Knowledge-based Systems, 5*(04), 411–427. doi:10.1142/S0218488597000312

González-Caballero, E., Leyva-Vázquez, M., & Smarandache, F. (n.d.). On Neutrosophic Uninorms. *Neutrosophic Sets and Systems.*

González-Caballero, E., Smarandache, F., & Leyva-Vázquez, M. (2019). On Neutrosophic Offuninorms. *Symmetry, 11*(9), 1136. doi:10.3390ym11091136

González-Hidalgo, M., Massanet, S., Mir, A., & Ruiz-Aguilera, D. (2014). A new edge detector based on uninorms. In *15th International Conference on Information Processing and Management of Uncertainty in Knowledge-Based Systems*. Springer International Publishing. 10.1007/978-3-319-08855-6_19

Hajék, P. (1985). Combining functions for certainty degrees in consulting systems. *International Journal of Man-Machine Studies, 22*(1), 59–76. doi:10.1016/S0020-7373(85)80077-8

Hart, P. E., Duda, L. O., & Einaudi, M. T. (1978). PROSPECTOR—A Computer-Based Consultation System for Mineral Exploration. *Mathematical Geology, 10*(5), 589–610. doi:10.1007/BF02461988

Isaacs, I. M. (2008). *Finite Group Theory: Graduate Studies in Mathematics* (Vol. 92). American Mathematical Society.

Kandasamy, V., Kandasamy, I., & Smarandache, F. (2020). NeutroAlgebra of Neutrosophic Triplets using {Zn, ×} NeutroAlgebra of Neutrosophic Triplets using {Zn, ×}. *Neutrosophic Sets and Systems*, *38*, 509–523.

Marík, V., & Kouba, Z. (1991). Some knowledge-acquisition methods for Prospector-like systems. *Knowledge-Based Systems*, *4*(4), 225–230. doi:10.1016/0950-7051(91)90052-4

Mohammadzadeh, E., & Rezaei, A. (2020). On NeutroNilpotent Groups. *Neutrosophic Sets and Systems*, *38*, 33–40.

Pís, P., & Mesiar, R. (1989). Fuzzy model of inexact reasoning in medicine. *Computer Methods and Programs in Biomedicine*, *30*(1), 1–8. doi:10.1016/0169-2607(89)90117-X PMID:2582743

Shortliffe, E. H. (1976). *Computer-Based Medical Consultations: MYCIN*. Elsevier.

Silva-Jiménez, D., Valenzuela-Mayorga, J. A., Roja-Ubilla, M. E., & Batista-Hernández, N. (2021). NeutroAlgebra for the evaluation of barriers to migrants' access in Primary Health Care in Chile based on PROSPECTOR function. *Neutrosophic Sets and Systems*, *39*, 1–9.

Smarandache, F. (2020a). NeutroAlgebra is a Generalization of Partial Algebra. *International Journal of Neutrosophic Science, 2*(1), 8-17.

Smarandache, F. (2020b). Introduction to NeutroAlgebraic Structures and AntiAlgebraic Structures (revisited). *Neutrosophic Sets and Systems*, *31*, 1–16.

Smarandache, F. (2020c). *Neutroalgebra is a generalization of partial algebra*. Infinite Study.

Smarandache, F. A. (2005). *Unifying Field in Logics: Neutrosophic Logic. Neutrosophy Set, Neutrosophic Probability and Statistics*. American Research Press.

Sussner, P., & Schuster, T. (2018). Interval-valued fuzzy morphological associative memories: Some theoretical aspects and applications. *Information Sciences*, *438*, 127–144. doi:10.1016/j.ins.2018.01.042

Tsadiras, A. K., & Margaritis, K. G. (1998). The MYCIN certainty factor handling function as uninorm operator and its use as a threshold function in artificial neurons. *Fuzzy Sets and Systems*, *93*(3), 263–274. doi:10.1016/S0165-0114(96)00185-6

Yager, R. R. (2001). Uninorms in fuzzy system modeling. *Fuzzy Sets and Systems*, *122*(1), 167–175. doi:10.1016/S0165-0114(00)00027-0

Yager, R. R., & Kreinovich, V. (2003). Universal approximation theorem for uninorm-based fuzzy systems modeling. *Fuzzy Sets and Systems*, *140*(2), 331–339. doi:10.1016/S0165-0114(02)00521-3

Yager, R. R., & Rybalov, A. (1996). Uninorm aggregation operators. *Fuzzy Sets and Systems*, *80*(1), 111–120. doi:10.1016/0165-0114(95)00133-6

Zhang, C., & Luo, X. (1999). Isomorphic Transformations of Uncertainties for Incorporating EMYCIN-Style and PROSPECTOR-Style Systems into a Distributed Expert System. *Journal of Computer Science and Technology*, *14*(4), 386–392. doi:10.1007/BF02948741

KEY TERMS AND DEFINITIONS

Abelian Group: In group theory it is a commutative group.

Combining function: It is a function used to aggregate assessments in Mycin and Prospector expert systems.

Group: Classical algebraic structure formed by a nonempty set of elements and an operation satisfying associativity, existence of a neutral element, and existence of an inverse element for every element in the set.

Mycin: This is a well-known expert system coded in Lisp by Edward Shortliffe at the University of Stanford in the 1970s to diagnosing infectious blood diseases.

NeutroAlgebra: An extension of classical algebra and partial algebra, where at least one axiom is satisfied by some elements and is not satisfied by others. If 100% of the elements do not satisfy one axiom it is called an AntiAlgebra. The NeutroAlgebra does not have any AntiAxiom, or an axiom that is not satisfied for any element.

NeutroGroup: NeutroAlgebra where the algebraic structure corresponds to a group.

Prospector: Expert system created for consulting mineral exploration.

Uninorm: Extends the notion of t-norm and t-conorm in fuzzy logic. Preserve the axioms of commutativity, associativity, and increasing, but in the boundary condition admits the neutral element lying in (0, 1).

APPENDIX 1

Table 2. Cayley table of ⊙

$x \odot y$	-1	-0.9	-0.8	-0.7	-0.6	-0.5	-0.4	-0.3	-0.2	-0.1	0
-1	-1	-1	-1	-1	-1	-1	-1	-1	-1	-1	-1
-0.9	-1	-1	-1	-1	-1	-1	-0.9	-0.9	-0.9	-0.9	-0.9
-0.8	-1	-1	-1	-0.9	-0.9	-0.9	-0.9	-0.9	-0.8	-0.8	-0.8
-0.7	-1	-1	-0.9	-0.9	-0.9	-0.9	-0.8	-0.8	-0.8	-0.7	-0.7
-0.6	-1	-1	-0.9	-0.9	-0.8	-0.8	-0.8	-0.7	-0.7	-0.6	-0.6
-0.5	-1	-1	-0.9	-0.9	-0.8	-0.8	-0.7	-0.6	-0.6	-0.5	-0.5
-0.4	-1	-0.9	-0.9	-0.8	-0.8	-0.7	-0.6	-0.6	-0.5	-0.5	-0.4
-0.3	-1	-0.9	-0.9	-0.8	-0.7	-0.6	-0.6	-0.5	-0.4	-0.4	-0.3
-0.2	-1	-0.9	-0.8	-0.8	-0.7	-0.6	-0.5	-0.4	-0.4	-0.3	-0.2
-0.1	-1	-0.9	-0.8	-0.7	-0.6	-0.6	-0.5	-0.4	-0.3	-0.2	-0.1
I	-1	-0.9	-0.8	-0.7	-0.6	-0.5	-0.4	-0.3	-0.2	-0.1	0
0	-1	-0.9	-0.8	-0.7	-0.6	-0.5	-0.4	-0.3	-0.2	-0.1	0
0.1	-0.8	-0.7	-0.6	-0.5	-0.5	-0.4	-0.3	-0.2	-0.1	0	0.1
0.2	-0.7	-0.6	-0.5	-0.4	-0.3	-0.3	-0.2	-0.1	0	0.1	0.2
0.3	-0.5	-0.5	-0.4	-0.3	-0.2	-0.2	-0.1	0	0.1	0.2	0.3
0.4	-0.4	-0.4	-0.3	-0.2	-0.1	-0.1	0	0.1	0.2	0.3	0.4
0.5	-0.3	-0.3	-0.2	-0.1	-0.1	0	0.1	0.2	0.3	0.4	0.5
0.6	-0.3	-0.2	-0.1	-0.1	0	0.1	0.1	0.2	0.3	0.5	0.6
0.7	-0.2	-0.1	-0.1	0	0.1	0.1	0.2	0.3	0.4	0.5	0.7
0.8	-0.1	-0.1	0	0.1	0.1	0.2	0.3	0.4	0.5	0.6	0.8
0.9	-0.1	0	0.1	0.1	0.2	0.3	0.4	0.5	0.6	0.7	0.9
1	I	0.1	0.1	0.2	0.3	0.3	0.4	0.5	0.7	0.8	1

Source: Authors

Table 3. Cayley table of ⊙ (Continued)

$x \odot y$	I	0.1	0.2	0.3	0.4	0.5	0.6	0.7	0.8	0.9	1
-1	-1	-0.8	-0.7	-0.5	-0.4	-0.3	-0.2	-0.2	-0.1	-0.1	I
-0.9	-0.9	-0.7	-0.6	-0.5	-0.4	-0.3	-0.2	-0.1	-0.1	0	0.1
-0.8	-0.8	-0.6	-0.5	-0.4	-0.3	-0.2	-0.1	-0.1	0	0.1	0.1
-0.7	-0.7	-0.5	-0.4	-0.3	-0.2	-0.1	-0.1	0	0.1	0.1	0.2
-0.6	-0.6	-0.5	-0.3	-0.2	-0.1	-0.1	0	0.1	0.1	0.2	0.3
-0.5	-0.5	-0.4	-0.3	-0.2	-0.1	0	0.1	0.1	0.2	0.3	0.3
-0.4	-0.4	-0.3	-0.2	-0.1	0	0.1	0.1	0.2	0.3	0.4	0.4
-0.3	-0.3	-0.2	-0.1	0	0.1	0.2	0.2	0.3	0.4	0.5	0.5
-0.2	-0.2	-0.1	0	0.1	0.2	0.3	0.3	0.4	0.5	0.6	0.7
-0.1	-0.1	0	0.1	0.2	0.3	0.4	0.5	0.5	0.6	0.7	0.8
I	I	I	I	I	I	I	I	I	I	I	I
0	I	0.1	0.2	0.3	0.4	0.5	0.6	0.7	0.8	0.9	1
0.1	I	0.2	0.3	0.4	0.5	0.6	0.6	0.7	0.8	0.9	1
0.2	I	0.3	0.4	0.4	0.5	0.6	0.7	0.8	0.8	0.9	1
0.3	I	0.4	0.4	0.5	0.6	0.6	0.7	0.8	0.9	0.9	1
0.4	I	0.5	0.5	0.6	0.6	0.7	0.8	0.8	0.9	0.9	1
0.5	I	0.6	0.6	0.7	0.7	0.8	0.8	0.9	0.9	1	1
0.6	I	0.6	0.7	0.7	0.8	0.8	0.8	0.9	0.9	1	1
0.7	I	0.7	0.8	0.8	0.8	0.9	0.9	0.9	0.9	1	1
0.8	I	0.8	0.8	0.9	0.9	0.9	0.9	0.9	1	1	1
0.9	I	0.9	0.9	0.9	0.9	1	1	1	1	1	1
1	I	1	1	1	1	1	1	1	1	1	1

Source: Authors

APPENDIX 2

```
function offuninG = offuninG(L,x)
%Code of the function to calculate the combining function in
%Prospector in the interval [L(1),L(2)]
if((x(1)==L(1) && x(2)==L(2)) || (x(1)==L(2) && x(2)==L(1)))
offuninG = (L(2)-L(1))*.5+L(1);
else
offuninG = (x(1)-L(1))*(x(2)-L(1))/((L(2)-x(1))*(L(2)-
x(2))+(x(1)-L(1))*(x(2)-L(1)));
offuninG = (L(2)-L(1))*offuninG+L(1);
end
function munin = munin(X,Y)
%Code .m to calculate the product of matrices using uninorms
    munin(1,1) = offunin(1,[X(1,1)*Y(1,1) X(1,2)*Y(2,1)]);
    munin(1,2) = offunin(1,[X(1,1)*Y(1,2) X(1,2)*Y(2,2)]);
    munin(2,1) = offunin(1,[X(2,1)*Y(1,1) X(2,2)*Y(2,1)]);
    munin(2,2) = offunin(1,[X(2,1)*Y(1,2) X(2,2)*Y(2,2)]);
%-----------------------------------------------------------------
--------------
function offunin = offunin(n,x)
%Code .m of the function o calculate the combining function in
Prospector in %the interval [-n, n]
  if(x(1)*x(2) == -n^2)
    offunin = 0;
  else
    offunin = (n^2)*(x(1)+x(2))/(n^2+x(1)*x(2));
  end
```

Chapter 11
On Neutro-Topological Spaces and Their Properties

Bhimraj Basumatary
Bodoland University, India

Jeevan Krishna Khaklary
Central Institute of Technology, Kokrajhar, India

Nijwm Wary
Bodoland University, India

Florentin Smarandache
https://orcid.org/0000-0002-5560-5926
University of New Mexico, USA

ABSTRACT

There is a lot of ambiguous information in the real world that crisp values can't manage. The fuzzy set theory was proposed by Zadeh. It is an age-old and excellent tool for dealing with uncertain information. As a result, intuitionistic fuzzy set theory was suggested. However, these theories are incapable of dealing with all forms of uncertainty, such as indeterminate and inconsistent data in various decision-making situations. To address this shortfall, Smarandache proposed the neutrosophic set theory by introducing a degree of indeterminacy as an independent component. In this current decade, neutrosophic environments are mainly interested by different fields of researchers. Recently, Smarandache introduced theNeutroAlgebra andAntiAlgebras. NeutroAlgebras and AntiAlgebras represent a new research subject that is based on real-world examples. In this chapter, some properties of NeutroTopological space are introduced and studied with examples. Several definitions of NeutroInterior, NeutroClosure, and NeutroBoundary are defined, and the authors also studied its properties with examples.

DOI: 10.4018/978-1-6684-3495-6.ch011

INTRODUCTION

The concept of a fuzzy set was first developed by Zadeh (1965), where the concept of the membership function is defined and discussed the concept of uncertainty using a fuzzy set. Atanassov (1986) proposed the intuitionistic fuzzy set by generalizing the concept of fuzzy sets and introducing the degree of non-membership as a component. Chang (1968) was the first to introduce fuzzy topology, while Coker (1997) defined intuitionistic fuzzy topological space. Salama et al. (2014) investigated topology using neutrosophic sets. Kelly (1963) introduced the concept of bitopological space. The concept of fuzzy bitopological space was investigated by Kandil et al. (1995). Florentin Smarandache (1998) defined the idea of neutrosophic logic and the concept of neutrosophic set. After that, the concepts of the neutrosophic set have been applied in many branches of sciences and technology. The concept of neutrosophic topological space was introduced by Salama and Alblowi (2012). Devi *et al.* (2017) discussed separation axioms in ordered neutrosophic bitopological space. Mwchahary et al. (2020) did their work in neutrosophic bitopological space. After defining the neutrosophic group, Sumathi et al. (2016) defined the concept of the topological group structure of the neutrosophic set and also, Sumathi and Arockiarani (2015) studied the fuzzy neutrosophic group.

In recent years, there has been a surge in academic interest in neutrosophic set theory. The concept of neutro-structures and anti-structures was first defined by Florentin Smarandache (2019, 2020). Şahin et al. (2021) discussed the idea of neutro-topological space and anti-topological space. Smarandache (2020) studied NeutroAlgebra as a generalization of partial algebra. Agboola (2020) investigated the idea of NeutroRings, NeutroGroups, and finite NeutroGroups of type-NG. Smarandache (2020) proposed the generalizations and alternatives of Classical Algebraic Structures to NeutroAlgebraic Structures and AntiAlgebraic Structures. Al-Tahan et al. (2021) studied the NeutroOrderedAlgebra, NeutroHyper structures, and their properties.

Blizard (1989) investigated the concept of multiset theory and neutrosophic multi groups and applications were studied by Bakbak (2019). Basumatary et al. (2020) studied on interval-valued triangular neutrosophic linear programming problem also Basumatary et al. (2021) investigated some properties of the neutrosophic multi topological group. Yager (1986) introduced the concept of fuzzy multiset (fuzzy bag) to generalize the fuzzy set, and Miyamoto (2001) studied Fuzzy Multisets and Their Generalizations. A fuzzy multiset element can appear more than once, with the same or different membership values. Onasanya et al. (2018) investigated the Algebraic properties of alpha-level subsets topology of a fuzzy subset, and the authors Onasanya et al. (2019) and Al Tahan et al. (2020) studied fuzzy multi-polygroups. Al Tahan et al. (2019) studied the fuzzy multi-Hv-ideals of Hv-rings fuzzy multi-Hv-ideals and their properties. Sebastian et al. (2011) investigated a novel type of fuzzy set (fuzzy multiset). The concept of multi-fuzzy complex numbers and multi-fuzzy complex sets was initially introduced by Dey et al. (2015). Yong et al. (2013) recently suggested the notion of the multi-fuzzy soft set for its application to decision making, which is a more broad fuzzy soft set. Basumatary (2016) investigated the concept of fuzzy closure and the fuzzy boundary based on reference function.

Smarandache (2019, 2020) founded and studied the concept of neutro-structures and anti-structures. From the concepts of NeutroAlgebra, he showed that if a statement (theorem, lemma, consequence, property, etc.) is totally true in a classical Algebra, it does not mean that it is also totally true in a NeutroAlgebra or an AntiAlgebra. It depends on the operations and axioms (if they are totally true, partially true, totally false, or partially or totally indeterminate) it is based upon.

For Example

1. Let $(A,*)$ be a NeutroAlgebra (it has NeutroOperations or NeutroAxioms while the others are classical Operations and classical Axioms, and no AntiOperation and no AntiAxiom).

Statement: If x, y in A, then $x * y$ in A.

This statement is true for classical Algebra.

But for a NeutroAlgebra we have:

a. The Statement is true if the operation * is a classical Operation (totally true).
b. The Statement is true if the operation * is a NeutroAxiom, but x, y both belong to the partially true subset;
c. The Statement is false if the operation * is a NeutroAxiom, and at least one of x or y belongs to the partially false subset.

2. Similarly, for the NeutroGroup.

Let A be a NeutroGroup, and x in A. Then its inverse x^{-1} is also in A.

This is true for the classical Group.

For the NeutroGroup:

a. This is true if the inverse element axiom is totally true;
b. This is true if the NeutroInverse element axiom is partially true, and x belongs to the true subset;
c. This is false otherwise.

By observing the above concepts, the properties of NeutroTopological space will be observed. Several definitions of NeutroInterior, NeutroClosure, and NeutroBoundary are defined and also, some properties with examples are studied.

BACKGROUND

Definition 2.1: [Smarandache, (2020c)] **The NeutroSophication of the Law**

1. Let X be a non-empty set and * be binary operation. For some elements $(a,b) \in (X,X)$, $(a*b) \in X$ (degree of well defined (T)) and for other elements $(x,y),(p,q) \in (X,X)$; [$x*y$ is indeterminate (degree of indeterminacy (I)), or $p*q \notin X$ (degree of outer-defined (F)], where (T, I, F) is different from (1,0,0) that represents the Classical Law, and from (0,0,1) that represents the AntiLaw.
2. In NeutroAlgebra, the classical well-defined for * binary operation is divided into three regions: degree of well-defined (T), degree of indeterminacy (I) and degree of outer-defined (F) similar to neutrosophic set and neutrosophic logic.

Definition 2.2: [Şahin et al. (2021)] Let X be the non-empty set and τ be a collection of subsets of X. Then τ is said to be a NeutroTopology on X and the pair (X,τ) is said to be a NeutroTopological space, if at least one of the following conditions hold good:

1. $\left[\left(\varnothing_N \in \tau, X_N \notin \tau\right) or \left(X_N \in \tau, \varnothing_N \notin \tau\right)\right]$ or $\left[\varnothing_N, X_N \in_{\sim} \tau\right]$.

2. For some n elements

$$a_1, a_2, \ldots, a_n \in \tau, \bigcap_{i=1}^{n} a_i \in \tau$$

[degree of truth T] and for other n elements

$$b_1, b_2, \ldots, b_n \in \tau, p_1, p_2, \ldots, p_n \in \tau; [\left(\bigcap_{i=1}^{n} b_i \notin \tau\right)$$

[degree of falsehood F] or ($\bigcap_{i=1}^{n} p_i$ is indeterminate (degree of indeterminacy I)], where n is finite; where (T, I, F) is different from (1,0,0) that represents the Classical Axiom, and from (0,0,1) that represents the AntiAxiom.)].

3. For some n elements

$$a_1, a_2, \ldots, a_n \in \tau, \bigcup_{i=1} a_i \in \tau$$

[degree of truth T] and for other n elements

$$b_1, b_2, \ldots, b_n \in \tau, p_1, p_2, \ldots, p_n \in \tau; [\left(\bigcup_{i=I} b_i \notin \tau\right)$$

[degree of falsehood F] or ($\bigcup_{i=I} p_i$ is indeterminate (degree of indeterminacy I)], where n is finite; where (T, I, F) is different from (1,0,0) that represents the Classical Axiom, and from (0,0,1) that represents the AntiAxiom.)].

Example 2.1: Let $X = \{p, q, r, s\}$ be a set and

$$\tau = \{\varnothing, \{p\}, \{p, q\}, \{r, s\}, \{q, r\}\}$$

be a collection of subsets of X. Then,

1. It is clear that $X \notin \tau$ and $\varnothing \in \tau$.
2. Let $m_1 = \{p\}$, $m_2 = \{p, q\}$, $n_1 = \{r, s\}$, $n_2 = \{q, r\}$. We obtain that $\left(\bigcap_{i=1}^{2} m_i \in \tau\right)$ and $\left(\bigcap_{i=1}^{2} n_i \notin \tau\right)$.
3. Let $m_1 = \{p, q\}$, $m_2 = \{r, s\}$, $n_1 = \{p\}$, $n_2 = \{q, r\}$. We obtain that $\left(\bigcup_{i=1,2} m_i \in \tau\right)$ and $\left(\bigcup_{i=I} n_i \notin \tau\right)$.

Thus, (X, τ) satisfies the conditions $\{(i), (ii), (iii)\}$ in Definition 2.3. Hence, (X, τ) is a NeutroTopological space.

Remark 2.1: [Şahin et al. (2021)] The symbol "$\in_{\sim}$" will be used for situations where it is an unclear appurtenance (not sure if an element belongs or not to a set). For example, if it is not certain whether "a" is a member of the set P, then it is denoted by a $\in_{\sim} P$.

Theorem 2.1: [Smarandache, (2020c)] Let $\left(X,\tau\right)$ be a classical topological space. Then $\left(X,\tau-\varnothing\right)$ is a NeutroTopological space.

Theorem 2.2: [Smarandache, (2020c)] Let $\left(X,\tau\right)$ be a classical topological space. Then $\left(X,\tau-X\right)$ is a NeutroTopological space.

MAIN FOCUS OF THE CHAPTER[1]

Some Properties of NeutroTopological Space

Proposition 3.1: In a NeutroTopological space. The conditions (i), (ii), and (or) (iii) are satisfied.

1. Empty set or X is not NeutroOpen.
2. There exist two or more NeutroOpen sets in which their union is NeutroOpen and two or more NeutroOpen sets in which their union is not NeutroOpen.
3. There exist two or more NeutroOpen sets in which their intersection is NeutroOpen and two or more NeutroOpen sets in which their intersection is not NeutroOpen.

Examples 3.1: Let $X=\left\{a,b,c,d\right\}$ and

$$\tau=\left\{\varnothing,\left\{a\right\},\left\{a,b\right\},\left\{c,d\right\},\left\{b,c\right\}\right\}.$$

Then $\left(X,\tau\right)$ is neutroTopological space.

1. Here X is not NeutroOpen.
2. From the above NeutroTopology let $A=\left\{a\right\}$, $B=\left\{a,b\right\}$, $P=\left\{a,b\right\}$ and $Q=\left\{b,c\right\}$, then clearly, $A\cup B=\left\{a,b\right\}$ which is NeutroOpen in $\left(X,\tau\right)$ and $P\cup Q=\left\{a,b,c\right\}$ which is not NeutroOpen in $\left(X,\tau\right)$.

3. From the above NeutroTopology let $A=\{a\}$, $B=\{a,b\}$, $P=\{c,d\}$ and $Q=\{b,c\}$, then clearly, $A\cap B=\{a\}$ which is NeutroOpen in (X,τ) and $P\cap Q=\{c\}$ which is not NeutroOpen in (X,τ).

Definition 3.1: Let (X,τ) be a NeutroTopological space over X and A is subset on X. Then, the NeutroInterior of A is the union of all NeutroOpen subsets of A. Clearly, NeutroInterior of A is the biggest NeutroOpen set over X which is contained in A.

Proposition 3.2: If A is Neutro Open then $NeutroInt(A)=A$.

Proof: If A is NeutroOpen then $NeutroInt(A)=A$.

But the converse part may not be true.

For this cite an example. Let $X=\{a,b,c,d\}$ and

$$\tau=\{X,\{a\},\{a,b\},\{c,d\},\{b,c\}\},$$

then clearly (X,τ) is NeutroTopological space and $NeutroInt(A)=A$, where $A=\varnothing$.

But here $\varnothing$ is not NeutroOpen set.

Remark 3.1: The converse of the proposition 3.2 may not be true.

For this we cite an example:

Let $X=\{a,b,c,d\}$ and

$$\tau=\{X,\{a\},\{a,b\},\{c,d\},\{b,c\}\},$$

then clearly (X,τ) is NeutroTopological space. Now, let $A=\{c\}$, then $NeutroInt(A)=\varnothing$, so $NeutroInt(A)\subseteq A$. But A is not NeutroOpen set.

Theorem 3.1: Let (X,τ) be NeutroTopological space. Then

1. $NeutroInt(A)\subseteq A$
2. $A\subseteq B\Rightarrow NeutroInt(A)\subseteq NeutroInt(B)$
3. $NeutroInt(A\cap B)\subseteq NeutroInt(A)\cap NeutroInt(B)$

Proof of the theorem is simple.

Remark 3.2: Equality does not hold in Theorem 3.1 (iii).

For this we cite an example:

Let $X = \{a,b,c,d\}$ and

$$\tau = \{\varnothing, \{a\}, \{a,b\}, \{c,d\}, \{b,c\}\},$$

then clearly (X,τ) is NeutroTopological Space. Let $A = \{c,d\}$ and $B = \{b,c\}$, then $A \cap B = \{c\}$, then $NeutroInt(A \cap B) = \phi$.

Also, $NeutroInt(A) = \{c,d\}$ and $NeutroInt(B) = \{b,c\}$, so,

$$NeutroInt(A) \cap NeutroInt(B) = \{c\}.$$

Hence,

$$NeutroInt(A \cap B) \neq NeutroInt(A) \cap NeutroInt(B).$$

Definition 3.2: Let (X,τ) be a NeutroTopological space and a subset $\mathcal{F}$ of X is said to be τ-NeutroClosed set if and only if its complement $\mathcal{F}^C$ is NeutroOpen set.

Proposition 3.3: In a NeutroTopological space. The conditions (i), (ii), and (or) (iii) are satisfied.

1. Empty set or X may not be NeutroClosed.
2. There exist two or more NeutroClosed sets in which their intersection is NeutroClosed and two or more NeutroClosed sets in which their intersection may not be NeutroClosed.
3. There exist two or more NeutroClosed sets in which their union is NeutroClosed and two or more NeutroClosed sets in which their union may not be NeutroClosed.

Definition 3.3: Let (X,τ) be a NeutroTopological space over X and A is subset on X. Then, the NeutroClosure of A is the intersection of all NeutroClosed super sets of A. Clearly, NeutroClosure of A is the smallest NeutroClosed set over X conatining A.

Proposition 3.4: Let (X,τ) be NeutroTopological space. Then

1. $Cl(A)$ is the smallest NeutroClosed set containing A.

2. If A is NeutroClosed then $A = \mathrm{Neutro}Cl(A)$.

Proof of the theorem is simple.

Remark 3.3: The converse of the Proposition 3.4 (ii) may not be true.

For this we cite an example:

Let $X = \{a,b,c,d\}$ and

$$\tau = \{\varnothing, X, \{a\}, \{b\}, \{c\}, \{c,d\}, \{b,d\}\}.$$

Here the NeutroClosed sets are

$$\varnothing, X, \{b,c,d\}, \{a,c,d\}, \{a,b,d\}, \{a,b\}, \{a,c\}$$

Let $A = \{c,d\}$, then

$$NeutroCl(A) = \{b,c,d\} \cap \{a,c,d\} = \{c,d\} = A.$$

Hence $NeutroCl(A) = A$ but A is not NeutroClosed set.

Theorem 3.2: Let (X,τ) be NeutroTopological space and let $A, B \subseteq X$. Then

1. $A \subseteq NeutroCl(A)$
2. $A \subseteq B \Rightarrow NeutroCl(A) \subseteq NeutroCl(B)$
3. $NeutroCl(A) \cup NeutroCl(B) \subseteq NeutroCl(A \cup B)$
4. $NeutroCl(A \cap B) \subseteq NeutroCl(A) \cap NeutroCl(B)$
5. $NeutroCl(NeutroCl(A)) = NeutroCl(A)$

Proof of the theorem is simple.

Remark 3.4: Equality does not hold in Theorem 3.2 (iii).

For this we cite an example:

$$NeutroCl(A \cup B) \neq NeutroCl(A) \cup NeutroCl(B)$$

Let $X = \{a,b,c,d\}$ and

$\tau = \{\phi, \{a\}, \{a,b\}, \{c,d\}, \{b,c\}\}$,

Let $A = \{a\}$ and $B = \{c,d\}$, Then $NeutroCl(A) = \{a\}$ and $NeutroCl(B) = \{c,d\}$,

$NeutroCl(A) \cup NeutroCl(B) = \{a,c,d\}$.

Now, $A \cup B = \{a,c,d\}$, so $NeutroCl(A \cup B) = X$.
Hence,

$NeutroCl(A \cup B) \neq NeutroCl(A) \cup NeutroCl(B)$.

Theorem 3.3: Let (X, τ) be NeutroTopological space and let $A \subseteq X$. Then

1. $NeutroInt(A) = \left(NeutroCl(A^C)\right)^C$
2. $NeutroCl(A^C) = \left(NeutroInt(A)\right)^C$
3. $NeutroCl(A) = \left(NeutroInt(A^C)\right)^C$.

Prove of the theorem is simple.
Example 3.2: Let $X = \{a,b,c,d\}$ and

$\tau = \{\varnothing, \{a\}, \{a,b\}, \{c,d\}, \{b,c\}\}$,

Let

$A = \{a,c\}$, $A^C = \{b,d\}$, $NeutroInt(A) = \{a\}$, $NeutroCl(A^C) = \{b,c,d\}$

and

$\left(NeutroCl(A^C)\right)^C = \{a\}$.

Hence,

$NeutroInt(A) = (NeutroCl(A^C))^C$.

Similarly, from the example, we can show

$NeutroCl(A^C) = (NeutroInt(A))^C$ and $NeutroCl(A) = (NeutroInt(A^C))^C$.

Definition 3.4: Let (X, τ) be a NeutroTopological space over X and A is subset on X. Then NeutroBoundary of A is defined as

$NeutroBd(A) = NeutroCl(A) \cap NeutroCl(A^C)$.

Theorem 3.4: Let (X, τ) be NeutroTopological space and let $A, B \subseteq X$. Then

1. $NeutroCl(A) \cap A^C - NeutroInt(A) \subseteq NeutroBd(A)$
2. $NeutroInt(A) = A - NeutroBd(A)$
3. $NeutroInt(A) \cup NeutroInt(A^C) = [NeutroBd(A)]^C$
4. $NeutroBd(NeutroInt(A)) \subseteq NeutroBd(A)$
5. $NeutroBd(NeutroCl(A)) \subseteq NeutroBd(A)$
6. $NeutroBd(A \cup B) \subseteq NeutroBd(A) \cup NeutroBd(B)$
7. $NeutroBd(A \cap B) \subseteq NeutroBd(A) \cup NeutroBd(B)$.

Proof:

1. 1. Let $x \in NeutroCl(A) \cap A^C - NeutroInt(A)$

$\Rightarrow x \in NeutroCl(A) \cap A^C \, and \, x \notin NeutroInt(A)$

$\Rightarrow x \in NeutroCl(A) and \, x \in A^C$

$\Rightarrow x \in NeutroCl(A) and \, x \in NeutroCl(A^C)$

$\Rightarrow x \in NeutroCl(A) \cap NeutroCl(A^C)$

$\Rightarrow x \in NeutroBd(A)$.

Hence,

$NeutroCl(A) \cap A^C - NeutroInt(A) \subseteq NeutroBd(A)$.

2. 2. Let $x \in NeutroInt(A)$

$\Rightarrow x \in A and x \notin A^C$

$\Rightarrow x \in A and x \in NeutroCl(A) and \ x \notin NeutroCl(A^C)$

$\Rightarrow x \in A and x \notin NeutroBd(A)$

$\Rightarrow x \in A - NeutroBd(A)$

Therefore,

$NeutroInt(A) \subseteq A - NeutroBd(A)$.

Conversely, we can show

$A - NeutroBd(A) \subseteq NeutroInt(A)$.

Hence

$NeutroInt(A) = A - NeutroBd(A)$.

3. From Theorem 3.2 (i) and Theorem 3.3 (i), (ii), we have

$$NeutroBd(A) = NeutroCl(A) \cap NeutroCl(A^C)$$

$$\Rightarrow [NeutroBd(A)]^C = [NeutroCl(A) \cap NeutroCl(A^C)]^C$$

$$\Rightarrow [NeutroBd(A)]^C = [NeutroCl(A)]^C \cup [NeutroCl(A^C)]^C$$

$$\Rightarrow [NeutroBd(A)]^C \supseteq NeutroInt(A^C) \cup NeutroInt(A)$$

Hence,

$$\text{Neutro}Int(A) \cup \text{Neutro}Int(A^C) \subseteq [\text{Neutro}Bd(A)]^C.$$

Conversely, we can show

$$[NeutroBd(A)]^C \subseteq NeutroInt(A) \cup NeutroInt(A^C)$$

Thus

$$[NeutroBd(A)]^C = NeutroInt(A) \cup NeutroInt(A^C)$$

4. 4. $NeutroBd(NeutroInt(A))$

$$= NeutroCl(NeutroInt(A)) \cap NeutroCl\left[(NeutroInt(A))^C\right]$$ [by (i)]

$$= NeutroCl\left(NeutroInt\left(A\right)\right) \cap NeutroCl\left[\left\{\left(NeutroCl\left(A^C\right)\right)^C\right\}^C\right]$$

$$\left[as\left(\mathrm{NeutroCl}\left(\mathrm{A}^{\mathrm{C}}\right)\right)^{\mathrm{C}} = \mathrm{NeutroInt}\left(A\right)\right]$$

$$= NeutroCl\left(NeutroInt\left(A\right)\right) \cap NeutroCl\left(A^C\right)$$

$$\left[as\ (P^C)^C = P\, and\, NeutroCl\left(NeutroCl\left(P\right)\right) = NeutroCl\left(P\right) \text{ for any set } P\right]$$

$$\subseteq NeutroCl\left(A\right) \cap NeutroCl\left(A^C\right)$$

$$\left[as\, NeutroInt\left(A\right) \subseteq A \Rightarrow NeutroCl\left(NeutroInt\left(A\right)\right) \subseteq NeutroCl\left(A\right)\right]$$

$$= NeutroBd\left(A\right) \text{[by (i)]}$$

Hence,

$$NeutroBd\left(NeutroInt\left(A\right)\right) \subseteq NeutroBd\left(A\right).$$

5. 5. $NeutroBd\left(NeutroCl\left(A\right)\right)$

$$= NeutroCl\left(NeutroCl\left(A\right)\right) \cap NeutroCl\left(\left(NeutroCl\left(A\right)\right)^C\right) \text{[by (i)]}$$

Now,

$$A \subseteq NeutroCl\left(A\right) \Rightarrow \left(NeutroCl\left(A\right)\right)^C \subseteq A^C$$

$$\Rightarrow NeutroCl\left[\left(NeutroCl\left(A\right)\right)^{C}\right] \subseteq NeutroCl\left(A^{C}\right)$$

$[A \subseteq B \Rightarrow NeutroCl\left(A\right) \subseteq NeutroCl\left(B\right)]$

Hence,

$$NeutroBd\left(NeutroCl\left(A\right)\right) \subseteq NeutroCl\left(A\right) \cap NeutroCl\left(A^{C}\right) = NeutroBd\left(A\right)$$

i.e.,

$$NeutroBd\left(NeutroCl\left(A\right)\right) \subseteq NeutroBd\left(A\right).$$

6. 6. $NeutroBd\left(A \cap B\right) = NeutroCl\left(A \cap B\right) \cap NeutroCl\left(A \cap B\right)^{C}$

$$\subseteq \left[NeutroCl\left(A\right) \cap NeutroCl\left(B\right)\right] \cap NeutroCl\left(A \cap B\right)^{C}$$

$$= \left[NeutroCl\left(A\right) \cap NeutroCl\left(B\right)\right] \cap \left[NeutroCl\left(A^{C}\right) \cup NeutroCl\left(B^{C}\right)\right]$$

$$= \left[\left\{NeutroCl\left(A\right) \cap NeutroCl\left(B\right)\right\} \cap NeutroCl\left(A^{C}\right)\right] \cup \left[\left\{NeutroCl\left(A\right) \cap NeutroCl\left(B\right)\right\} \cap NeutroCl\left(A^{C}\right)\right]$$

$$= \left[\left\{NeutroCl\left(A\right) \cap NeutroCl\left(A^{C}\right)\right\} \cap NeutroCl\left(B\right)\right] \cup \left[NeutroCl\left(A\right) \cap \left\{NeutroCl\left(B\right) \cap NeutroCl\left(B^{C}\right)\right\}\right]$$

$$= \left[NeutroBd\left(A\right) \cap NeutroCl\left(B\right)\right] \cup \left[NeutroCl\left(A\right) \cap NeutroBd\left(B\right)\right]$$

$$\subseteq NeutroBd\left(A\right) \cup NeutroBd\left(B\right).$$

Hence,

$$NeutroBd\left(A\cap B\right)\subseteq NeutroBd\left(A\right)\cup NeutroBd\left(B\right).$$

7. $7.\ NeutroBd\left(A\cup B\right)=NeutroCl\left(A\cup B\right)\cap NeutroCl\left(A\cup B\right)^{C}$

$$\subseteq\left[NeutroCl\left(A\right)\cup NeutroCl\left(B\right)\right]\cap\left[NeutroCl(A^{C})\cap NeutroCl(B^{C})\right]$$

$$=\left[NeutroCl\left(A\right)\cap\left\{NeutroCl(A^{C})\cap NeutroCl(B^{C})\right\}\right]\cup$$
$$\left[NeutroCl\left(B\right)\cap\left\{NeutroCl\left(A^{C}\right)\cap NeutroCl(B^{C})\right\}\right]$$

$$=\left[\left\{NeutroCl\left(A\right)\cap NeutroCl(A^{C})\right\}\cap NeutroCl(B^{C})\right]\cup$$
$$\left[\left\{NeutroCl\left(B\right)\cap NeutroCl\left(B^{C}\right)\right\}\cap NeutroCl\left(A^{C}\right)\right]$$

$$=\left[NeutroBd\left(A\right)\cap NeutroCl\left(B^{C}\right)\right]\cup\left[NeutroBd\left(B\right)\cap A^{C}\right]$$

$$\subseteq NeutroBd\left(A\right)\cup NeutroBd\left(B\right)$$

Hence,

$$NeutroBd\left(A\cup B\right)\subseteq NeutroBd\left(A\right)\cup NeutroBd\left(B\right).$$

Remark 3.5: Equality hold good in general topology for Theorem 3.4 (i), whereas Equality does not hold good in NeutroTopological space.

For this we cite an example:

Let $X=\left\{a,b,c,d\right\}$ and

$$\tau=\left\{\phi,\left\{a\right\},\left\{a,b\right\},\left\{c,d\right\},\left\{b,c\right\}\right\}$$

be a NeutroTopological space.

1. 1. $A=\{a,c\}, A^C=\{b,d\}$

$NeutroInt(A)=\{a\}$

Therefore,

$NeutroCl(A)=X$, $Cl(A^C)=\{b,c,d\}$.

Therefore,

$NeutroBd(A)=\{b,c,d\}$.

Now,

$NeutroCl(A)\cap A^C = X\cap\{b,d\}=\{b,d\}$

Therefore,

$NeutroCl(A)\cap A^C - NeutroInt(A)\subseteq NeutroBd(A)$.

Hence

$NeutroBd(A)\neq NeutroCl(A)\cap A^C - NeutroInt(A)$.

Theorem 3.5: Let (X,τ) be NeutroTopological space and let $A\subseteq X$. Then

If A is NeutroOpen then $NeutroCl(A)-A\subseteq NeutroBd(A)$
If A is NeutroOpen then $NeutroBd(A)\subseteq NeutroCl(A)$.
If A is NeutroClosed then $NeutroBd(A)\subseteq A$.

Proof: Since A is NeutroOpen, therefore $NeutroInt(A)\subseteq A$ [from Theorem 3.1 (i)] and

$$NeutroInt(A) \subseteq \left(NeutroCl\left(A^C\right)\right)^C$$

[from Theorem 3.3 (i)]

1. 1. $NeutroCl(A) - A \subseteq NeutroCl(A) - NeutroInt(A)$

$$\subseteq NeutroCl(A) - \left(NeutroCl\left(A^C\right)\right)^C$$

$$\subseteq NeutroCl(A) \cap \left\{\left(NeutroCl\left(A^C\right)\right)^C\right\}^C$$

$$= NeutroCl(A) \cap NeutroCl\left(A^C\right)$$

$$= NeutroBd(A).$$

Hence

$NeutroCl(A) - A \subseteq NeutroBd(A)$.

2. 2. $NeutroInt(A) \subseteq A$ and

$NeutroBd(A) = NeutroCl(A) \cap NeutroCl\left(A^C\right) \subseteq NeutroCl(A)$.

3. Similarly, we can prove following same steps as in part (ii) above.

Remark 3.6: The converse of the Theorems 3.5 may not be true. For this we cite an example:

1. 1. Let $X = \{a, b, c, d\}$ and

$\tau = \{\phi, \{a\}, \{a,b\}, \{c,d\}, \{b,c\}\}$

Let $A = \{c\}$, $NeutroCl(A) = \{c,d\}$

$A^C = \{a,b,d\}$, $Cl(A^C) = X$.

Therefore, $NeutroBd(A) = \{c,d\}$

$NeutroCl(A) - A = \{c,d\} - \{c\} = \{d\}$.

Here,

$NeutroCl(A) - A \subseteq NeutroBd(A)$

but A is not NeutroOpen.

Similarly, we can show for (ii) and (iii).

CONCLUSION

In this study, it is observed that many properties of NeutroTopological space are not the same as general topological space. Then we have investigated the properties of the interior, closure, and boundary of NeutroTopological spaces. Hope our work will help in further study of NeutroTopological space. This may lead to a new beginning for further research on the study of generalized closed sets in NeutroTopological space.

REFERENCES

Agboola, A. A. A. (2020a). Introduction to NeutroGroups. *International Journal of Neutrosophic Science*, *6*, 41–47.

Agboola, A. A. A. (2020b). Introduction to NeutroRings. *International Journal of Neutrosophic Science*, *7*(2), 62–73.

Agboola, A. A. A. (2020c). On Finite NeutroGroups of Type-NG. *International Journal of Neutrosophic Science*, *10*(2), 84–95.

Al-Tahan, M. (2021). NeutroOrderedAlgebra: Theory and Examples. In *International Workshop on Advanced Topics in Dynamical Systems*. University of Kufa.

Al-Tahan, M., Davvaz, B., Smarandache, F., & Anis, O. (2021). On Some NeutroHyperstructures. *Symmetry*, *13*(535), 1–12.

Al Tahan, M., Hoskova-Mayerova, S., & Davvaz, B. (2019). Some results on (generalized) fuzzy multi-Hv-ideals of Hv-rings. *Symmetry*, *11*(11), 1376. doi:10.3390ym11111376

Al Tahan, M., Hoskova-Mayerova, S., & Davvaz, B. (2020). Fuzzy multi-polygroups. *Journal of Intelligent & Fuzzy Systems*, *38*(2), 2337–2345. doi:10.3233/JIFS-191262

Al-Tahan, M., Smarandache, F., & Davvaz, B. (2021). NeutroOrderedAlgebra: Applications to Semigroups. *Neutrosophic Sets and Systems*, *39*, 133–147.

Atanassov, K. T. (1986). Intuitionistic fuzzy sets. *Fuzzy Sets and Systems*, *20*(1), 87–96. doi:10.1016/S0165-0114(86)80034-3

Bakbak, D., Uluçay, V., & Sahin, M. (2019). Neutrosophic Multigroups and Applications. *Mathematics*, *7*(1), 95. doi:10.3390/math7010095

Basumatary, B. (2016). Towards forming the field of fuzzy closure with reference to fuzzy boundary. *Journal of Process Management. Nuclear Technology*, *4*(1), 30–40.

Basumatary, B., & Broumi, S. (2020). Interval-Valued Triangular Neutrosophic Linear Programming Problem. *International Journal of Neutrosophic Science*, *10*(2), 105–115.

Basumatary, B., Wary, N., Mwchahary, D. D., Brahma, A. K., Moshahary, J., Basumatary, U. R., & Basumatary, J. (2021). A Study on Some Properties of Neutrosophic Multi Topological Group. *Symmetry*, *13*(9), 1689. doi:10.3390ym13091689

Blizard, W. (1989). Multiset theory. *Notre Dame Journal of Formal Logic*, *30*, 36–66.

Chang, C. L. (1968). Fuzzy Topological Space. *Journal of Mathematical Analysis and Applications*, *24*(1), 182–190. doi:10.1016/0022-247X(68)90057-7

Coker, D. (1997). An introduction to intuitionistic fuzzy topological spaces. *Fuzzy Sets and Systems*, *88*(1), 81–89. doi:10.1016/S0165-0114(96)00076-0

Devi, R., Dhavaseelan, R., & Jafari, S. (2017). On Separation Axioms in an Ordered Neutrosophic Bitopological Space. *Neutrosophic Sets and Systems*, *18*, 27–36.

Dey, A., & Pal, M. (2015). Multi-fuzzy complex numbers and multi-fuzzy complex sets. *International Journal of Fuzzy System Applications*, *4*(2), 15–27. doi:10.4018/IJFSA.2015040102

Kandil, A., Nouth, A. A., & El-Sheikh, S. A. (1995). On fuzzy bitopological spaces. *Fuzzy Sets and Systems*, *74*(3), 353–363. doi:10.1016/0165-0114(94)00333-3

Kelly, J. C. (1963). Bitopological spaces. *Proceedings of the London Mathematical Society*, *3*(1), 71–89. doi:10.1112/plms3-13.1.71

Miyamoto, S. (2001). Fuzzy Multisets and Their Generalizations. In *Multiset Processing* (pp. 225–235). Springer. doi:10.1007/3-540-45523-X_11

Mwchahary, D. D., & Basumatary, B. (2020). A note on Neutrosophic Bitopological Space. *Neutrosophic Sets and Systems*, *33*, 134–144.

Onasanya, B. O., & Hoskova-Mayerova, S. (2018). Some Topological and Algebraic Properties of alpha-level Subsets Topology of a Fuzzy Subset. *An. St. Univ. Ovidius Constanta*, *26*, 213–227.

Onasanya, B. O., & Hoskova-Mayerova, S. (2019). Multi-fuzzy group induced by multisets. *International Journal of Pure and Applied Mathematics*, *41*, 597–604.

Şahin, M., Kargın, A., & Yücel, M. (2021). Neutro-Topological Space and Anti-Topological Space. *NeutroAlgebra Theory*, *I*, 16.

Salama, A. A., & Alblowi, S. A. (2012). Neutrosophic Set and Neutrosophic Topological Spaces. *IOSR Journal of Mathematics, 3*(4), 31-35.

Salama, A. A., Smarandache, F., & Kroumov, V. (2014). Closed sets and Neutrosophic Continuous Functions. *Neutrosophic Sets and Systems*, *4*, 4–8.

Sebastian, S., & Ramakrishnan, T. V. (2011). Multi-fuzzy sets: An extension of fuzzy sets. *Fuzzy Information and Engineering*, *3*(1), 35–43. doi:10.100712543-011-0064-y

Smarandache, F. (1998). Neutrosophy. Neutrosophic Probability, Set, and Logic, 105.

Smarandache, F. (2019). *Introduction to NeutroAlgebraic Structures and AntiAlgebraic Structures. In Advances of Standard and Nonstandard Neutrosophic Theories*. Pons Publishing House Brussels.

Smarandache, F. (2020a). Generalizations and Alternatives of Classical Algebraic Structures to NeutroAlgebraic Structures and AntiAlgebraic Structures. *Journal of Fuzzy Extension and Applications*, *1*(2), 85–87.

Smarandache, F. (2020b). Introduction to NeutroAlgebraic Structures and AntiAlgebraic Structures (revisited). *Neutrosophic Sets and Systems*, *31*, 1–16.

Smarandache, F. (2020c). NeutroAlgebra is a Generalization of Partial Algebra. *International Journal of Neutrosophic Science*, *2*, 8–17. doi:10.54216/IJNS.020103

Smarandache, F. (2020d). Structure, NeutroStructure, and AntiStructure in Science. *International Journal of Neutrosophic Science*, *13*(1), 28–33.

Sumathi, I. R., & Arockiarani, I. (2015). Fuzzy Neutrosophic Groups. *Advanced in Fuzzy Mathematics*, *10*(2), 117–122.

Sumathi, I. R., & Arockiarani, I. (2016). Topological Group Structure of Neutrosophic set. *Journal of Advanced Studies in Topology*, *7*(1), 12–20. doi:10.20454/jast.2016.1003

Yager, R. R. (1986). On the theory of bags. *International Journal of General Systems*, *13*(1), 23–37. doi:10.1080/03081078608934952

Yang, Y., Tan, X., & Meng, C. (2013). The multi-fuzzy soft set and its application in decision making. *Applied Mathematical Modelling*, *37*(7), 4915–4923. doi:10.1016/j.apm.2012.10.015

Zadeh, L. A. (1965). Fuzzy sets. *Information and Control*, *8*(3), 338–353. doi:10.1016/S0019-9958(65)90241-X

Chapter 12
An Introduction to NeutroHX-Groups

Sonea Andromeda
University of Life Sciences, Romania

Saba Al-Kaseasbeh
https://orcid.org/0000-0003-4590-310X
Tafila Technical University, Jordan

ABSTRACT

The chapter aims to introduce the concept of NeutroHX-Group. Starting with HX-groups and using the newly defined concept of NeutroAlgebra, the authors define the NeutroHX-Group. Using the HX-groups associated with the dihedral group, they obtain the Chinese hypergroup associated with them. They present some examples to illustrate situations when the NeutroHX-Groups and AntiHX-Groups are obtained. Also, they determine a way to obtain commutative hyperstructure, considering the set of all HX-Groups with the dihedral group as support. The important result of this chapter is given by Theorem 1, where it is establishing a connection between NeutroHX-Groups and NeutroSemihypergroups.

INTRODUCTION

Hongxing Li et. al (Hongxing L., Qinzi D., Peizhuang W., 1985) introduced the notion of HX-groups. They considered a new operation on the family of non-empty sets of a group and if this new family support together with the new operation formed a group, which is called an HX-group. More precisely, it is consider (G,.) a group and $P^*(G) = P(G) \setminus \{\varnothing\}$ is the set of nonempty subsets of G, then the $P^*(G)$ endowed

DOI: 10.4018/978-1-6684-3495-6.ch012

with hypeoperation "$*$": P *(G)× P *(G) → P *(G) is also a group. Recently, Corsini (Corsini, P., 2016) determined the HX-groups associated to group Z/nZ and Sonea (Sonea A., 2019) found the HX-groups with D_n as group support. Also, starting with HX-groups, Corsini defined a hyperoperation on G, $\oplus$:G×G→ P *(G) such that

$$x \oplus y = \bigcup_{\substack{x \in A \\ y \in B}} \{x.y \mid x \in A, y \in B\}$$

is called Chinese hypergroupoid in the honor of Chinese scholar Hongxing Li. In 2021, Cristea presented a link between HX-groups and hypergroups.

Neutrosophy, the study of neutralities, is a new branch of philosophy initiated by Smarandache (Smarandache, F., 1998) and it has connections with different domains in mathematics. One of these connections is Algebra. As a generalization of partial algebra (Smarandache, F., 2020), NeutroAlgebra was introduced whose operations and axioms are partially true, partially indeterminate, and partially false. On the other hand, algebraic hyperstructures are a generalization of algebraic structures and they were introduced in 1934 by Marty (Marty, F., 1934). As a combination between the two notions: algebraic hypertsructures and NeutroAlgebra, NeutroHyperstructure theory was introduced (Ibrahim, M.A. & Agboola, A.A.A, 2020) where Ibrahim and Agboola defined NeutroHypergroups and studied a special type of it.

This chapter introduces NeutroHX-Groups and we establish a connection between NeutroHX-Groups and Chinese hypergroupoid associated. Also, we present examples to illustrate this connection.

PRELIMINARIES

Hyperstructure theory is a generalization of classical algebraic structures and it was introduced in 1934 at the eighth Congress of Scandinavian Mathematicians by Marty (Marty, F. 1934). Marty generalized the notion of groups by defining hypergroups. The class of algebraic hyperstructures is larger than that of algebraic structures where the operation on two elements in the latter is again an element, whereas the hyperoperation of two elements in the first class is a non-void set. This theory has many applications in various fields of science like Cryptography, Automata theory, Fuzzy sets, Biology, Chemistry, Physics, Neutrosophy, Coding theory. Some of the applications can be found in the book of Vougiouklis (Vougiouklis, T., 1994).

Let H be a nonempty set and $P^*(H)$ be the set of all nonempty subsets of H. Then a hyperoperation on H is a mapping $\bigcirc$: $H \times H \to P^*(H)$. The couple $(H,\bigcirc)$ is called a hypergroupoid.

A hypergroupoid $(H,\bigcirc)$ is called semihypergroup if it satisfies the associative axiom. i.e. $a\bigcirc(b\bigcirc c) = (a\bigcirc b)\bigcirc c$ for all $a,b,c \in H$. (Corsini, P., 1993).

The following definitions and results of HX-groups appear in (Corsini P., 2016) and (Corsini P., 2017). The results refer to the cardinality of elements from HX-groups and necessary conditions to obtain a hypergroup structure.

Definition 1: An HX-group is a nonempty subset Γ of P*(G) which is a group with respect to the operation " $*$" defined *by:*

$$\forall A, B \in \Gamma, A * B = \{a.b \mid a \in A, b \in B\} \tag{1}$$

We say that Γ has group G as support.

Definition 2: An HX-group with support G is said to be regular if the identity element e of the group G belongs to the set of identities in Γ. That is if e ∈ E.

Proposition 1: Let Γ be a HX-group with support G and let E be the identities of Γ. Then for any *A,B* ∈ Γ we have

|A| = |E|;

A ∩ B ≠∅ ⇒ |A ∩ B| = |E|.

Definition 3: Let (Γ, $*$) be an HX-group with the support (G, .) and let E be the identity of group Γ. A Chinese hypergroupoid is a hyperstructure (H, ⊕), where $H = \bigcup_{A \in G} A$ and for any

$$(x,y) \in H \times H, x \oplus y = \bigcup_{\substack{x \in A \\ y \in B \\ \{A,B\} \subset \Gamma}} A * B.$$

The following proposition provides us with the condition that makes the hypergroupoid a hypergroup.

Proposition 2: If Γ is an HX-group such that for any A, B ∈ *G*, A ∩ B ≠∅ implies A=B, then (H, ⊕) is a hypergroup.

Definition 4: Let G be a nonempty set and let $*$: $G \times G \rightarrow G$ be a binary operation on G. The couple (G,$*$) is called a NeutroGroup if it satisfies at least one of the following conditions and it has no AntiOperation nor AntiAxioms.

Definition of the Neutrosophic Triplet <Axiom, NeutroAxiom, AntiAxiom>

1. A classical **Axiom** defined on a nonempty set is an axiom that is totally true (i.e., true for all set's elements).
2. A **NeutroAxiom** (or Neutrosophic Axiom) defined on a nonempty set is an axiom that is true for some set's elements [degree of truth (T)], indeterminate for other set's elements [degree of indeterminacy (I)], and false for the other set's elements [degree of falsehood (F)], where $T,I,F \in [0,1]$, with $(T,I,F) \neq (1,0,0)$ that represents the classical Axiom, and $(T,I,F) \neq (0,0,1)$ that represents the AntiAxiom.
3. An **AntiAxiom** defined on a nonempty set is an axiom that is false for all set's elements.

Therefore we have formed the neutrosophic triplet:

<Axiom(1,0,0), NeutroAxiom(T,I,F), AntiAxiom(0,0,1)>,

where (T, I, F) $\notin$ {(1, 0, 0), (0, 0, 1)}.

NeutroAssociativity

1. there are some elements a, b, c such that a*(b*c) = (a*b)*c (this is degree of truth T);
2. there are some elements d, e, f such that d*(e*f) or (d*e)*f are indeterminate (this is degree of indeterminacy I);
3. there are some elements g, h, j such that g*(h*j) is different from (g*h)*j (this is degree of falsehood F).

NeutroNeutral

1. there is an element p such that a*p = p*a (this is degree of truth T);
2. there is an element such that a*q = q*a is indeterminate (this is degree of indeterminacy I);

3. there is an element r such that a*r is different from r*a (this is degree of falsehood F).

NeutroInverse

1. for every element *a* there is an inverse b such that a*b = b*a=p (this is degree of truth T);
2. for every element *a* there is an inverse b such that a*b = b*a=p is indeterminate (this is degree of indeterminacy I);
3. for every element *a* there is an inverse b such that a*b is different from p (this is degree of falsehood F).

NeutroCommutative

1. There are some elements a,b such that a*b = b *a (this is degree for truth T);
2. There are some elements c,d such that c*d or d*c are indeterminate (this is degree of indeterminacy I);
3. There are some elements e,f such that e∗f is different from f ∗e (this is degree of falsehood F).

Then (G,∗) is called a NeutroCommutativeGroup or a NeutroAbelianGroup.

In 2021, Al-Tahan, Davvaz, Smarandache, and Anis studied NeutroSemihypergroups and they defined some new related concepts such as NeutroSubsemihypergroups, NeutroHyperideals, etc. and they proved some related properties.

Definition 5: Let A be any non-empty set and "$\bigcirc: A \to P^*\left(A\right)$" be a map. Then "$\bigcirc$" is called a NeutroHyperoperation on A if some (or all) of the following conditions hold in a way that

$$\left(T,I,F\right) \notin \left\{\left(1,0,0\right),\left(0,0,1\right)\right\}.$$

1. There exist $x, y \in A$ with $x \bigcirc y \subseteq A$. (This condition is called degree of truth, "T").
2. There exist $x, y \in A$ with $x \bigcirc y \subsetneq A$. (This condition is called degree of falsity, "F").
3. There exist $x, y \in A$ with $x \bigcirc y$ is indeterminate in A. (This condition is called degree of indeterminacy, "I").

The couple $(A,\bigcirc)$ is called a NeutroHypergroupoid.

Definition 6: Let A be any non-empty set and "$\bigcirc: A \to P^*(A)$" be a map. Then "$\bigcirc$" is called an AntiHyperoperation on A if $x \bigcirc y \subsetneq A$ for all $x, y \in A$. The couple $(A,\bigcirc)$ is called an AntiHypergroupoid.

Definition 7: Let A be any nonempty set and "$\bigcirc: A \to P^*(A)$" be a map. Then "$\bigcirc$" is called a NeutroAssociative on A if some (or all) of the following conditions hold in a way that

$$(T, I, F) \notin \{(1,0,0),(0,0,1)\}.$$

1. There exist $x, y, z \in A$ with $x \bigcirc (y \bigcirc z) = (x \bigcirc y) \bigcirc z$. (This condition is called degree of truth, "T").
2. There exist $x, y, z \in A$ with $x \bigcirc (y \bigcirc z) \neq (x \bigcirc y) \bigcirc z$. (This condition is called degree of falsity, "F").
3. There exist $x, y, z \in A$ with $x \bigcirc (y \bigcirc z)$ is indeterminate or $(x \bigcirc y) \bigcirc z$ is indeterminate or we can not find if $x \bigcirc (y \bigcirc z)$ and $(x \bigcirc y) \bigcirc z$ are equal. (This condition is called degree of indeterminacy, "I").

Definition 8: Let A be any nonempty set and "$\bigcirc: A \to P^*(A)$" be a map. Then "$\bigcirc$" is called an AntiAssociative on A if

$$x \bigcirc (y \bigcirc z) \neq (x \bigcirc y) \bigcirc z$$

for all $x, y, z \in A$.

Definition 9: Let A be any non-empty set and "$\bigcirc$" be a map. Then $(A,\bigcirc)$ is called NeutroSemihypergroup if "$\bigcirc$" is a NeutroAssociative but not an AntiHyperoperation.

NEUTROHX-GROUPS AND ANTIHX-GROUPS

In this section, we introduce for the first time, the concept of NeutroHX-Groups, using the new theory of NeutroGroups (Agaboola A.A.A., 2020) .

Definition 10 Let (G, .) be a group and P*(G) is the set of nonempty subsets of G.

Table 1.

*	A_0	A_1
A_0	A_0	A_1
A_1	A_1	A_2

Let H ⊆ P*(G) and " ∗" be defined as follows:

$$A * B = \{a.b \mid a \in A, b \in B\}. \quad (1)$$

Then:

1. (H, ∗) is a NeutroHX-Group if (H, ∗) is a NeutroGroup.
2. (H, ∗) is an AntiHX-Group if (H,, ∗) is an AntiGroup.

Sonea (2019) presented the form of HX-groups with dihedral group D_n as support. Let $G^{p_1} = \{A_i | 0 \leq i \leq 2p_2 - 1\}$, then ($G^{p_1}$, ∗) is an HX-group associated to dihedral group D_{n}.

Example 1 Let G= D_n and H={A_0, A_1}with Table 1.

1. A NeutroOperator, because $A_0 * A_0 = A_0 \in H$ and $A_1 * A_1 = A_2 \notin H$.
2. Not AntiAssociative:

$$(A_0 * A_0) * A_0 = A_0 * A_0 = A_0$$

and

$$A_0 * (A_0 * A_0) = A_0 * A_0 = A_0.$$

So, we have $(A_0 * A_0) * A_0 = A_0 * (A_0 * A_0)$.

3. Not AntiIdentity: $A_0 * A_1 = A_1 * A_0 = A_1.$
4. Not AntiInverse: $A_0 * A_0 = A_0.$

Example 2 Let G= D_n and H={A_1, A_2 }with Table 2.

(H, ∗) is an AntiHX-Groups, because we have AntiIdentity as it follows:

$$A_1 * A_1 = A_2 \neq A_1;$$

Table 2.

*	A_1	A_2
A_1	A_2	A_3
A_2	A_3	A_4

$A_1 * A_2 = A_2 * A_1 = A_3 \neq A_1 \neq A_2$;

$A_2 * A_2 = A_4 \neq A_2.$

Example 3: *The* set of all HX-groups with dihedral group D_6 as support formed a NetruoGroup

The dihedral group is generated by rotation ρ and axial symmetry σ, so

$D_6 = \{e,\rho,\rho^2,\rho^3,\rho^4,\rho^5,\sigma, \rho\sigma, \rho^2\sigma, \rho^3\sigma, \rho^4\sigma, \rho^5\sigma\}$,

where the rotation ρ and the symmetry σ are satisfying the next conditions:

$\rho^6 = \sigma^2 = e; \rho^k\sigma = \sigma\rho_{n-k}, 0 \leq k \leq 6.$

For $n = 6 = 2 \cdot 3$, we consider $p_1 = 2$ and $p_2 = 3$, so we get an HX -group formed by six sets, each one having the cardinal equal with two. So, it has the following form

$$G_3^2 = \{(e,\rho^3);(\rho,\rho^4);(\rho^2,\rho^5);(\sigma,\rho^3\sigma);(\rho\sigma,\rho^4\sigma);(\rho^2\sigma,\rho^5\sigma);\}$$

Also, n=3·2, so $p_1 = 3$ and $p_2 = 2$ we obtain an HX -group with four elements, each one having cardinal equal with three.

$$G_2^3 = \{(e,\rho^2,\rho^4);(\rho,\rho^3,\rho^5);(\sigma,\rho^2\sigma,\rho^4\sigma);(\rho\sigma,\rho^3\sigma,\rho^5\sigma);\}$$

For $n = 6 \cdot 1$, we have $p_1 = 6$ and $p_2 = 1$. We get an HX -group with two elements, each one of cardinal equal with six, so it has the next form:

$$G_1^6 = \{(e,\rho,\rho^2,\rho^3,\rho^4,\rho^5);(\sigma,\rho\sigma,\rho^2\sigma,\rho^3\sigma,\rho^4\sigma,\rho^5\sigma)\}$$

In last case, when n=1·6, we consider $p_1 = 1$ and $p_2 = 6$. The corresponding HX -group has twelve elements, each element having cardinal 1; in fact, it is represented by the dihedral group D_6.

$$G_6^1 = \{(e);(\rho);(\rho^2);(\rho^3);(\rho^4);(\rho^5);(\sigma);(\rho\sigma);(\rho^2\sigma);(\rho^3\sigma);(\rho^4\sigma);(\rho^5\sigma);\}$$

We denote by $\bar{G}=\{G_2^3, G_3^2, G_1^6, G_6^1\}$the set of all HX groups associated to dihedral group $(D_6,.)$. On $\bar{G}$ we consider the following hyperoperation "$\odot$":$\bar{G}\times \bar{G}\rightarrow P^*(\bar{G})$ as it follows

$$G_a^b \odot G_{a'}^{b'} = \{C_{i,j} = A_i^b * B_j^{b'} \mid A_i^b \subseteq G_a^b, B_j^{b'} \subseteq G_{a'}^{b'}, 0 \le i \le 2a-1, 0 \le j \le 2a'-1\}$$

where n=a·b=a'·b'.

For n=6, we have

$$G_2^3 \odot G_3^2 = \{C_{i,j} = A_i^3 * B_j^2 \mid A_i^3 \subseteq G_2^3, B_j^2 \subseteq G_3^2, 0 \le i \le 3, 0 \le j \le 5\}$$

So, for i=0, j=0, we have

$$C_{0,0} = A_0^3 * B_0^2 = \{e,\rho^2,\rho^4\} * \{e,\rho^3\} = e.e \cup e.\rho^3 \cup \rho^2.e \cup \rho^2.\rho^3 \cup \rho^4.e \cup \rho^4.\rho^3$$
$$= \{e,\rho^3,\rho^2,\rho^5,\rho^4,\rho\} = \{e,\rho,\rho^2,\rho^3,\rho^4,\rho^5\}$$

For i=0, j=1, we get

$$C_{0,1} = A_0^3 * B_1^2 = \{e,\rho^2,\rho^4\} * \{\rho,\rho^4\} = e.\rho \cup e.\rho^4 \cup \rho^2.\rho \cup \rho^2.\rho^4 \cup \rho^4.\rho \cup \rho^4.\rho^4$$
$$= \{\rho,\rho^4,\rho^3,e,\rho^5,\rho^2\} = \{e,\rho,\rho^2,\rho^3,\rho^4,\rho^5\}$$

In a similar way we obtain

$$C_{0,0} = C_{0,1} = C_{0,2} = C_{0,3} = C_{0,4} = C_{0,5} = C_{1,0} = C_{1,1} = C_{1,2}$$

For i=1, j=3, we get

$$C_{1,3} = A_1^3 * B_3^2 = \left\{\rho, \rho^3, \rho^5\right\} * \left\{\sigma, \rho^3, \sigma\right\} = \rho.\sigma \cup \rho.\rho^3\sigma \cup \rho^3.\sigma \cup \rho^3.\rho^3\sigma \cup \rho^5.\sigma \cup \rho^5.\rho^3\sigma$$
$$= \left\{\rho\sigma, \rho^4\sigma, \rho^3\sigma, \sigma, \rho^5\sigma, \rho^2\sigma\right\} = \left\{\sigma, \rho\sigma, \rho^2\sigma, \rho^3\sigma, \rho^4\sigma, \rho^5\sigma\right\}$$

Analogous, we obtain

$$C_{1,3} = C_{1,4} = C_{1,5} = C_{2,0} = C_{2,1} = C_{2,2} = C_{3,0} = C_{3,1} = C_{3,2}$$

and

$$C_{1,5} = C_{3,3} = C_{3,4} = C_{3,5}$$

So, we notice that

$$G_2^3 \odot G_3^2 = G_1^6 .$$

Proceeding in a similar way, we obtain the following relations

$$\begin{array}{l} G_3^2 \odot G_2^3 = G_1^6; G_3^2 \odot G_1^6 = G_1^6 = G_1^6 \odot G_3^2; G_3^2 \odot G_6^1 = G_3^2 = G_6^1 \odot G_3^2; \\ G_6^1 \odot G_2^3 = G_2^3 = G_2^3 \odot G_6^1; G_6^1 \odot G_1^6 = G_1^6 = G_1^6 \odot G_6^1; \\ G_1^6 \odot G_2^3 = G_2^3 \odot G_1^6 = G_1^6 \end{array} .$$

Also, we obtain

$$G_1^6 \odot G_1^6 = G_1^6; G_2^3 \odot G_2^3 = G_2^3; G_3^2 \odot G_3^2 = G_3^2; G_6^1 \odot G_6^1 = G_6^1 .$$

Remark 1: If n=6, we have

$$G_a^b \odot G_{a'}^{b'} = G_{(a,a')}^{[a,a']},$$

where n=a·b=a'·b'.

According with remark 1, we can affirm that G_6^1 is the identity for Ḡ, because

$$G_a^b \odot G_6^1 = G_{(a,6)}^{[a,6]} = G_a^b$$

for any a, b natural numbers with property that 6= a·b and (a,a') represents the greatest common factor of a and a' and [a,a'] means the least common multiple of a and a'.

Also, we show that G_6^1 and G_1^6 has inverse.

$$G_a^b \odot G_{a'}^{b'} = G_{a'}^{b'} \odot G_a^b = G_6^1,$$

equivalent with $(a, a') = 6$, but a and a' are divisors of 6 which implies a=1 and a'=6 or a=6 and a'=1. If a=1 then b=6 and the inverse of G_1^6 is G_6^1 and the inverse of G_6^1 is G_1^6, according to Definition 4, "$\odot$" satisfies condition iii), i.e. exists a NeutroInverse element.

In conclusion, ($\bar{G}$, $\odot$) is a NeutroGroup.

In the following, we will consider that for any A, B $\in \Gamma$, A $\cap$ B $\neq \varnothing$ implies A=B.

Theorem 1: Let (H, $\oplus$) be the Chinese hypergroupoid associated to the HX-group Γ if (Γ,*) is a NeutroHX-Group, then (H, $\oplus$) is a NeutroSemihypergroup.

Proof: Let (Γ,*) be a NeutroHX-Group. So, there is A,B,C$\in \Gamma$ such that

$$(A * B) * C \neq A * (B * C)$$

and there is A',B',C'$\in \Gamma$ such that

$$(A' * B') * C' = A' * (B' * C')$$

According to the Proposition 2 results that (H, $\oplus$) is a Chinese hypergroup associated to the HX-group Γ.

$$H = \bigcup_{A \subseteq \Gamma} A$$

and

$$x \oplus y = \bigcup_{\substack{x \in A \\ y \in B}} A * B.$$

But, using the condition A, B $\in \Gamma$, A $\cap$ B $\neq \varnothing$ implies A=B, we assert that

$$x \oplus y = A * B, x \in A, y \in B.$$

To show that (H, $\oplus$) is a NeutroSemihypergrop, we will prove that "$\oplus$" is a NeutroAssociative and not an AntiHyperoperation, according to Definition 9.

We consider

$$x = a, a \in A;$$
$$y = b, b \in B;$$
$$z = c, c \in C.$$

We can affirm that

$$A \bigcap B = \varnothing, B \bigcap C = \varnothing, C \bigcap A = \varnothing.$$

Therefore,

$$a \oplus b = \bigcup_{\substack{a \in A \\ b \in B}} A * B = A * B.$$

We denote by $D = A * B$. So,

$$\left(a \oplus b\right) \oplus c = \bigcup_{\alpha \in D} \left(\alpha \oplus c\right) = \bigcup_{\substack{\alpha \in D \\ c \in C}} D * C = \bigcup_{\substack{\alpha \in D \\ c \in C}} \left(A * B\right) * C = \left(A * B\right) * C$$

Now, we denote by $E = B * C$

$$a \oplus \left(b \oplus c\right) = \bigcup_{\alpha \in E} \left(a \oplus \alpha\right) = \bigcup_{\substack{\alpha \in A \\ \alpha \in E}} A * E = \bigcup_{\substack{\alpha \in A \\ \alpha \in E}} A * \left(B * C\right) = A * \left(B * C\right)$$

In conclusion,

$$\left(a \oplus b\right) \oplus c \neq a \oplus \left(b \oplus c\right),$$

Because we choose the sets A,B,C such that $\left(A * B\right) * C \neq A * \left(B * C\right)$.

We consider

$$x = a', a' \in A';$$
$$y = b', b' \in B';$$
$$z = c', c' \in C',$$

with

$$A' \cap B' = \varnothing, B' \cap C' = \varnothing, C' \cap A' = \varnothing$$

and similarly we prove that

$$\left(a' \oplus b'\right) \oplus c' = a' \oplus \left(b' \oplus c'\right).$$

In conclusion, "$\oplus$" is NeutroAssociative and "$\oplus$" it is not AntiHyperoperation. So, (H, $\oplus$) is a NeutroSemihypergroup.

Example 4: Let $G = \left(\mathbb{Z}_{10}, \bigcirc\right)$, where $a \bigcirc b = a + 2b \left(\text{mod}\, 10\right)$ a NeutroGroup and

$$\left(\Gamma = \left\{A_0, A_1, A_2, A_3, A_4\right\}, *\right)$$

a NeutroHX-Group, where

$$A_0 = \left\{\widehat{0}, \widehat{5}\right\}; A_1 = \left\{\widehat{1}, \widehat{6}\right\}; A_2 = \left\{\widehat{2}, \widehat{7}\right\}; A_3 = \left\{\widehat{3}, \widehat{8}\right\}; A_4 = \left\{\widehat{4}, \widehat{9}\right\}$$

Let $H = \bigcup_{A_i \subseteq \Gamma} A_i$ the Chinese hypergroup and $x \oplus y = \bigcup_{\substack{x \in A_i \\ y \in A_j}} A_i * A_j$.

We will show that $\left(\widehat{0} \oplus \widehat{1}\right) \oplus \widehat{6} \neq \widehat{0} \oplus \left(\widehat{1} \oplus \widehat{6}\right)$.

$$\widehat{0} \oplus \widehat{1} = A_0 * A_1 = \left\{\widehat{0}, \widehat{5}\right\} * \left\{\widehat{1}, \widehat{6}\right\} = \widehat{0} \bigcirc \widehat{1} \cup \widehat{0} \bigcirc \widehat{6} \cup \widehat{5} \bigcirc \widehat{1} \cup \widehat{5} \bigcirc \widehat{6}$$
$$= \left\{\widehat{2}\right\} \cup \left\{\widehat{2}\right\} \cup \left\{\widehat{7}\right\} \cup \left\{\widehat{7}\right\} = \left\{\widehat{2}, \widehat{7}\right\} = A_2.$$
$$\left(\widehat{0} \oplus \widehat{1}\right) \oplus \widehat{6} = \bigcup_{x \in A_2} \left(x \oplus \widehat{6}\right) = \widehat{2} \oplus \widehat{6} \cup \widehat{7} \oplus \widehat{6} = \left\{\widehat{4}, \widehat{9}\right\} = A_4.$$

$$\widehat{1} \oplus \widehat{6} = A_1 * A_1 = \left\{\widehat{1}, \widehat{6}\right\} * \left\{\widehat{1}, \widehat{6}\right\} = \left\{\widehat{3}, \widehat{8}\right\} = A_3.$$
$$\widehat{0} \oplus \left(\widehat{1} \oplus \widehat{6}\right) = \bigcup_{x \in A_3} \left(\widehat{0} \oplus \widehat{x}\right) = \widehat{0} \oplus \widehat{3} \cup \widehat{0} \oplus \widehat{8} = A_0 * A_3$$
$$= \left\{\widehat{0}, \widehat{5}\right\} * \left\{\widehat{3}, \widehat{8}\right\} = \widehat{0} \circ \widehat{3} \cup \widehat{0} \circ \widehat{8} \cup \widehat{5} \circ \widehat{3} \cup \widehat{5} \circ \widehat{8} - \left\{\widehat{1}, \widehat{6}\right\} = A_1.$$

So, we have $\left(\widehat{0} \oplus \widehat{1}\right) \oplus \widehat{6} \neq \widehat{0} \oplus \left(\widehat{1} \oplus \widehat{6}\right)$.

Also, in a similar way, we obtain $\left(\widehat{0} \oplus \widehat{0}\right) \oplus \widehat{0} = \widehat{0} \oplus \left(\widehat{0} \oplus \widehat{0}\right)$ and it is obvious that "$\oplus$" is not AntiHyperoperation.

In conclusion, $(H, \oplus)$ is a NeutroSemihypergroup.

Example 5: We will consider the Chinese hypergroup associated to NeutroHX-Group by Example 1.

$$H = \left\{A_0, A_1\right\},$$

where

$$A_0 = \left\{e, \rho^{p_2}, \rho^{2p_2}, \ldots \rho^{(p_1-1)p_2}\right\};$$
$$A_1 = \left\{\rho, \rho^{p_2+1}, \rho^{2p_2+1}, \ldots, \rho^{(p_1-1)p_2+1}\right\},$$

where the dihedral group is generated by the rotation ρ and axial symmetry σ.

We have

$$\rho^{p_2} \oplus \rho^{p_2} = A_0 * A_0 = A_0 \subseteq H;$$
$$\rho^{p_2+1} \oplus \rho^{p_2+1} = A_1 * A_1 = A_2 \not\subset H.$$

So, "$\oplus$" is a NeutroHyperoperation on H. Therefore, H is a NeutroHypergroupoid.

FUTURE RESEARCH DIRECTIONS

For future research, one can investigate more properties about NeutroHX-Groups and AntiHX-Groups.

CONCLUSION

This chapter introduced NeutroHX-Groups and AntiHX-Groups by presenting some examples on them studying some of their properties.

FUNDING

This research received no specific grant from any funding agency in the public, commercial, or not-for-profit sectors.

ACKNOWLEDGMENT

The authors are grateful to Dr. Madeleine Al-Tahan for advice offering a valuable interdisciplinary view. Also, we would like to thank the referees for their very constructive suggestions and for their valuable comments.

REFERENCES

Agaboola, A. A. A. (2020). On Finite NeutroGroups of Type-NG [1,2,4]. *International Journal of Neutrosophic Science*, *10*(2), 84–95.

Al-Tahan, M., Davvaz, B., Smarandache, F., & Anis, O. (2021). On some NeutroHyperstructures. *Symmetry*, *13*(4), 535. doi:10.3390ym13040535

Al-Tahan, M., Smarandache, F., & Davvaz, B. (2021). NeutroOrderedAlgebra: Applications to semigroups. *Neutrosophic Sets and System*, *39*, 133–147.

Corsini, P. (1993). Prolegomena of Hypergroup Theory. Udine Aviani Editore.

Corsini, P. (2016). HX-Hypergroups associated with the direct product of some Z/nZ. *Journal of Algebraic Structures and Their Applications*, (3), 1–15.

Corsini P. (2016). HX-groups and Hypergroups. *Analele Univer sitatii Ovidius Constanta*, *24*(3), 101-121.

Corsini, P. (2016). On Chinese hyperstructures Z/nZ. *Journal of Discrete Mathematical Science Cryptography*, (6), 133–137.

Corsini, P. (2017). Hypergroups associated with HX-groups. *Analele Universitații Ovidius Constanța*, *25*(2), 49–64. doi:10.1515/auom-2017-0020

Cristea, I., Novák, M., & Onasanya, B. (2021). Link between HX-Groups and Hypergroups. *Algebra Colloquium*, *28*(3), 441–452. doi:10.1142/S1005386721000341

Heidari, D., & Davvaz, B. (2011). On ordered hyperstructures. *Politehn. Univ. Bucharest Sci. Bull. Ser. A Appl. Mathematics and Physics*, *73*(2), 85–96.

Hongxing L., Qinzi D., & Peizhuang W. (1985), *Hypergroup (I), Busefal*, (23). Academic Press.

Hongxing L., Qinzi D., & Peizhuang W. (1986), *Hypergroup (I), Busefal*, (25). Academic Press.

Ibrahim, M. A., & Agboola, A. A. A. (2020). Introduction to NeutroHyperGroups. *Neutrosophic Sets and System*, *38*, 15–32.

Marty, F. (1934). Sur une generalization de la notion de groupe. *Proceedings of the 8th Congress on Mathmatics, Scandinaves*, 45–49.

Smarandache, F. (1998). Neutrosophy. Neutrosophic Probability, Set, and Logic, 105.

Smarandache, F. (2020). NeutroAlgebra is a generalization of partial algebra. *International Journal of Neutrosophic Science*, *2*, 8–17. doi:10.54216/IJNS.020103

Smarandache, F. (2020). Introduction to NeutroAlgebraic structures and AntiAlgebraic structures (revisited). *Neutrosophic Sets and System*, *31*, 1–16.

Sonea A. (2019). HX-Groups associated to the dihedral group D_n. *Journal of Multiple Valued Logic and Soft Computing*.

Vougiouklis, T. (1994). *Hyperstructures and Their Representations*. Hadronic Press, Inc.

ADDITIONAL READING

Al-Tahan, M., & Davvaz, B. (2017). Algebraic Hyperstructures Associated to Biological Inheritance. *Mathematical Biosciences*, *285*, 112–118. doi:10.1016/j.mbs.2017.01.002 PMID:28108295

Al-Tahan, M., Davvaz, B., Smarandache, F., & Anis, O. (2021). Properties of Productional NeutroOrderedSemigroups. *Neutrosophic Sets and System*, *42*, 178–190.

Davvaz, B., & Leoreanu-Fotea, V. (2008). *Hyperring Theory and Applications*. International Academic Press.

Davvaz, B., & Omidi, S. (2016). Basic notions and properties of ordered semihyperrings. *Categ. Gen. Algebr. Struct. Appl.*, *4*(1), 43–62.

Fuchs, L. (1963). *Partially Ordered Algebraic Systems, Int. Ser. of Monographs on Pure and Appl. Math. 28*. Pergamon Press.

Hamidi, M., & Smarandache, F. (2020). Neutro-BCK-Algebra. *International Journal of Neutrosophic Science*, *8*, 110–117. doi:10.54216/IJNS.080204

Rezaei, A., & Smarandache, F. (2020). On Neutro-BE-algebras and Anti-BE-algebras. *International Journal of Neutrosophic Science*, *4*, 8–15.

Smarandache, F. (2020). Extension of HyperGraph to n-SuperHyperGraph and to Plithogenic n-SuperHyperGraph, and Extension of HyperAlgebra to n-ary (Classical-/ Neutro-/Anti-) HyperAlgebra. *Neutrosophic Sets and Systems*, *33*, 290–296.

Chapter 13
Neutro Ordered Ring

Ahmed Hatip
University of Gaziantep, Turkey

Necati Olgun
University of Gaziantep, Turkey

ABSTRACT

The objective of this chapter is to introduce the concept of Neutro Ordered Rings by considering three Neutro Axioms (Neutro Abelian Group [additive], Neutro Semi group [multiplicative], and Neutro Distributive [multiplication over addition]) and total ordered relation or partially ordered relation. The concepts of Neutro Ordered Rings, Neutro Ordered Subrings, Neutro Ordered Ideals, Neutro Ordered Quotient Rings, and Neutro Ordered Ring Homomorphisms are presented. Moreover, several interesting results were reached, and a lot of illustrative examples were added.

INTRODUCTION

Vagueness or uncertainty is a critical issue in the representation of incomplete knowledge in the fields of Computer Science and artificial intelligence. To deal with the uncertainty, the fuzzy set introduced by (Zadeh 1965) allows the uncertainty of a set with a membership degree between 0 and 1. Then, (Atanassov 1986) introduced an intuitionistic Fuzzy set (IFS) as a generalization of the Fuzzy set. The IFS represents the uncertainty with respect to both membership and non-membership. However, it can only handle incomplete information but not the indeterminate and inconsistent information, which exists commonly in real situations. Therefore, (Smarandache 1998) proposed a neutrosophic set. It can independently express truth-membership degree T, indeterminacy-membership degree I, and false

DOI: 10.4018/978-1-6684-3495-6.ch013

membership degree F and deal with incomplete, indeterminate, and inconsistent information. The indeterminate element I is such that ordinary multiplication. $I = I^2 = I$, I^{-1} the inverse of I is not defined and hence does not exist. Moreover, $I + I + \ldots I = nI : n \in N$. Also, several generalizations of the set theories made such as fuzzy multi-set theory (Sebastian 2010), intuitionistic fuzzy multi-set theory (Rajarajeswari 2014) and refined neutrosophic set theory (Deli et al 2016). Many research treating imprecision and uncertainty have been developed and studied. Since then, it is applied to various areas, such as decision-making problems (Pramanik 2015) and (Hatip 2019) machine learning (Sahin 2014), intelligent disease diagnosis (Shahzadi 2017) communication services (Chakraborty 2020) pattern recognition (Sahin 2017)social network analysis and e-learning systems (Lotfy 2015) physics (Yuhua 2015), … etc.

Smarandache (Agboola 2020) recently introduced new fields of research in neutrosophy called Neutro-Structures and Anti-Structures respectively. In, (Smarandache 2019) Smarandache introduced the concepts of Neutro-Algebras and Anti-Algebras and in (Smarandache 2020) he revisited the concept of Neutro-Algebras and Anti-Algebras where he studied Partial Algebras, Universal Algebras, Effect Algebras and Boole's Partial Algebras and he showed that Neutro-Algebras are generalization of Partial Algebras. Agboola (Agboola 2020) introduced the concept of Neutro-Group. Inspired by Neutro Algebra and ordered Algebra (Al-Tahan 2021) Introduced Neutro Ordered Algebra and some related terms such as Neutro Ordered Sub Algebra and Neutro Ordered Homomorphism.

As a special case of NeutroAxioms, i.e. when T>0,F>0, and I belongs to [0,1] the present research is devoted to the presentation of the concept of Neutro Ordered Ring. Several interesting results and examples Neutro Ordered Ring, Neutro Ordered Subring and Neutro Ordered Ring Homomorphisms are presented in this chapter.

BACKGROUND

In this section, we will give some background that will be useful in other sections of the research.

Neutrosophic Sets (Smarandache 1998)

Let $\mathcal{U}$ be a universe. A neutrosophic sets $\mathcal{A}$ over $\mathcal{U}$ is defined by

$$\mathcal{A} = \left\{ \prec u, \left(T_{\mathcal{A}}(u), I_{\mathcal{A}}(u), F_{\mathcal{A}}(u)\right) \succ : u \in \mathcal{U} \right\}$$

Where, $T_A(u)$, $I_A(u)$ and $F_A(u)$ are called truth-membership function, indeterminacy-membership function and falsity- membership function, respectively. They are defined by

$$T_A : \mathcal{U} \rightarrow]^-0,1^+[, I_A : \mathcal{U} \rightarrow]^-0,1^+[, F_A : \mathcal{U} \rightarrow]^-0,1^+[$$

Such that $0^- \leq T_A(u) + I_A(u) + F_A(u) \leq 3^+$.

Single Valued Neutrosophic Set (Smarandache et al 2010)

Let $\mathcal{U}$ be a universe. A single valued neutrosophic set (SVN-set) over $\mathcal{U}$ is a neutrosophic set over $\mathcal{U}$, but the truth-membership function T, indeterminacy-membership function I and falsity- membership function F are respectively defined by

$$T_A : \mathcal{U} \rightarrow]^-0,1^+[, I_A : \mathcal{U} \rightarrow]^-0,1^+[, F_A : \mathcal{U} \rightarrow]^-0,1^+[$$

Such that $0 \leq T_A(u) + I_A(u) + F_A(u) \leq 3$.

Neutro-Axiom, Anti-Axiom (Smarandache et al 2020)

1. A classical axiom defined on a non-empty set is an axiom that is totally true (i.e., true for all set's elements in the set).
2. A Neutro-Axiom (or Neutrosophic Axiom) defined on a non-empty set is an axiom that is true for some elements in the set [degree of truth (T)], indeterminate for other elements in the set. [degree of indeterminacy (I)], or false for the other elements in the set [degree of falsehood (F)], where $T, I, F \in [0,1]$,, such that $(T, I, F) \neq (1,0,0)$ that represents the classical axiom, and $(T, I, F) \neq (0,0,1)$ that represents the Anti-Axiom.
3. 3. An Anti-Axiom defined on a nonempty set is an axiom that is false for all elements in the set. Therefore, we have the neutrosophic triplet:

< Axiom,Neutro-Axiom,Anti-Axiom >.

Neutro-Group (Agboola 2020)

Let G be a non-empty set and let $* : G \times G \rightarrow G$ be a binary operation on G. The couple (G,*) is called a Neutro-Group if the following conditions are satisfied:

1. * is Neutro-Associative that is there exists at least one triplet $(a,b,c) \in G$ such that

$$a * (b * c) = (a * b) * c$$

and there exists at least one triplet $(x,y,z) \in G$ such that

$$x * (y * z) \neq (x * y) * z..$$

2. There exists a Neutro-Neutral element in G that is there exists at least an element $a \in G$ that has a single neutral element that is we have e Î G such that $b * e = e * b = b$ and for $b \in G$ there does not exist $e \in G$ such that $b * e = e * b = b$ or there exist $e_1, e_2 \in G$ such that $b * e_1 = e_1, *b = b$ or $b * e_2 = e_2 * b = b$ with $e_1 \neq e_2$.
3. There exists a Neutro-Inverse element that is there exists an element $a \in G$ that has an inverse $b \in G$ with respect to a unit element $e \in G$ that is $a * b = b * a = e$ or there exists at least one element $b \in G$ that has two or more inverses $c, d \in G$ with respect to some unit element $u \in G$ that is $b * c = c * b = u$, $b * d = d * b = u$. In addition, if * is Neutro-Commutative that is there exists at least a duplet $(a,b) \in G$ such that $a * b = b * a$ and there exists at least a duplet $(c,d) \in G$ such that $c * d \neq d * c$, , then (G,*) is called a Neutro-Commutative Group or a Neutro-Abelian Group.

If only condition I is satisfied, then (G,*) is called a Neutro-Semi Group and if only conditions I and II are satisfied, then (G,*) is called a Neutro-Monoid.

Neutro-Ring (Agboola 2020)

1. A Neutro-Ring (R,+,.) is a ring structure that has at least one Neutro-Operation among "+" or "•" at least one Neutro-Axiom.

2. Let R be a non-empty set and let $+,. : R \times R \rightarrow R$ be binary operations of ordinary addition and multiplication on R. The triple (R,+,.) is called a Neutro-Ring if the following conditions are satisfied:
   a. (R,+) is a Neutro-Abelian Group.
   b. (R,.) is a Neutro-Semi Group.
   c. 2. c. "•" is both left and right Neutro-Distributive over "+" that is there exists at least a triplet $(a,b,c) \in R$ and at least a triplet $(d,e,f) \in R$ such that

$$a.(b+c) = a.b + a.c$$

$$(b+c).a = b.a + c.a$$

$$d.(e+f) \neq d.e + d.f \; (e+f).d \neq e.d + f.d \, .$$

If"•" is Neutro-Commutative, then (R,+,.) is called a Neutro-Commutative Ring.

Ordered Algebra (Fuchs 1963)

Let A be an Algebra with n operations "$*_i$" and "$\leq$" be a partial order relation (reflexive, anti-symmetric, and transitive) on A. Then $(A, *_1, *_2, \ldots, *_n, \leq)$ is an Ordered Algebra if the following conditions hold.

If $x \leq y \in A$, then $z *_i x \leq z *_i y$ and $x *_i z \leq y *_i z$ for all $i = 1, \ldots, n$ and $z \in A$.

Neutro Ordered Algebra (Al-Tahan et al 2021)

Let A be a Neutro Algebra with n (Neutro) operations "i" and "$\leq$" be a partial order (reflexive, anti-symmetric, and transitive) on A. Then $(A, *_1, *_2, \ldots, *_n, \leq)$ is a Neutro Ordered Algebra if the following conditions hold.

1. There exist $x \leq y \in A$ with $x \neq y$ such that $z *_i x \leq z *_i y$ and $x *_i z \leq y *_i z$ for all $i = 1, \ldots, n$ and $z \in A$ (This condition is called degree of truth, "T".)

2. There exist $x \le y \in A$ and $z \in A$ such that $z *_i x\, z *_i y$ and $x *_i z\, y *_i z$ for some $i = 1,\ldots,n$. (This condition is called degree of falsity, "F".)
3. *There exist $x \le y \in A$ and $z \in A$ such that $z *_i x$ or $z *_i y$ or $x *_i z$ or $y *_i z$ are indeterminate, or the relation between $z *_i x$ and $z *_i y$, or the relation between $x *_i z$ and $y *_i z$ are indeterminate for some $i = 1,\ldots,n.$. (This condition is called degree of indeterminacy, "I".) Where (T,I,F) is different from (1,0,0) that represents the classical Ordered Algebra as well from (0,0,1) that represents the AntiOrderedAlgebra.

Neutro Total Ordered Algebra (Al-Tahan et al 2021)

Let $\left(A, *_1, *_2, \ldots, *_n, \le\right)$ be a Neutro Ordered Algebra. If "≤" is a total order on, then A is called Neutro Total Ordered Algebra.

NEUTRO ORDERED RING AND ITS PROPERTIES

In this section, we use the defined notion of Neutro Ordered Algebra and apply it to Neutro Ordered Ring. Moreover, we study some properties of Neutro Ordered Ring and, Neutro Ordered Ring Homomorphism.

Neutro Ordered Ring

Let R be a Neutro Ring with two Neutro operations and "≤" be a partial order (reflexive, anti-symmetric, and transitive) on R. Then $\left(R, *_1, *_2, \le\right)$ is a Neutro Ordered Ring if the following conditions hold.

1. There exist $x \le y \in R$ with $x \ne y$ such that $z *_i x \le z *_i y$ and $x *_i z \le y *_i z$ for all $i = 1,2$ and $z \in R$ (This condition is called degree of truth, "T".)
2. There exist $x \le y \in R$ and $z \in R$ such that $z *_i x\, z *_i y$ and $x *_i z\, y *_i z$ for some $i = 1,2$. (This condition is called degree of falsity, "F".)
3. There exist $x \le y \in R$ and $z \in R$ such that $z *_i x$ or $z *_i y$ or $x *_i z$ or $y *_i z$ are indeterminate, or the relation between that $z *_i x$ and $z *_i y$, or the relation between $x *_i z$ and $y *_i z$ are indeterminate for some $i = 1,2.$. (This condition is called degree of indeterminacy, "I".)

Table 1.

+	m	n	p	•	m	n	p
m	m	n	n	m	m	m	m
n	p	m	p	n	m	p	m
p	n	p	n	p	m	p	n

Where (T,I,F) is different from (1,0,0) that represents the classical Ordered Ring as well from (0,0,1) that represents the Anti-Ordered Ring.

Neutro Total Ordered Ring

Let $\left(R,*_1,*_2,\leq\right)$ be a Neutro Ordered Ring. If "≤" is a total order on R, then R is called Neutro Total Ordered Ring.

Notes:

1. R is called a finite Neutro Ordered Ring if the number of elements in R is n that is o(R) = n. If no such n exists, then *R* is called an infinite Neutro Ordered Ring and we write $o\left(R\right)=\infty$.
2. An element $x \in R$ is called a Neutro Ordered Idempotent element if $x^2 = x$.
3. An element $x \in R$ is called a Neutro Ordered INilpotent element if there exists some positive integer n, such that $x^n = e$ where e is Neutro-Neutral element in R.

EXAMPLE: Let X={m,n,p,q,t} be a universe of discourse and let R={m,n,p} be a subset of .

Let + and • be tow binary operations defined on R as shown in the Cayely tables in Table 1.

It is clear from the table that it (R,+,•) is a Neutro-Commutative Ring with Neutro-Unity and by defining the total order

$$\leq=\left\{\left(m,m\right),\left(n,n\right),\left(p,p\right),\left(m,n\right),\left(m,p\right),\left(n,p\right)\right\}$$

on R, we get that (R,+,•, ≤) is a Neutro Total Ordered Ring. This is easily seen as:

1. $m \leq p$ implies that $m \cdot x \leq p \cdot x$ and $x \cdot m \leq x \cdot p$ for all $x \in R$.

And having $n \leq p$ but $p + n \leq p + p$.

2. $m \leq n$ implies that $m \cdot xn \cdot x$ and $x \cdot mx \cdot n$ for all $x \in R$.

And having $n \leq p$ but $p * np * p$.
It follows that (R,+,•) is a Neutro Ring.

Neutro Ordered Sub Ring

Let $\left(R, *_1, *_2, \leq\right)$ be a Neutro Ordered Ring and $\varnothing \neq S \subseteq M$. Then S is a Neutro Ordered Sub Ring of R if $\left(S, *_1, *_2, \leq\right)$ is a Neutro Ordered Ring.

EXAMPLE: Let R be a Neutro Ordered Ring. R is a Neutro Ordered Sub Ring called a trivial Neutro Ordered Sub Ring.

THEOREM: Let $\left(R, *_1, *_2, \leq\right)$ be a Neutro Ordered Ring and let N be a nonempty subset of R.

N is Neutro Ordered Sub Ring of R if the following conditions hold:

1. there exists some $m, n \in N$ such that $m *_1 n \in N$.
2. there exists some $m, r \in N$ such that $r *_2 m \in N$.

THEOREM: Let $\left(R, *_1, *_2, \leq\right)$ be the Neutro Ordered Ring and let $\left\{S_i\right\}, i = 1,2,\ldots,n$ be a family of Neutro Ordered Sub Rings of R. Then

1. $\bigcap_{1}^{n} S_i$ is a Neutro Ordered Sub Ring of R.
2. $\prod_{1}^{n} S_i$ is a Neutro Ordered Sub Ring of R.

Left Neutro Ordered Ideal

Let (R,+, •, ≤) be a Neutro Ordered Ring. A nonempty subset I of R is called a left Neutro Ordered Ideal of R if the following conditions hold:

1. I is a Neutro Ordered Sub Ring of R.
2. For $x \in I$ there exists $xr \in I$ for some $r \in R$.

EXAMPLE: Let (R,+,•) be the Neutro Ordered Ring of Example3.3. and let N= {p,n}:

$p, n \in N, p + n = p \in N$ but $n + n = m$

$p, n \in N, p \cdot n = p \in N$ but $n \cdot p = m$

It follows that N is Neutro Ordered Sub Ring of R.

Right Neutro Ordered Ideal

Let (R,+, •, ≤) be a Neutro Ordered Ring. A nonempty subset I of R is called right Neutro Ordered Ideal of R if the following conditions hold:

1. I is a Neutro Ordered Subring of R.
2. For $x \in I$ there exists $rx \in I$ for some $r \in R$.

Neutro Ordered Ideal

Let (R,+, •, ≤) be a Neutro Ordered Ring. A nonempty subset I of R is called a Neutro Ordered Ideal of R if the following conditions hold:

I is a Neutro Ordered Sub Ring of R.

For $x \in I$ there exists $xr, rx \in I$ for some $r \in R$.

EXAMPLE: Let R = {m,n,p} be the Neutro ordered Ring of Example 3.3 and let I = {n,p}

1. I = {n,p} is Neutro Ordered Sub Ring of R as we showed in Example 3.9
2. p•p=n, m•p=m,n•p=m $\notin$ I,n•n=p,p•n=p,m•n =m $\notin$ I.

This shows that I is a left Neutro ordered Ideal of R.

Again, p•p=n, p•m=m $\notin$ I,p•n=p,n•n=p, m•n=m $\notin$ I.

This also shows that I is a right Neutro ordered Ideal of R. Hence, I is a Neutro ordered Ideal of R.

Neutro Ordered Quotient Ring

Let (R,+, •, ≤) be a Neutro Ordered Ring and let I be a Neutro ordered Ideal of R. The set R/I is defined by

$$R / I = \{x + I : x \in R\}.$$

For $x+I, y+I \in R/I$ with at least a pair $(x,y) \in R$, let $\oplus$ and $\otimes$ be binary operations on R/I defined as follows:

$$(x+I) \oplus (y+I) = (x+y)+I,$$

$$(x+I) \otimes (y+I) = x \cdot y + I.$$

Then it can be shown that the tripple (R/I, $\oplus$, $\otimes$, $\leq$) is a Neutro Ordered Ring which we call a Neutro Ordered Quotient Ring.

EXAMPLE: Let R = {m,n,p} be the Neutro ordered Ring of Example3.3 and let I = {n,p} be the Neutro Ordered Ideal of Example3.12 then:

$$m+I = \{n\},$$

$$n+I = \{m,p\},$$

$$p+I = \{n,p\} = I,$$

Now the Neutro Ordered Quotient Ring is the set

$$R/I = \{m+I, n+I, c+I\} = \{\{n\},\{m,p\},\{n,p\}\}.$$

Neutro Ordered Ring Homomorphism

Let $(R, *_1, *_2, \leq_1)$ and $(S, \square_1, \square_2, \leq_2)$ be any two Neutro Ordered Rings. The mapping $\varphi : R \to S$ is called a Neutro Ordered Ring Homomorphism if the following conditions hold: for some $(x,y) \in R$, we have:

- $\varphi(x *_1 y) = \varphi(x) \square_1 \varphi(y)$
- $\varphi(x *_2 y) = \varphi(x) \square_2 \varphi(y)$ and there exist $a \leq_1 b, a \neq b, \varphi(a) \leq_2 \varphi(b)$ φ is called Neutro Ordered Ring Isomorphism if φ is a bijective Neutro Ordered R-Module Homomorphism.

EXAMPLE: From the Example3.3 Let

$\varphi:(R,+,\cdot,\leq)\to(R,+,\cdot,\leq)$

be a mapping defined as:

$\varphi(m)=m*m$

It can be shown that φ is a Neutro Ordered Ring Homomorphism such that for $m,n,p\in R$, we have:

1. $\varphi(m+m)=\varphi(m)=m\cdot m=m$

$\varphi(m)+\varphi(m)=(m\cdot m)+(m\cdot m)=m+m=m$

but

$\varphi(m+n)=\varphi(n)=n\cdot n=p$.

$\varphi(m)+\varphi(n)=(m\cdot m)+(n\cdot n)=m+p=n$.

$\varphi(m\cdot n)=\varphi(m)=m\cdot m=m$

$\varphi(m)\cdot\varphi(n)=m\cdot p=m$

but

$\varphi(p\cdot n)=\varphi(p)=p\cdot p=n$

$\varphi(p)\cdot\varphi(n)=n\cdot p=m$

The kernel of φ is $Ker\varphi = \{x : \varphi(x) = e_M\} = \{m, p\}$ where $e_M \in R$ is Neutro-Neutral element in R.

The image of φ is $Im\varphi = \{y \in N : y = \varphi(x) \text{ for at least one } y \in N\} = \{m, n, p\}$

2. there exist $m \leq n$ such that $\varphi(m) \leq \varphi(n)$

CONCLUSION

We have for the first time introduced in this paper the concept of Neutro Ordered Rings by considering three Neutro Axioms (Neutro Abelian Group (additive), Neutro Semi group (multiplicative) and Neutro Distributive (multiplication over addition)). Several interesting results and examples on Neutro Ordered Rings, Neutro Ordered Subrings.

REFERENCES

Abobala, M. (2021). Semi Homomorphisms and Algebraic Relations Between Strong Refined Neutrosophic Modules and Strong Neutrosophic Modules. *Neutrosophic Sets and Systems*, *39*, 107–119.

Abobala, M., & Hatip, A. (2020, July). AH-Substructures in Strong Refined Neutrosophic Modules. *International Journal of Neutrosophic Science*, *18*, 110–116.

Abobala, M., Hatip, A., & Alhamido, R. K. (2019, October). A Contribution to Neutrosophic Groups. *International Journal of Neutrosophic Science*, *14*, 67–76. doi:10.54216/IJNS.000202

Agboola, A. A. A. (2020). Introduction to NeutroGroups. *International Journal of Neutrosophic Science*, *6*(1), 41–47.

Agboola, A. A. A. (2020). Introduction to NeutroRings. *International Journal of Neutrosophic Science*, *7*, 62–73. doi:10.5281/zenodo.3991389

Al-Tahan, . (2021). NeutroOrderedAlgebra: Applications to Semigroups. *Neutrosophic Sets and Systems*, *39*, 133–147.

Anuradha. (2020). Neutrosophic Fuzzy Hierarchical Clustering for Dengue Analysis in Sri Lanka. *Neutrosophic Sets and Systems*, *31*, 179-199.

Atanassov, K. (1986). Intuitionistic fuzzy sets. *Fuzzy Sets and Systems*, *20*(1), 87–96. doi:10.1016/S0165-0114(86)80034-3

Chakraborty, A., Banik, B., Mondal, S. P., & Alam, S. (2020). Arithmetic and Geometric Operators of Pentagonal Neutrosophic Number and its Application in Mobile Communication Service Based MCGDM Problem. *Neutrosophic Sets and Systems*, *32*, 61–79.

Deli, I. (2016). Refined Neutrosophic Sets and Refined Neutrosophic Soft Sets: Theory and Applications. Handbook of Research on Generalized and Hybrid Set Structures and Applications for Soft Computing, 321-343.

Deli, I., Broumi, S., & Ali, M. (2014). Neutrosophic Soft Multi-Set Theory and Its Decision Making. *Neutrosophic Sets and Systems*, *5*, 65–76.

Ejaita, O. A., & Asagba, P. (2017). An Improved Framework for Diagnosing Confusable Diseases Using Neutrosophic Based Neural Network. *Neutrosophic Sets and Systems*, *16*, 28–34.

Fathi, S., ElGhawalby, H., & Salama, A. A. (2016). A Neutrosophic Graph Similarity Measures. *New Trends in Neutrosophic Theories and Applications, 1*.

Fuchs, L. (1963). *Partially Ordered Algebraic Systems*. Pergamon Press.

Hatip, A. (2020). *The Special Neutrosophic Functions. International Journal of Neutrosophic Science*.

Karaaslan, F. (2016). Correlation Coefficient between Possibility Neutrosophic Soft Sets. *Math. Sci. Lett.*, *5*(1), 71–74. doi:10.18576/msl/050109

Karaaslan, F. (2016). Correlation coefficients of single-valued neutrosophic refined soft sets and their applications in clustering analysis. *Neural Computing & Applications*, 1–13.

Lotfy, M. M. (2015). Review of Recommender Systems Algorithms Utilized in Social Networks based e-Learning Systems & Neutrosophic System. *Neutrosophic Sets and Systems*, *8*, 32–41.

Olgun. (2020). *On Refined Neutrosophic R-module. International Journal of Neutrosophic Science*.

Rajarajeswari, P., & Uma, N. (2014). Correlation Measure for Intuitionistic Fuzzy Multi Sets. *International Journal of Research in Engineering and Technology*, *3*(1), 611–617. doi:10.15623/ijret.2014.0301104

Rajarajeswari, P., & Uma, N. (2014). Zhang and Fu's Similarity Measure on Intuitionistic Fuzzy Multi Sets, International Journal of Innovative Research in Science. *Engineering and Technology*, *3*(5), 12309–12317.

Sahin, M., Deli, I., & Ulucay, V. (2016). Jaccard vector similarity measure of bipolar neutrosophic set based on multi-criteria decision making. *International conference on natural science and engineering (ICNASE'16).*

Sahin, M., Deli, I., & Ulucay, V. (2016). Jaccard Vector Similarity Measure of Bipolar Neutrosophic Set Based on Multi-Criteria Decision Making. *International Conference on Natural Science and Engineering (ICNASE'16).*

Sahin, M., Deli, I., & Ulucay, V. (2017). Extension principle based on neutrosophic multi-sets and algebraic operations. *Journal of Mathematical Extension*, *12*(1), 69–90.

Sahin, M., Olgun, N., Uluçay, V., Kargın, A., & Smarandache, F. (2017). A New Similarity Measure Based on Falsity Value between Single Valued Neutrosophic Sets Based on the Centroid Points of Transformed Single Valued Neutrosophic Numbers with Applications to Pattern Recognition. *Neutrosophic Sets and Systems*, *15*, 31–48.

Sahin, M., Ulucay, V., & Menekse, M. (2018). Some New Operations of (α, β, γ) Interval Cut Set of Interval Valued Neutrosophic Sets. *Journal of Mathematical and Fundamental Sciences*, *50*(2), 103–120. doi:10.5614/j.math.fund.sci.2018.50.2.1

Sahin, R. (2014). Neutrosophic Hierarchical Clustering Algoritms. *Neutrosophic Sets and Systems*, *2*, 19–24.

Sahin, M., Olgun, N., Ulucay, V., Kargin, A., & Smarandache, F. (2017). *A new similarity measure based on falsity value between single valued neutrosophic sets based on the centroid points of transformed single valued neutrosophic numbers with applications to pattern recognition.* Infinite Study.

Sankari, H., & Abobala, M. (2020). n-Refined Neutrosophic Modules. *Neutrosophic Sets and Systems*, *36*, 1–11.

Sebastian, S., & Ramakrishnan, T. V. (2010). Multi-Fuzzy Sets. *International Mathematical Forum*, *5*(50), 2471–2476.

Shahzadi, G., Akram, M., & Saeid, A. B. (2017). An Application of Single-Valued Neutrosophic Sets in Medical Diagnosis. *Neutrosophic Sets and Systems*, *18*, 80–88.

Shinoj & John. (2012). Intuitionistic fuzzy multisets and its application in medical diagnosis. *World Academy of Science, Engineering and Technology*, *6*, 1–28.

Smarandache, F. (1998). *A Unifying Field in Logics Neutrosophy: Neutrosophic Probability, Set and Logic*. American Research Press.

Smarandache, F. (2013). n-Valued Refined Neutrosophic Logic and Its Applications in Physics. *Progress in Physics*, *4*, 143–146.

Smarandache, F. (2019). Introduction to NeutroAlgebraic Structures. In *Advances of Standard and Nonstandard Neutrosophic Theories* (pp. 240–265). Pons Publishing House.

Smarandache, F. (2020). Introduction to NeutroAlgebraic Structures and AntiAlgebraic Structures (revisited). *Neutrosophic Sets and Systems*, *31*, 1–16. doi:10.5281/zenodo.3638232

Smarandache, F., & Abobala, M. (2020). n-Refined Neutrosophic Vector Spaces. *International Journal of Neutrosophic Science*, *7*, 47–54. doi:10.54216/IJNS.070104

Smarandache, F. (2020). NeutroAlgebra is a Generalization of Partial Algebra. *International Journal of Neutrosophic Science, 2*(1), 8-17.

Ulucay, V., Sahin, M., & Hassan, N. (2015). Neutrosophic soft expert sets. *Applied Mathematics*, *6*(01), 116–127. doi:10.4236/am.2015.61012

Wang, H., Smarandache, F.Y., Zhang, Q., & Sunderraman, R. (2010). Single valued neutrosophic sets. *Multi Space and Multi Structure, 4*, 410-413.

Yuhua, F. (2013). Neutrosophic Examples in Physics. *Neutrosophic Sets and Systems*, *1*, 26–33.

Yuhua, F. (2015). Examples of Neutrosophic Probability in Physics. *Neutrosophic Sets and Systems*, *7*, 32–33.

Zadeh, L. A. (1965). Fuzzy Sets. *Information and Control*, *8*(3), 338–353. doi:10.1016/S0019-9958(65)90241-X

Chapter 14
On NeutroAlgebra Environment for Image Retrieval

Ahmed A. Salama
https://orcid.org/0000-0003-2300-615X
Port Said University, Egypt

Mariam Hariri
Zayed University, UAE

ABSTRACT

A NeutroStructure is a structure that has at least one NeutroRelation or one NeutroAttribute, and neither AntiRelation nor AntiAttribute. In this chapter, the authors review and study the NeutroAlgebra structure and introduce several operations and properties of algebraic equations for image retrieval, such as NeutroContent-based image retrieval. It is a generalization of classical and fuzzy algebraic structures for image retrieval. An explanation of the new particular processes is provided by several results.

INTRODUCTION

The Neutrosophic Fuzzy Sets (NFS) approach has been applied to image processing such as image thresholding (Cheng & Guo, 2008), image segmentation (Zhang et al., 2010) and image de-noising based on Neutrosophic median filtering. Neutrosophy of image segmentation is defined by T, I, F in image processing. T is the degree to be the object, I is the degree to be the boundary, and F is the degree to be the

DOI: 10.4018/978-1-6684-3495-6.ch014

background. Neutrosophy is used in image segmentation to increase noise-tolerant ability and to produce a superior blurry boundary image as opposed to other methods. The image in the Neutrosphic Fuzzy Domain (NFD) is represented as an array of neutrosophic fuzzy singletons by Salama et al. in (El-Nakeeb et al., 2017; Salama, Eisa, ElGhawalby et al, 2018; Salama, ElGhawalby, & Nasr, 2018; Salama, Smarandache, & ElGhawalby, 2018; Salama, 2017; Kahraman et al., 2019; Salama et al., 2014), and in (Salama & Samarandache, 2014) the concept of neutrosophic crisp sets was introduced to represent any event by a triple crisp structure. A crisp structure is a structure whose all elements are characterized by the same given Relationships and Attributes. A NeutroStructure is a structure that has at least one NeutroRelation or one NeutroAttribute, and neither AntiRelation nor AntiAttribute. In 2019 and 2020, Smarandache (Smarandache, 1991; Smarandache, 2019; Smarandache, 2020a; Smarandache, 2020b) generalized the classical Algebraic Structures to NeutroAlgebraic Structures. Salama introduced and studied the NeutroAlgebra structures and many operations and properties of the mathematical morphology (Salama, Eisa, ElGhawalby et al, 2018; Salama, ElGhawalby, & Nasr, 2018; Salama, Smarandache, & ElGhawalby, 2018; Salama, 2017; Kahraman et al., 2019). In our work, we study the NeutroAlgebra structure and many operations and properties of algebraic equations for image retrieval such as the NeutroContent-Based Image Retrieval. An explanation of the new given NeutroOperations is provided through several examples.

TERMINOLOGIES

We recollect some relevant basic preliminaries, and in particular, the work of Smarandache in (Smarandache, 1991; Smarandache, 2019; Smarandache, 2020a; Smarandache, 2020b), Salama et al. (Salama, Eisa, ElGhawalby et al, 2018; Salama, ElGhawalby, & Nasr, 2018; Salama, Smarandache, & ElGhawalby, 2018; Salama, 2017; Kahraman et al., 2019), and other references in ().

DIGITAL IMAGES IN THE NEUTROSOPHIC ALGEBRA ENVIRONMENT

Let U be a universe of discourse. A neutrosphic fuzzy domain W is a set in U which is composed of bright pixels. A NFI P_{NS} is characterized by three subsets T, I, F, which can be defined as T is the degree of membership, I is the degree of indeterminacy, and F is the degree of non-membership. In the image, a pixel P can be described as

P(T,I,F) which belongs to W such that t% is true in the bright pixel, i% is the ineutrosophic domain determinate, and f% is false. Where t varies in T, i varies in I, and f varies in F. In the image crisp domain, the pixel p(i,j) is transformed to the neutrosphic fuzzy domain $NFDP_{NS}(i,j)$ which can be defined as

$$NFDP_{NS}(i,j) = \{T(i,j), I(i,j), F(i,j)\},$$

where $T(i,j)$ belongs to white set, $I(i,j)$ belongs to determinate set and $F(i,j)$ belongs to non-white set. It can be defined as

$$P_{NS} = \{T(i,j), I(i,j), F(i,j)\} \tag{3.1}$$

$$T(i,j) = \frac{\overline{g(i,j)} - \overline{g}_{\min}}{\overline{g}_{\max} - \overline{g}_{\min}} \tag{3.2}$$

$$I(i,j) = 1 - \frac{H_0(i,j) - H_0}{H_{0_{\max}} - H_{0_{\min}}} \tag{3.3}$$

$$F(i,j) = 1 - T(i,j) \tag{3.4}$$

$$H_0(i,j) = abs\left(g(i,j) - \overline{g(i,j)}\right) \tag{3.5}$$

Where $\overline{g}(i,j)$ is the local mean value of the pixels of w in neutrosophic domain of w size, and $H_0(i,j)$ is the homogeneity value of T at (i,j), which is the absolute value of the difference between the intensity $g(i,j)$ and the local mean value $\overline{g}(i,j)$.

The second Neutrosophic transformation is

$$NFDP_{NS}(i,j) = \{T(i,j), I(i,j), F(i,j), \pi(i,j)\}, \tag{3.6}$$

where

$$\pi(i,j) = 3 - T(i,j) - I(i,j) - F(i,j)$$

Neutrosophic Texture Feature Extraction

Neutrosophic Entropy

In general, to represent a variable that can take one of n values, Shannon Entropy takes $\log_2(n)$ bits. The entropy is equal to the number of bits equality, in case these values are equally probable. When all outcomes are equally probable The Shannon Entropy can be held. The Shannon Entropy only considers the probability of observing a specific event that encapsulates information about the neutrosophic domain underlying probability distribution, not the meaning of the events themselves.

Moreover, when one outcome is certain, entropy is equal to zero. However, when a probability distribution of the source is known the Shannon entropy can quantify all these considerations exactly.

Although, the Neutrosophic Fuzzy Entropy was defined in one dimension in (Albergsten, 2008), we will define it in neutrosophic fuzzy to be as follows:

$$NeutroEn_{Ns} = En_T + En_I + En_F \tag{3.7}$$

$$En_T = \sum_i \sum_j P_T(i,j) \log P_T(i,j) \tag{3.8}$$

$$En_I = \sum_i \sum_j P_I(i,j) \log P_I(i,j) \tag{3.9}$$

$$En_F = \sum_i \sum_j P_F(i,j) \log P_F(i,j) \tag{3.10}$$

Where P contains the histogram counts. A Neutrosphic Fuzzy domain takes its values from the interval [0,1]. Hence, $\log P(i,j)$ may have negative values. So, we use the absolute value of Ent_T, Ent_I and Ent_F.

Neutrosophic Contrast

We will define the Neutrosophic Contrast to be as follows:

$$Neutrocont_{NS} = cont_T + cont_I + cont_F \tag{3.11}$$

$$cont_T = \sum_i \sum_j (i-j)^2 P_T(i,j) \tag{3.12}$$

$$cont_I = \sum_i \sum_j (i-j)^2 P_I(i,j) \tag{3.13}$$

$$cont_F = \sum_i \sum_j (i-j)^2 P_F(i,j) \tag{3.14}$$

Neutrosophic Energy

We will define the Neutrosophic Energy to be as follows:

$$NeutroEny_{NS} = En_T + En_I + En_F \tag{3.15}$$

$$En_T = \sum_i \sum_j P_T^2(i,j) \tag{3.16}$$

$$En_I = \sum_i \sum_j P_I^2(i,j) \tag{3.17}$$

$$En_F = \sum_i \sum_j P_F^2(i,j) \tag{3.18}$$

Neutrosophic Homogeneity

The properties of database can be described by using Homogeneity. For example, homoscedasticity considerations examine how much the variability of data values changes throughout a dataset. Nevertheless, homogeneity questions can be applied to all aspects of the statistical distributions, including the location parameter.

We will define the Neutrosophic Homogeneity to be as follows:

$$\text{Homo}_{NS} = \text{Homo}_T + \text{Homo}_I + \text{Homo}_F \tag{319}$$

$$\text{Homo}_T = \sum_i \sum_j \frac{P_T(i,j)}{1+|i-j|} \tag{3.20}$$

$$\text{Homo}_I = \sum_i \sum_j \frac{P_I(i,j)}{1+|i-j|} \tag{3.21}$$

$$\text{Homo}_F = \sum_i \sum_j \frac{P_F(i,j)}{1+|i-j|} \tag{3.22}$$

Neutrosophic Angular Seco Moment

Angular Neutrosphic Fuzzy domain Moment (ASM) is used to measure the homogeneity of an image. A homogeneous scene which contains only a few gray levels can give only a few but relatively high values of $p(i,j)$. So the sum of squares will be high, which can be defined as in [54]:

$$ASM = \sum_i \sum_i \left(p(i,j)\right)^2 \tag{3.23}$$

We will define the Neutrosophic ASM to be as follows:

$$NeutroASM_{NS} = ASM_T + ASM_I + ASM_F \tag{3.24}$$

$$ASM_T = \sum_i \sum_i \left(p_T\left(i,j\right)\right)^2 \tag{3.25}$$

$$ASM_I = \sum_i \sum_i \left(p_I\left(i,j\right)\right)^2 3.26()$$

$$ASM_F = \sum_i \sum_i \left(p_F\left(i,j\right)\right)^2 \tag{3.27}$$

Neutrosophic Inverse Difference Moment

The homogeneity of the image is also affected by Inverse Difference Moment (IDM). IDM will get small contributions from inhomogeneous areas because of the weight of the factor $\left(1+\left(i-j\right)^2\right)^{-1}$ when $i \neq j$. The result will be a low IDM value for inhomogeneous images, a Neutrosphic Fuzzy domain relatively higher value for homogeneous images. IDM can be defined in Neutrosphic Fuzzy domains as [54]:

$$IDM = \sum_i \sum_j \frac{1}{1+\left(i-j\right)^2} P\left(i,j\right) \tag{3.28}$$

We will define the IDM to be as follows:

$$NeutroIDM_{NS} = IDM_T + IDM_I + IDM_F \tag{3.29}$$

$$IDM_T = \sum_i \sum_j \frac{1}{1+\left(i-j\right)^2} P_T\left(i,j\right) \tag{3.30}$$

$$IDM_I = \sum_i \sum_j \frac{1}{1+(i-j)^2} P_I(i,j) \tag{3.31}$$

$$IDM_F = \sum_i \sum_j \frac{1}{1+(i-j)^2} P_F(i,j) \tag{3.32}$$

Neutrosophic Correlation Measure

Correlation measures the gray level linear deep neutrosophic dominance between the pixels at the speciðed positions relative to each other. It can be defined as [54]:

$$COR = \sum_i \sum_j \frac{\{i \times j\} \times P(i,j) - \{\mu_x \times \mu_y\}}{\sigma_x \times \sigma_y} \tag{3.33}$$

We will define the Correlation to be as follows:

$$NeutroCOR_{NS} = COR_T + COR_I + COR_F \tag{3.34}$$

$$COR_T = \sum_i \sum_j \frac{\{i \times j\} \times P_T(i,j) - \{\mu_x \times \mu_y\}}{\sigma_x \times \sigma_y} \tag{3.35}$$

$$COR_I = \sum_i \sum_j \frac{\{i \times j\} \times P_I(i,j) - \{\mu_x \times \mu_y\}}{\sigma_x \times \sigma_y} \tag{3.36}$$

$$COR_F = \sum_i \sum_j \frac{\{i \times j\} \times P_F(i,j) - \{\mu_x \times \mu_y\}}{\sigma_x \times \sigma_y} \tag{3.37}$$

Neutrosophic Variance

This feature can relatively put high weights on the elements which differ from the average value of $P(i,j)$. Variance can be defined as [54]:

$$VAR = \sum_i \sum_j (i-\mu)^2 P(i,j) \tag{3.38}$$

We will define the Neutrosophic Variance to be as follows:

$$NeutroVAR_{NS} = VAR_T + VAR_I + VAR_F \tag{3.39}$$

$$VAR_T = \sum_i \sum_j (i-\mu)^2 P_T(i,j) \tag{3.40}$$

$$VAR_I = \sum_i \sum_j (i-\mu)^2 P_I(i,j) \tag{3.41}$$

$$VAR_F = \sum_i \sum_j (i-\mu)^2 P_F(i,j) \tag{3.42}$$

Neutrosophic Sum Average

Sum average (mean) is the sum of all values divided by the total number of values which can be defined as [54]:

$$AVER = \sum_{k=0}^{2G-2} k \sum_i \sum_j P(i,j) \tag{3.43}$$

We will define the Neutrosophic Sum Average to be as follows:

$$NeutroAVER_{NS} = AVER_T + AVER_I + AVER_F \tag{3.44}$$

$$AVER_T = \sum_{k=0}^{2G-2} k \sum_i \sum_j P_T(i,j) \tag{3.45}$$

$$AVER_I = \sum_{k=0}^{2G-2} k \sum_i \sum_j P_I(i,j) \tag{3.46}$$

$$AVER_F = \sum_{k=0}^{2G-2} k \sum_i \sum_j P_F(i,j) \tag{3.47}$$

Neutrosophic Sum Entropy

Sum entropy can be defined as [58]:

$$SENT = -\sum_{i=0}^{2G-2} P_{x+y}(i) \log\left(P_{x+y}(i)\right) \tag{3.48}$$

We will define the Neutrosphic Fuzzy Sum Entropy to be as follows:

$$NeutroSENT_{NS} = SENT_T + SENT_I + SENT_F \tag{3.49}$$

$$SENT_T = -\sum_{i=0}^{2G-2} P_{x_T+y_T}(i) \log\left(P_{x_T+y_T}(i)\right) \tag{3.50}$$

$$SENT_I = -\sum_{i=0}^{2G-2} P_{x_I+y_I}(i) \log\left(P_{x_I+y_I}(i)\right) \tag{3.51}$$

$$SENT_F = -\sum_{i=0}^{2G-2} P_{x_F+y_F}(i) \log\left(P_{x_F+y_F}(i)\right) \tag{3.52}$$

Neutrosophic Difference Entropy

Difference entropy can be defined as [58]:

$$DENT = -\sum_{i=0}^{G-1} P_{x+y}(i)\log\left(P_{x+y}(i)\right) \tag{3.53}$$

We will define the Neutrosophic Difference entropy to be as follows:

$$NeutroDENT_{NS} = DENT_{T} + DENT_{I} + DENT_{F} \tag{3.54}$$

$$DENT_{T} = -\sum_{i=0}^{G-1} P_{x_T+y_T}(i)\log\left(P_{x_T+y_T}(i)\right) \tag{3.55}$$

$$DENT_{I} = -\sum_{i=0}^{G-1} P_{x_I+y_I}(i)\log\left(P_{x_I+y_I}(i)\right) \tag{3.56}$$

$$DENT_{F} = -\sum_{i=0}^{G-1} P_{x_F+y_F}(i)\log\left(P_{x_F+y_F}(i)\right) \tag{3.57}$$

Neutrosophic Cluster Shade

Cluster shade can be defined as [54]:

$$SHADE = \sum_{i}\sum_{j}\left\{i + j - \mu_x - \mu_y\right\}^3 \times P(i,j) \tag{3.58}$$

We will define the Neutrosophic Fuzzy Cluster shade to be as follows:

$$NeutroSHADE_{NS} = SHADE_{T} + SHADE_{I} + SHADE_{F} \tag{3.59}$$

$$SHADE_{T} = \sum_{i}\sum_{j}\left\{i + j - \mu_x - \mu_y\right\}^3 \times P_T(i,j) \tag{3.60}$$

$$SHADE_I = \sum_i \sum_j \left\{i + j - \mu_x - \mu_y\right\}^3 \times P_I\left(i, j\right) \tag{3.61}$$

$$SHADE_F = \sum_i \sum_j \left\{i + j - \mu_x - \mu_y\right\}^3 \times P_F\left(i, j\right) \tag{3.62}$$

Neutrosophic Cluster Prominence

Cluster prominence can be defined as [54]:

$$PROM = \sum_i \sum_j \left\{i + j - \mu_x - \mu_y\right\}^4 \times P\left(i, j\right) \tag{3.63}$$

We will define the Neutrosophic Cluster prominence to be as follows:

$$NeutroPROM_{NS} = PROM_T + PROM_I + PROM_F \tag{3.64}$$

$$PROM_T = \sum_i \sum_j \left\{i + j - \mu_x - \mu_y\right\}^4 \times P_T\left(i, j\right) \tag{3.65}$$

$$PROM_I = \sum_i \sum_j \left\{i + j - \mu_x - \mu_y\right\}^4 \times P_I\left(i, j\right) \tag{3.66}$$

$$PROM_F = \sum_i \sum_j \left\{i + j - \mu_x - \mu_y\right\}^4 \times P_F\left(i, j\right) \tag{3.67}$$

Neutrosophic Shape Feature Extraction

Neutrosophic Geometric Moment Invariants

A set of moment invariants, which is usually called geometric moment, has the desirable properties of being invariant neutrosophic domain translation, scaling a rotation, and derived by using nonlinear combinations of the lower order moments. Using higher order moments for pattern analysis has not been addressed. A normalization process, such as z-score normalization, is needed in the implementation when the values of the acquired moment invariants are usually very small.

For a binary image which can be considered a gray-level image with $Z\left(x,y\right)=1$ for print pixels, $Z\left(x_i,y_i\right)=1$ for background pixels by Neutrosophic Summing over the N print pixels only, which can be represented as:

$$m_{pq}=\sum_{i=1}^{N}(x_i)^p(y_i)^q \tag{3.68}$$

$$\mu_{pq}=\sum_{i=1}^{N}(x_i-\overline{x})^p(y_i-\overline{y})^q \tag{3.69}$$

Where

$$\overline{x}=\frac{m_{1,0}}{m_{0,0}},\overline{y}=\frac{m_{0,1}}{m_{0,0}} \tag{3.70}$$

In Neutrosophic Geometric moment invariants, we will apply Eq. (3.68), (3.69) and (3.70) in T, I, F in order to get geometric moment invariants for T, I, F. After that, construct a vector of geometric moment invariants.

Neutrosophic Discrete Wavelet Transform

In Neutrosphic Fuzzy Domain, adjacent pixel values are highly correlated, hence red. These existing among pixels need to be eliminated in order to compress images. Discrete Wavelet Transform (DWT) processor the Neutrosphic Fuzzy domain pixels transform into frequency domain information by DWT processor that are represented in multiple sub-a neutrosophic domains, representing different time scale points. It can be classified as Hear Transform Daubechies wavelet. The DWT [56] is characteristics of frequency or time used in various application fields compared to other transforms. In Fig. (3.2) the DWT Sub flow chart for digital image is presented. In the flow chart, L represents the component of low frequency, H represents the high frequency, the numbers 1 and 2 represent the decomposition level of the DWT. The sub image LL is the element of low frequency, which is the approximate usual image sub image. HL sub image is the component of low

Figure 1. Neutrosophic DWT sub- decomposition.

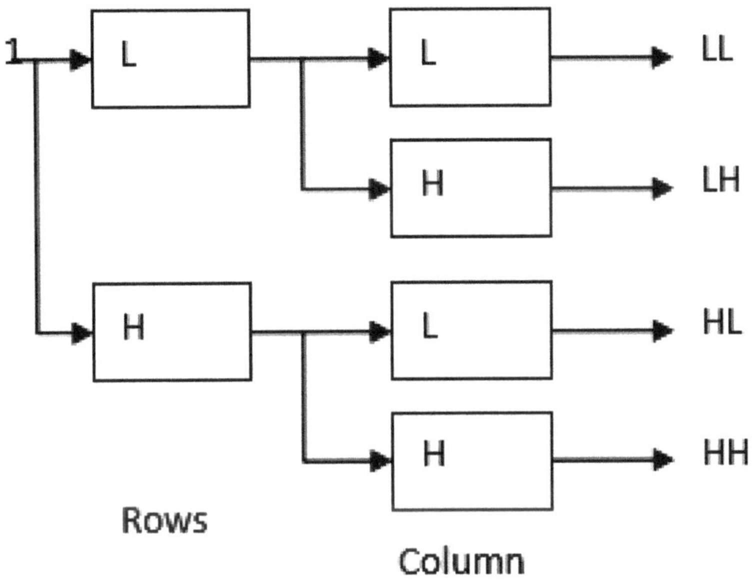

frequency in horizontal course high in vertical direction, which manifests the long established image horizontal side. LH sub image is the factor of high frequency in horizontal course a vertical direction which manifests the original image vertical edge. Sub image HH is the component of high frequency. It manifests the original image oblique edge, which presents that original image energy is contained in the LL2 low frequency region. In NeutroDWT, we will apply this method in T, I, F in order to get DWT for T, I, F. After that, we will construct a vector of DWT [56].

Neutrosophic Color Feature Extraction

Neutrosophic Color Co-Occurrence Matrix

A Colour Co-occurrence Matrix (CCM) is used to represent the traversal of adjacent pixel colour difference in an image, each pixel corresponding neutrosophic domains to four adjacent pixel colours. Each image can be presented by four images of motifs of scan pattern that can be constructed into four two-dimensional matrices of the image size. The attribute of the image will then be computed with the motifs of scan pattern based on these four matrices a Neutrosphic Fuzzy domain CCM can be obtained. As presented in Fig. (3.3), for each pixel $G(x,y)$, a 3×3 convolution mask can be generated, where this 3×3 convolution mask can be divided into four blocks of 2×2 grids (pixels) with each including pixel $G(x,y)$. Generally, there are 25 different scan patterns in a grid. If the traversal goes from four angles, the scan starting from the top left corner pixel p1 as shown in Fig. (3.3) (a) can be considered. Because of the representation of a complete family of space filling curve, we reduce the number of patterns to only 7 as shown in Fig. (3.3). Due to the equivalence, the situation wherein a motif cannot be formed among them motif number 0 signifies. Then, these 2 × 2 grids are replaced by motifs of scan pattern which would traverse the grid in an optimal sense, where the optimality of the scan is related to the incremental difference in intensity along the scan line minimizing the variation in the intensities in a local neighbourhood.

Figure 2. 3×3 convolution mask are divided in four 2×2 grids
(Kannan & Andres, 2012).

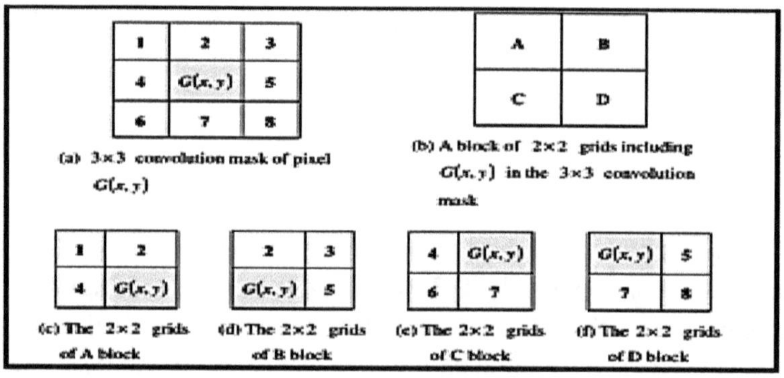

Figure 3. The seven scan patterns
(Kannan & Andres, 2012).

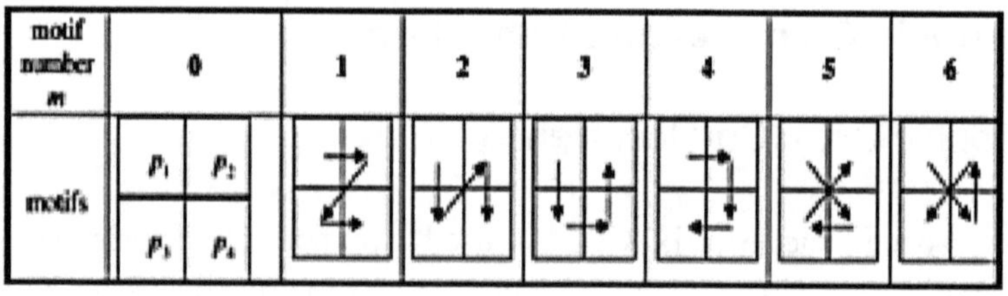

Let $G(x,y) :\rightarrow Z$ be the gray levels of an $N_x \times N_y$ image I for $Z = \{0,1,\ldots,255\}$. To form four different motifs shown by number m of motif, pixel $G(x,y)$ is divided into four blocks, each of which has 2×2 grids (pixels). These four motifs will be saved in four $N_x \times N_y$ two-dimensional motifs of scan pattern matrix $P_i[x,y]$, where

$$i = 1,2,3,4, x = 1,2,\ldots,N_x, y = 1,2,\ldots,N_y,$$

a Neutrosphic Fuzzy domain

$$Pi[x,y] : N_x \times N_y \rightarrow W$$

denotes an $N_x \times N_y$ matrix for $W = \{0,1,\ldots,6\}$. CCM calculates the distribution within the two-dimensional motifs of scan pattern matrix $P_i[N_x, N_y]$, where it takes into account the probability of the co-occurrence between the two motifs neutrosophic domain corresponding to (x,y) its adjacent $(x+\sigma_x, y+\sigma_y)$ respectively. Then, the total number of co-occurring motifs of scan pattern pairs (u,v), where $u = (0,1,\ldots,6$, $v = 0,1,\ldots,6)$. By dividing (u,v) by the total number of counts across all v, the co-occurring probabilities of the number i motifs of scan pattern matrix are determined. As a result, in total there will be a 7×7 two-dimensional CCM grids. In Neutrosophic CCM, we will apply this method in T, I, F in order to get CCM for T, I, F. After that, we will construct a vector of CCM.

NEUTROSOPHIC HESITANCY DEGREE IMAGE RETRIEVAL

Neutrosophic Hesitancy Degree

We will now extract the concepts of distances to the case of neutrosophic hesitancy degree by taking into account the four parameters characterization of NS

$$A = \{\langle \mu_A(x), v_A(x), \gamma_A(x), \pi_A(x)\rangle, x \in X\}$$

Definition 4.1.1 (Kr & Mazumdar, 2010):
Let

$$A = \{\langle \mu_A(x), v_A(x), \gamma_A(x)\rangle, x \in X\},$$

$$B = \{\langle \mu_B(x), v_B(x), \gamma_B(x)\rangle, x \in X\}$$

on $X = \{x_1, x_2, x_3, \ldots\ldots, x_n\}$. For a NFS

$$A = \{\langle \mu_A(x), v_A(x), \gamma_A(x) \rangle, x \in X\}$$

in X, We can call,

$$\pi_A(x) = 3 - \mu_A(x) - v_A(x) - \gamma_a(x),$$

x of X in A. It is a Neutrosophic Hesitancy degree of x to A. It is obvious that $0 \leq \pi_A(x) \leq 3$. It can be defined as Eq. (3.1), (3.2), (3.3), (3.4), (3.5) where

$$\pi(i,j) = 3 - T(i,j) - I(i,j) - F(i,j) \tag{3.71}$$

NeutoTexture Features With Hesitancy Degree

Neutrosophic Entropy With Hesitancy Degree

We will define the Neutrosophic Entropy with hesitancy degree as Eq.(3.7), (3.8), (3.9), (3.10) $NeutroEnt_\pi = 3 - (Ent_T + Ent_I + Ent_F)$ (3.72)

Where P will contain the histogram counts.

Neutrosophic Contrast With Hesitancy Degree

We will define the Neutrosophic Contrast with hesitancy degree as Eq. (3.11),(3.12), (3.13),(3.14) a Neutrosphic Fuzzy domain(3.73).

$$\text{Neutrocont}_{\text{A}} = 3 - (\text{cont}_{\text{T}} + \text{cont}_{\text{I}} + \text{cont}_{\text{F}}) \tag{3.73}$$

Neutrosophic Energy With Hesitancy Degree

We will define the Neutrosophic Energy with hesitancy degree as Eq. (3.15), (3.16), (3.17), (3.18) a Neutrosphic Fuzzy domain (3.74).

$$NeutroEn_{\pi} = 3 - \left(En_T + En_I + En_F\right) \tag{3.74}$$

Neutrosophic Homogeneity With Hesitancy Degree

We will define the Neutrosophic Homogeneity with hesitancy degree as Eq. (3.19), (3.20), (3.21), (3.22) and (3.75).

$$\text{Homo}_{\text{A}} = 3 - \left(\text{Homo}_{\text{T}} + \text{Homo}_{\text{I}} + \text{Homo}_{\text{F}}\right) \tag{3.75}$$

Neutrosophic Euclidean Distances

Recently, the Euclidean distance was calculated as the distance between the query image in Neutrosphic Fuzzy domain the first image in the database in an array. This process is repeated for each image in the database then storing their values respectively. After that, the array is stored in a neutrosophic domaining order displayed the first *N* closest image matches. Let

$$A = \left\{\left(T_A(x), I_A(x), F_A(x)\right), x \in X\right\}$$

and

$$B = \left\{\left(T_B(x), T_B(x), F_B(x)\right), x \in X\right\}$$

in $X = \left\{x_1, x_2, x_3, \ldots, x_n\right\}$ then the Euclidean distance is equal to

$$d(A,B) = \sqrt{\sum_{i=1}^{n}\left[\left(T_A(x_i) - T_B(x_i)\right)^2 + \left(I_A(x_i) - I_B(x_i)\right)^2 + \left(F_A(x_i) - F_B(x_i)\right)^2\right]} \tag{6.2}$$

Fig. (4.1) represents the Neutrosophic Algebra image retrieval system which is a generalization of Classical Algebra, Fuzzy Algebra and Intuitionistic Fuzzy Alegebra image retrieval.

Figure 4. Algebra relations for image retrieval structure.

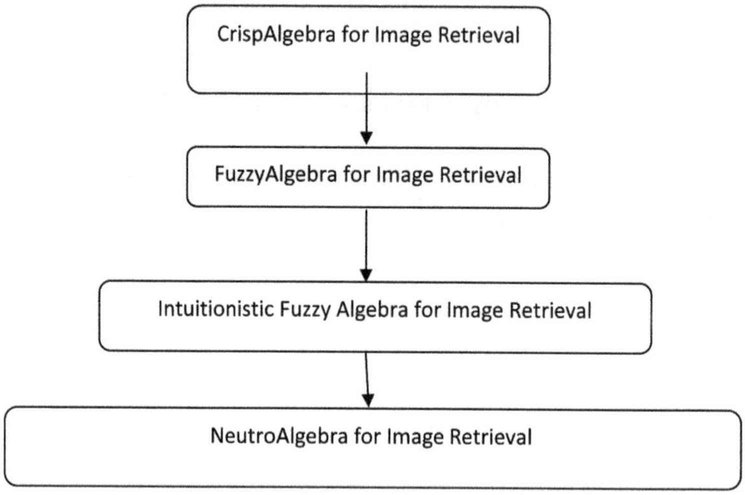

CONCLUSION

Conclusions for Neutrosophic Algebra CBIR

- Neutrosophic Content-Based Image Retrieval is introduced.
- Convert each image in the database from spatial domain to NFD.
- Extract Neutrosophic Texture feature, Neutrosophic Shape feature, Neutrosophic color feature of each image in the database.
- Construct a combined feature vector for T, I, F and stored in another database called Featured Database.
- Euclidean distance was presented in order to commute similarity measure.
- Find the distance between feature vectors of query images and that of featured databases.
- Sort the Neutrosophic distance and Retrieve the N-top most similar images.
- The proposed Neutrosophic technique is better than other techniques.
- Creating new Neutrosophic Content-Based Image Retrieval system.
- Apply Neutrosophic Hesitancy degree Image Retrieval.

REFERENCES

Adiga, & Deepak. (2008). A Universal Model for Content-Based Image Retrieval. *World Academy of Science, Engineering and Technology*, 46.

Ajorloo, H., & Lakdashti, A. (2012). IRFUM: Image retrieval via fuzzy modeling. *Computer Information*, *30*(5), 913–941.

Al-Tahan, M., Davvaz, B., Smarandache, F., & Anis, O. (2021a). On Some NeutroHyperstructures. *Symmetry*, *13*(4), 535. doi:10.3390ym13040535

Al-Tahan, M., Davvaz, B., Smarandache, F., & Anis, O. (2021b). Properties of Productional NeutroOrderedSemigroups. *Neutrosophic Sets and Systems*, *42*, 178–190.

Al-Tahan, M., Smarandache, F., & Davvaz, B. (2021). NeutroOrderedAlgebra: Applications to Semigroups. *Neutrosophic Sets and Systems*, *39*, 133–147.

Albergsten, F. (2008). *Statistical texture measures computed from gray level co-occurrence matrices.* Image Processing Laboratory, Department of Informatics, University of Oslo.

Alhasan, Salama, & Smarandache. (2021). Introduction to Neutrosophic Reliability Theory. *International Journal of Neutrosophic Science*, *15*(1), 52–61.

Amin, Salama, El-Henawy, Mahfouz, & Gafar. (2021). Intelligent Neutrosophic Diagnostic System for Cardiotocography Data. Computational Intelligence and Neuroscience.

Amin, B., & Gamall, M. (2019). Classifying Cardiotocography Data based on Rough Neural Network. *International Journal of Advanced Computer Science and Applications*, *10*(8), 352–356. doi:10.14569/IJACSA.2019.0100846

Atanassov, K. (1989). More on intuitionistic fuzzy sets. *Fuzzy Sets and Systems*, *33*(1), 37–46. doi:10.1016/0165-0114(89)90215-7

Banerjee, M., & Kundu, M. K. (2003). Content Based Image Retrieval With Fuzzy Geometrical Features. *The 12th IEEE International Conference on Fuzzy Systems*, 932-937. 10.1109/FUZZ.2003.1206556

Birgale, L., Kokare, M., & Doye, D. (2006). Colour and Texture Features for Content Based Image Retrieval. *Proceedings of the International Conference on Computer Graphics, Imaging and Visualisation.* 10.1109/CGIV.2006.30

Chan, Y. K., & Chen, C. Y. (2004). Image Retrieval System Based on Color Complexity and Color Spatial Features. *Journal of Systems and Software, 71*(1), 65–70. doi:10.1016/S0164-1212(02)00140-1

Cheng, H. D., & Guo, Y. (2008). A new neutrosophic approach to image thresholding. *New Mathematics and Natural Computation*, *4*(3), 291–308. doi:10.1142/S1793005708001082

Danish, M., Rawat, R., & Sharma, R. (2013). A Survey: Content –Based Image Retrieval Based on Color, Texture, Shape & Neuro Fuzzy. *Int. Journal of Engineering Research and Applications*, *3*(5), 839-844.

El-Nakeeb, ElGhawalby, Salama, & El-Hafeez. (2017). Neutrosophic Crisp Mathematical Morphology. *Neutrosophic Sets and Systems, 16*, 57-69.

ElWahsh, H., Gamal, M., Salama, A., & El-Henawy, I. (2018a). A novel approach for classifying MANETs attacks with a neutrosophic intelligent system based on genetic algorithm. *Security and Communication Networks*, *2018*, 1–7. doi:10.1155/2018/5828517

ElWahsh, H., Gamal, M., Salama, A., & El-Henawy, I. (2018b). Intrusion detection system and neutrosophic theory for MANETs:A comparative study. *Neutrosophic Sets and Systems*, *23*, 16–22.

ElWahsh, H., Gamal, M., Salama, A. A., & El-Henawy, I. M. (2017). Modeling Neutrosophic Data by Self-Organizing Feature Map: MANETs Data Case Study. *Procida Computer*, *121*, 152–157. doi:10.1016/j.procs.2017.11.021

Gonzalez, R. C., & Woods, R. E. (2002). *Digital Image Processing, seconded.* Prentice-Hall.

Gupta, D., & Choubey, S. (2013, March). Discrete wavelet transform for image processing. *International Journal of Emerging Technology and Advanced Engineering*, *4*(3), 598–602.

Hu, K. M. (1962). Visual pattern recognition by moment invariants. *I.R.E. Transactions on Information Theory*, *8*(2), 179–187. doi:10.1109/TIT.1962.1057692

Hwang, K. H., Lee, H., & Choi, D. (2012). Medical Image Retrieval: Past and Present, Healthc. *Information Research*, *18*(1), 3–9. PMID:22509468

Ingle, D., & Bhatia, S. (2012). Content Based Image Retrieval using Combined Features. *International Journal of Computer Applications, 44*(17).

Jarrah, K., Kyan, M., Krishnan, S., & Guan, L. (2006). Computational Intelligence Techniques and Their Applications in Content-Based Image Retrieval. *IEEE Int. Conf. on Multimedia & Expo*, 33-36.

Ji, R., Yao, H., & Liang, D. (2008). DRM: Dynamic Region Matching for Image Retrieval Using Probabilistic Fuzzy Matching and Boosting Feature Selection. *Springer Journal of Signal, Image and Video Processing*, *2*(1), 59–71. doi:10.100711760-007-0037-0

Kang, S. J., Woo, C. H., & Hwang, H. S. (2000). Evolutionary Design of Fuzzy Rule Base for Nonlinear System Modeling and Control. *IEEE Transactions on Fuzzy Systems*, *8*(1), 37–45. doi:10.1109/91.824766

Kannan, R., & Andres, F. (2012). Content Based Image Retrieval Using Various Distance Metrics. *ICDEM*, 154–161.

Kchang, S., & Hsu, A. (1992). Image Information System: Where Do We Go From Here? *IEEE Trans. on Knowledge and Data Engineering*. doi:10.1109/69.166986

Kekre, H.B., & Sudeep, D. (2008a). Creating the Color Panoramic View using Medley of Grayscale and Color Partial Images. *WASET International Journal of Electrical Computer and System Engineering,* 2-3.

Kekre, H. B., & Sudeep, D. (2008b). *Scaling Invariant Fusion of Image Pieces in Panorama Making and Novel Image Blending Technique. International Journal on Imaging, 1(A08).*

Kekre, H. B., & Sudeep, D. (2009). Improving the Performance of Image Retrieval using Partial Coefficients of Transformed Image. *International Journal of Information Retrieval, Serials Publications*, *2*(1).

Kekre, H. B., & Sudeep, D. (2010). *Performance Comparison for Face Recognition using PCA, DCT & Walsh Transform of Row Mean and Column Mean. International Journal on Graphics, Vision and Image Processing.*

Kokare, M., Chatterji, B. N., & Biswas, P. K. (2003). Comparison of Similarity Metrics for Texture Image Retrieval. *TENCON 2003. Conference on Convergent Technologies for the Asia-Pacific Region*, *2*, 571-575. 10.1109/TENCON.2003.1273228

Kothyari, P., Dwivedi, S., & Mandoria, H. L. (2016, June). Content-Based Image Retrieval Using Statistical Feature and Shape Extraction. *IUP Journal of Information Technology*, *12*(3), 62–69.

Kr, S. B., & Mazumdar, B. (2010). Content Retrieval from X-RAY Images Using Color &Texture Features. *International Journal of Electronics Engineering, 2*(1), 25–28.

Kurian, J., & Karunakaran, V. (2012, October). A Survey on Image Classification Methods. *International Journal of Advanced Research in Electronics and Communication Engineering, 1*(4).

Laaksonen, J., Oja, E., Koskela, M., & Brandt, S. (2000). Analyzing low-level visual features using content-based image retrieval. *Proc Int'l Conf. Neural Information Processing*.

Lakdashti, A., & Moin, M. S. (2007). A New Content-Based Image Retrieval Approach Based on Pattern Orientation Histogram. In *LNCS 4418* (pp. 587–595). Springer-Verlag. doi:10.1007/978-3-540-71457-6_54

Lakdashti, A., Moin, M. S., & Badie, K. (2008a). Reducing Feature Space Dimensionality for Image Retrieval. *The 3rd International Symposium on Communications, Control and Signal Processing (ISCCSP),* 305–308. 10.1109/ISCCSP.2008.4537239

Lakdashti, A., Moin, M. S., & Badie, K. (2008b). A Novel Semantic-Based Image Retrieval Method. *The 10th International Conference on Advanced Communication Technology (ICACT),* 17–20. 10.1109/ICACT.2008.4493928

Li, J., Liang, H., Wang, L., & Zhang, J. (2007). The Medical Image Retrieval Based on the Fuzzy Feature. *International Conference on Mechatronics and Automation (ICMA)*, 1245–1250. 10.1109/ICMA.2007.4303727

Lin, H. C., Chiu, C. Y., & Yang, S. N. (2003). Finding textures by textual descriptions, visual examples and relevance feedback. *Pattern Recognition Letters, 24*(14), 2255–2267. doi:10.1016/S0167-8655(03)00052-7

Long, F., Zhang, H., & Feng, D. D. (2003). Fundamentals of content-based image retrieval. In *Multimedia Information Retrieval and Management Technological Fundamentals andApplications*. Springer-Verlag. doi:10.1007/978-3-662-05300-3_1

Manjunath, B. S., Salembier, P., & Sikora, T. (2003). *Introduction to MPEG-7 – Multimedia Content Description Interface*. Wiley.

Puviarasan, N., Bhavani, R., & Vasanthi, A. (2014, March). Image Retrieval Using Combination of Texture and Shape Features. *International Journal of Advanced Research in Computer and Communication Engineering, 3*(3), 5873–5877.

Ramamurthy, B. (2012). CBMIR: Content Based Medical Image Retrieval System Using Texture and Intensity for Dental Images. In *Eco-friendly Computing and Communication Systems*. Springer Berlin Heidelberg.

Rehman, M., Iqbal, M., Sharif, M., & Raza, M. (2012). Content Based Image Retrieval: Survey. *World Applied Sciences Journal*, *19*(3), 404–412.

Salama, Eisa, ElGhawalby, & Fawzy. (2018). Neutrosophic Image Retrieval with Hesitancy Degree. In *New Trends in Neutrosophic Theories and Applications*. Pons Editions.

Salama, ElGhawalby, & Nasr. (2018). Retract neutrosophic crisp system for gray scale image. *Asian Journal of Mathematics and Computer Research*, *3*, 104–117.

Salama, A., & Samarandache, F. (2014). *Neutrosophic Crisp Set theory*. Educational Publisher Columbus.

Salama, A., Smarandache, F., & ElGhawalby, H. (2018). Neutrosophic approach to grayscale images domain. *Neutrosophic Sets and Systems*, *21*, 13–19.

Salama, A. A. (2017, July). A Neutrosophic Image Retrieval Classifier. *International Journal of Computers and Applications*, *170*(9), 1–6. doi:10.5120/ijca2017914798

Salama, A. A. (2020). Study on the Suez Canal Port, Egypt. *Neutrosophic Sets and Systems*, *35*, 22–44.

Salama, A. A., Eisa, M., ElGhawalby, H., & Fawzy, A. E. (2019). A New Approach in Content-Based Image Retrieval Neutrosophic Domain. In C. Kahraman & İ. Otay (Eds.), *Fuzzy Multi-criteria Decision-Making Using Neutrosophic Sets. Studies in Fuzziness and Soft Computing* (Vol. 369, pp. 361–369). Springer. doi:10.1007/978-3-030-00045-5_14

Salama, A. A., Smarandache, F., & Eisa, M. (2014). Introduction to Image Processing via Neutrosophic Techniques. *Neutrosophic Sets and Systems*, *5*, 59–64.

Saxena, K., Richaria, V., & Trivedi, V. (2012, September). A Survey on Content Based Image Retrieval using BDIP, BVLC and DCD. *Journal of Global Research in Computer Science*, *3*(9), 38–41.

Sifuzzaman, M., Islam, M. R., & Ali, M. Z. (2009). Application of Wavelet Transform and Its Advantage Compared to Fourier Transform. *The Journal of Physiological Sciences; JPS*, *13*, 121–134.

Smarandache, F. (1991). *A Unifying field in logic: Neutrosophic Logic, Neutrosophy, Neutrosophic Set, Neutrosophic Probability*. American Research press.

Smarandache, F. (2003). *A Unifying Field in Logics, Neutrosophic Logic, Neutrosophy, Neutrosophic Set, Neutrosophic Probability and Statistics*. American Research Press.

Smarandache, F. (2019). Introduction to NeutroAlgebraic Structures and AntiAlgebraic Structures. In *Advances of Standard and Nonstandard Neutrosophic Theories* (pp. 240–265). Pons Publishing House Brussels.

Smarandache, F. (2020a). NeutroAlgebra is a Generalization of Partial Algebra. *International Journal of Neutrosophic Science*, *2*, 8–17. doi:10.54216/IJNS.020103

Smarandache, F. (2020b). Introduction to NeutroAlgebraic Structures and AntiAlgebraic Structures. *Neutrosophic Sets and Systems*, *31*, 1–16.

Sridevi, Y., & Babu, R. M. (2011). Content Based Image Retrieval Using Color and Texture Features. *International Journal of Computer Science and Information Technologies*, *2*(5), 2384–2389.

Sudhakar, P., Prasad, P., & Edara, S. R. (2013). Content Based Image Retrieval Using Fuzzy Texton and Shearlet Transform. In *Industrial Engineering, Management Science and Applications* (pp. 791–80). Springer.

Tripathy, S. S., Shekhar, R., & Kumar, R. S. (2011). Texture retrieval system using intuitionistic fuzzy set theory. In *Devices and Communications (ICDeCom), International Conference*. IEEE. 10.1109/ICDECOM.2011.5738490

Tseng, V. S., Su, J. H., & Huang, W. J. (2007). FCBIR: A Fuzzy Matching Technique for Content-Based Image Retrieval. In Theoretical Advances and Applications of Fuzzy Logic and Soft Computing. Springer Berlin Heidelberg.

Wang, J. Z., Li, J., & Wiederhold, G. (2001, September). SIMPLIcity: Semantics-Sensitive Integrated Matching for Picture Libraries. *IEEE Transactions on Pattern Analysis and Machine Intelligence*, *23*(9), 947–963. doi:10.1109/34.955109

Wu, K., & Yap, K. H. (2007). Content-Based Image Retrieval Using Fuzzy Perceptual Feed-back. *Springer Journal of Multimedia Tools and Applications*, *32*(3), 235–251. doi:10.100711042-006-0050-2

Xiaoling, W., & Kanglin, X. (2003, April). Application of the Fuzzy Logic in Content-based Image Retrieval. *JCS&T*, *5*(1), 19–24.

Yasser, El-Khalek, Twakol, Abo-Elsoud, Salama, & Khalifa. (n.d.). A hybrid Automated Intelligent COVID-19 Classification System Based on Neutrosophic Logic and Machine Learning Techniques Using Chest X-ray Images, Studies in Systems, Decision and Control. *Advances in Data Science and Intelligent Data Communication Technologies for COVID-19, 378*.

Yasser, I., & Twakol, A. (2020). COVID-X: Novel Health-Fog Framework Based on Neutrosophic Classifier for Confrontation Covid-19. *Neutrosophic Sets and Systems*, *35*, 1–21.

Younes, A. A., Truck, I., & Akdag, H. (2007). Image Retrieval Using Fuzzy Representation of Colors. *Springer Journal of Soft Computing – A Fusion of Foundations, Methodologies and Applications*, *11*(3), 287–298.

Zadeh, L. A. (1963). Fuzzy Sets. *Information and Control*, *8*(3), 338–353. doi:10.1016/S0019-9958(65)90241-X

Zhang, G., Ma, Z. M., & Cai, Z. P. (2007). Directed Filter for Dominant Direction Fuzzy Set in Content-based Image Retrieval. In *Proceedings of the 22th ACM Symposium on Applied Computing (SAC 2007)*. ACM Press. 10.1145/1244002.1244019

Zhang, M. (2010). *Novel approaches to image segmentation based on neutrosophic logic*. Academic Press.

Zhang, M., Zhang, L., & Cheng, H. D. (2010, May). A Neutrosophic approach to image segmentation based on Watershed method. *Signal Processing*, *90*(5), 1510–1517. doi:10.1016/j.sigpro.2009.10.021

Chapter 15
NeutroAlgebra of Ideals in a Ring

Ilanthenral Kandasamy
Vellore Institute of Technology, India

Vasantha W. B.
Vellore Institute of Technology, India

Florentin Smarandache
https://orcid.org/0000-0002-5560-5926
University of New Mexico, USA

ABSTRACT

In this chapter, the authors define operations '+' and '×' on the substructure ideals of the ring. They find under what conditions the collection of ideals are NeutroAlgebra under + and ×. The study is carried out in the case of Z and Z_n, the ring of integers and ring of modulo integers, respectively, where n is a composite number. This study is extended to the case of polynomial rings R[x], Q[x], Z[x], and Z_n [x] where R and Q are real and rational, respectively. Some open problems based on this study are proposed.

INTRODUCTION

To deal with uncertainty, inconsistency and indeterminacy present in any real-world data, Neutrosophic theory proposed by Smarandache (1999) comes handy. It is well-known that Neutrosophy deals mainly with neutralities and indeterminacies present in real-world problems. The new notion of NeutroAlgebraic structures and AntiAlgebraic structures was analysed in 2019 by Smarandache. In this chapter, the

DOI: 10.4018/978-1-6684-3495-6.ch015

authors study the new notion of NeutroAlgebra defined/introduced by Smarandache (2019). Several interesting results have been obtained in this direction in Smarandache (2020), Smarandache and Rezaei (2020) and Smarandache and Hamidi (2020); these authors have introduced Neutro BE Algebra and Neutro BCK Algebra and so on. NeutroAlgebras and AntiAlgebras in classical number systems have been studied by Agboola and et al (2020).

Kandasamy et al (2020) have analysed and NeutroAlgebra of Neutrosophic triplets in $\{\mathbb{Z}_n, \times\}$. They have proved interesting results related with the NeutroAlgebra of neutrosophic triplets and extended neutrosophic triplets. Here for the first-time, the authors study about the NeutroAlgebra of ideals in $\mathbb{Z}$ and $\mathbb{Z}_n$ $(2 \le n < \infty)$; under the operations + and ×, then the study is extended to the case NeutroAlgebra of ideals in polynomial rings. In fact, this study is very new and innovative.

This chapter has six sections. Section one is introductory. Basic concepts are recalled in section two to make this chapter a self-contained one. In section three NeutroAlgebra of ideals in $\mathbb{Z}$ (the ring of integers) is analyzed. Study of NeutroAlgebra of ideals in $\mathbb{Z}_n$ (n a composite number $4 \le n < \infty$) is carried out in section four. Section five analyses the NeutroAlgebra of ideals in the polynomial rings $\mathbb{Z}[x], \mathbb{R}[x], \mathbb{Q}[x]$ and $\mathbb{Z}_n[x]$. The final section gives conclusions based on this study.

BACKGROUND

Here some of the basic concepts are recalled to make this chapter a self-contained one. Mainly the definition of NeutroAlgebra are recalled and described as this concept is very new.

A NeutroAlgebra is an algebra with at least one Neutro operation or one Neutro axiom (axiom that is true for some elements, indeterminate or false for the other elements) Smarandache (2020) and Smarandache and Hamidi (2020). A partial algebra has at the minimum one partial operation, and all of its axioms are classical. Smarandache and Rezaei (2020) proved that NeutroAlgebra is a generalization of partial algebra and has given illustrations of NeutroAlgebras that are partial algebras. Partial algebra has some elements for which operation is undefined (not outer defined). Similarly, an AntiAlgebra is a non-empty set endowed with at least one anti operation (or anti operations) or at least one anti axiom. Examples of NeutroAlgebra and AntiAlgebra can be found in Smarandache (2020) and Smarandache and Hamidi (2020).

Examples of NeutroAlgebra and AntiAlgebra are provided in the following to make this chapter a self-contained one.

Table 1. Cayley table of W

+	3	4
3	od	od
4	od	od

Example 1: *Let $S = \{\mathbb{Z}_6, \times\}$ be the semigroup under product modulo 6. The non-trivial idempotents of S are $\{3,4\} = W$ (as $3 \times 3 = 3(\bmod 6)$ and $4 \times 4 = 4(mod 6)$). The Cayley table for W under + is given in Table 1.where od means outer defined. Hence W is an AntiAlgebra and is not a partial algebra. Now let $V = \{0,1,3,4\}$ where V denotes the collection of all idempotents of $\{\mathbb{Z}_6, \times\}$ both trivial and non-trivial. The Cayley table of V under + mod 6 is given in Table 2.*

V is a NeutroAlgebra under + with several undefined terms.

Next, the authors describe, define and develop the notion of NeutroAlgebra of ideals of the ring of integers $\mathbb{Z}$.

NEUTROALGEBRA OF IDEALS IN THE RING OF INTEGERS Z

In this section the study of ideals of $\mathbb{Z}$ and their NeutroAlgebra of ideals of $\mathbb{Z}$, (where $\mathbb{Z}$ denotes the set of positive and negative integers including zero) is carried out. The term ideals mean only non-trivial ideals of $\mathbb{Z}$.

Example 2: *Consider the collection of ideals J of $\mathbb{Z}$, where $J = \{\langle t \rangle$, where $\langle t \rangle$ denotes the ideal of Z generated by the integer t, $t \neq 1$ and $t \neq 0$, $t \in \mathbb{Z}^+ \setminus \{0,1\}\}$. Clearly (0), the zero ideal and Z the whole ring does not belong to J. The sum of two ideals is defined if it results in non-trivial ideal of $\mathbb{Z}$ in J otherwise the sum of the two ideals is undefined.*

Table 2. Cayley table of V

+	0	1	3	4
0	0	1	3	4
1	1	od	4	od
3	3	4	0	1
4	4	od	1	od

Table 3. Cayley table of S

$+$	I_1	I_2
I_1	I_1	od
I_2	od	I_2

Let the two ideals be generated by $I_1 = \langle 3 \rangle$ and $I_2 = \langle 5 \rangle$ in $\mathbb{Z}$, the sum of $I_1 + I_2 = \langle 3 \rangle + \langle 5 \rangle$ generated is in $\mathbb{Z}$. Thus $\mathbb{Z}$ is not a non-trivial ideal of $\mathbb{Z}$ so the set $S = \left\{ I_1, I_2 \right\}$ is a NeutroAlgebra of ideals of $\mathbb{Z}$ under $+$. The Cayley table of S is given by Table 3.

Hence there exists NeutroAlgebra of ideals of $\mathbb{Z}$ under $+$.

In view of this the following theorem is proved.

Theorem 3: *Let $\mathbb{Z}$ be the ring of integers. J be the collection of all non-trivial ideals of $\mathbb{Z}$, that is J does not contain the zero ideal (0) and $\mathbb{Z}$ the ideal, is the total ring. J under + is an infinite NeutroAlgebra.*

Proof: Given J is the set of all non-trivial ideals of $\mathbb{Z}$ that is $J = \{I_j \mid I_j$ is a proper ideal of $\mathbb{Z}, j \in N\}$ that is (0) and $\mathbb{Z} \notin J$. Now if I_1 and I_2 are ideals in general and if I_1 and I_2 is generated by odd primes then $I_1 + I_2$ is $\mathbb{Z}$ so the collection J under + is a NeutroAlgebra of ideals. Clearly J is of infinite order, hence J is an infinite NeutroAlgebra of ideals under $+$.

Next authors proceed onto define NeutrosubAlgebra of a NeutroAlgebra.

Definition 4: *Let N be a NeutroAlgebra defined by some Neutro operation or Neutro axiom. A subset M of N is said to be a NeutrosubAlgebra of N if the following conditions are true*

- *M is proper non empty subset of N.*
- *M under the operations of N has some elements in M which are undefined.*

Some examples of NeutrosubAlgebras of a NeutroAlgebra is provided.

Example 5: *Let $S = \{$ collection of all ideals of $\mathbb{Z}$ generated by primes$\}$. Now S under + is a NeutroAlgebra of ideals of infinite order. Consider $P = \left\{ \langle 2 \rangle, \langle 5 \rangle, \langle 13 \rangle \right\} \subseteq S$ be the subset of ideals of S. The Cayley table of P under + is given by Table 4.*

Table 4. Cayley table of P

$+$	$\langle 2\rangle$	$\langle 5\rangle$	$\langle 13\rangle$
$\langle 2\rangle$	$\langle 2\rangle$	od	od
$\langle 5\rangle$	od	$\langle 5\rangle$	od
$\langle 13\rangle$	od	od	$\langle 13\rangle$

Thus P is a NeutrosubAlgebra of S of ideals of $\{\mathbb{Z},+,\times\}$. One can have infinite NeutrosubAlgebra of S by taking an infinite subset of S and defining the $+$ operation on ideals. Thus S can have finite and infinite NeutrosubAlgebras of the NeutroAlgebra S.

Synonymously instead of NeutrosubAlgebra the phrase subNeutroAlgebra is used and both convey the same meaning.

Example 6: *Let $S = \{$ collection of all non-trivial prime ideals generated by $p \in \mathbb{Z} \setminus \{1\}$, p a prime of the ring of integers $\{\mathbb{Z},+,\times\}\}$. Clearly S under $\times$ is not a NeutroAlgebra of ideals for if $\langle p\rangle$ and $\langle q\rangle$ are two principal ideals generated by the two primes p and q then the product $\langle p\rangle \times \langle q\rangle = \langle pq\rangle$ is the ideal generated by the composite number so $\langle pq\rangle \notin S$ so $\langle pq\rangle$ is undefined and does not belong to the set $P = \{\langle p\rangle, \langle q\rangle\}$. Hence S the collection of all ideals generated by primes in $\mathbb{Z}$ under $\times$ is not a NeutroAlgebra only a AntiAlgebra. The Cayley table is given in Table 5.*

Hence the claim. In view of all these the following examples give non-trivial examples of AntiAlgebra and NeutroAlgebra which are not partial algebras.

This situation is illustrated by an example.

Table 5. Table of $\{p,q\}$

$\times$	p	q
p	od	od
q	od	od

Table 6. Cayley table of B under +

+	$\langle 3\rangle$	$\langle 5\rangle$	$\langle 7\rangle$
$\langle 3\rangle$	$\langle 3\rangle$	od	od
$\langle 5\rangle$	od	$\langle 5\rangle$	od
$\langle 7\rangle$	od	od	$\langle 7\rangle$

Example 7: *Let* $\{Z,+,\times\}$ *be the ring of integers. Let* $B=\{I_1=\langle 3\rangle, I_2=\langle 5\rangle$ *and* $I_3=\langle 7\rangle\}$ *be 3 prime ideals of* $\mathbb{Z}$*. Now find* B *under the + operation which is described by Table 6.*

Thus $\{B,+\}$ is a NeutroAlgebra of prime ideals of $\mathbb{Z}$.

Consider $\{B,\times\}$, it is represented by Table 7.

Thus $\{B,\times\}$ is the AntiAlgebra of ideals of $\mathbb{Z}$.

Theorem 8: *Let* $\{\mathbb{Z},\times,+\}$ *be the ring of integers.* J = {*collection of all proper ideals of* $\{Z,+,\times\}$ *generated by singleton element* $n\in Z\setminus\{1\}$} *and* $S=\{$ *collection of ideals of* $\{Z,+,\times\}$ *generated by the primes* $p, p\in Z\setminus\{1\}\}$.

1. $\{J,+\}$ is NeutroAlgebra of ideals of $\{Z,+,\times\}$ which is not a partial algebra under axiom being an ideal of $\{Z,+,\times\}$.

Table 7. Cayley table of B under ×

×	$\langle 3\rangle$	$\langle 5\rangle$	$\langle 7\rangle$
$\langle 3\rangle$	od	od	od
$\langle 5\rangle$	od	od	od
$\langle 7\rangle$	od	od	od

2. $\{J,\times\}$ is not a NeutroAlgebra of ideals under $\times$.
3. $\{S,+\}$ is a NeutrosubAlgebra of ideals under +.
4. $\{S,\times\}$ is not a NeutrosubAlgebra of ideals under $\times$, in fact an AntiAlgebra.

Proof: Given J is a collection of ideals of $\{Z,+,\times\}$ different from $\mathbb{Z}$ and $\{0\}$. Every ideal of $\mathbb{Z}$ is a principal ideal.

Now consider two ideals $I_1 = \langle p \rangle$ and $I_2 = \langle q \rangle \in J$, where specifically p and q to be distinct prime and say $p = 2$ is choosen.

Now $\langle I_1 + I_2 \rangle$ generates the ideal which is $\mathbb{Z}$ but J does not contain $\mathbb{Z}$. In J under + ideal $I_1 + I_2$ is undefined. Hence under the axiom of being a proper ideal $\{J,+\}$ is a NeutroAlgebra of infinite order. Hence the claim in (i).

Proof of (ii). Now J under $\times$ is well defined for I_1 and I_2 are two ideals in J generated by $\langle p \rangle$ and $\langle q \rangle$ respectively; $I_1 \times I_2 = \langle p \rangle \times \langle p \rangle = \langle pq \rangle$ generated by pq hence once again an ideal in J. So $\{J,\times\}$ is well defined under product for ideals so $\{J,\times\}$ is not a NeutroAlgebra of ideals.

Proof of (iii). Let S be the collection of all proper ideals of $\mathbb{Z}$ generated by primes. Now $S \subseteq J$ and S has ideals I_1, I_2 such that $I_1 + I_2 \notin S$ hence S is a NeutrosubAlgebra of ideals of J.

Proof of (iv) is obvious as $S \subseteq J$ is such that $\{S,\times\}$ every product is not in S but only in J so $\{S,\times\}$ is not a NeutrosubAlgebra of J. This is in fact a very opposite case of J, for any $I_1, I_2 \in S, I_1 \times I_2 \notin S$. Hence an AntiAlgebra and not a NeutrosubAlgebra of ideals under $\times$.

Next authors proceed onto study the existence of NeutroAlgebra of ideals in case of $\mathbb{Z}_n$, n a composite number.

NEUTROALGEBRA OF IDEALS USING $\mathbb{Z}_n$

In this section, consider the modulo integer rings $\{Z_n,+,\times\}$, n a composite number and study the property of ideals under + and $\times$. This is illustrated by some examples.

Example 9: *Let $R = \{Z_6,+,\times\}$ be the ring of modulo integers of order 6. The only non-trivial ideals of R are $I_1 = \{0,2,4\}$ and $I_2 = \{0,3\}$. The Cayley Table under $\times$ for I_1 and I_2 is given by the Cayley Table 8.*

Table 8. Cayley table of P

$\times$	I_1	I_2
I_1	I_1	od
I_2	od	I_2

So under the axiom of ideals $P = \{I_1, I_2\}$ is a NeutroAlgebra under $\times$. Consider the sum in $P = \{I_1, I_2\}$ given by Table 9.

So P is a NeutroAlgebra of proper ideals under $+$.

Example 10: *Consider the modulo integer ring $R = \{Z_{30}, \times, +\}$. The ideals (proper) of R are $T = \{I_1 = \{0, 15\}, I_2 = \{0, 10, 20\}, I_3 = \{0, 5, 10, 15, 20, 25\}, I_4 = \{0, 6, 12, 18, 24\}, I_5 = \{0, 3, 6, 9, \ldots, 27\}, I_6 = \{0, 2, 4, 6, 8, \ldots, 28\}\}$. Now the Cayley table under + and $\times$ can be verified to have undefined terms as $\mathbb{Z}_{30}$ and (0) are not proper ideals of R.*

Thus the following theorem proves the existence of NeutroAlgebra of ideals which are not partial algebra or AntiAlgebra under the axiom of classical operations on ideals.

Theorem 11: *Let $R = \{Z_n, +, \times\}$ be the ring of modulo integers n; n a composite number $6 \leq n < \infty$. Let B denote the set of non-trivial ideals of R (that is (0) and ($\mathbb{Z}_n$) are not included as ideals in B).*

1. *$\{B, \times\}$ is a NeutroAlgebra under the axiom of product of ideals of R which is not a partial algebra or a AntiAlgebra.*

Table 9. Cayley table of P

+	I_1	I_2
I_1	I_1	od
I_2	od	I_2

2. $\{B,+\}$ *under the axiom of sum of ideals is a NeutroAlgebra which is not a partial algebra or AntiAlgebra.*

Proof: Given $B = \{$collection of all non-trivial ideals of $\mathbb{Z}_n$, that is the zero ideal and the whole ideals $\mathbb{Z}_n$ are not included$\}$.

Let $n = p_1^{\alpha_1} p_2^{\alpha_2} \dots p_m^{\alpha_m}$ where p_i s are primes with $1 \le \alpha_j < \infty$ and $1 \le i \le m$. Now the non-trivial ideals of R would be generated by p_i 's or $p_i^{\alpha_i}$'s or $p_{t_1}^{\alpha_{t_1}} \dots p_{t_r}^{\alpha_{t_r}}$; $1 \le t_r \le m$. For every ideal I_1 of B generated by $p_{t_1}^{\alpha_{t_1}} \dots p_{t_r}^{\alpha_{t_r}}$ there exists a unique ideal $J_1 \in B$ such that J_1 is generated by $p_{s_1}^{\alpha_{s_1}} \dots p_{s_q}^{\alpha_{s_q}}$ with

$$\left(p_{t_1}^{\alpha_{t_1}} \dots p_{t_r}^{\alpha_{t_r}}\right) \times \left(p_{s_1}^{\alpha_{s_1}} \dots p_{s_q}^{\alpha_{s_q}}\right) = 0(\text{mod } n).$$

Thus $I_1 \times I_2 = (0)$ and $(0) \notin B$. Hence $\{B,\times\}$ is a NeutroAlgebra of ideals under the axiom of being an ideal in B, so B which is not a partial algebra. Hence (1) is proved.

Proof of (2): Consider B the collection of ideals of R excluding $\{0\}$ and $\mathbb{Z}_n$. Now the ideals I_1 and I_2 in R are such that when the sum is taken it gives $1 \in I_1 + I_2$ there by making $\langle I_1 + I_2 \rangle = Z_n$. Hence $\{B,+\}$ is a NeutroAlgebra of ideals. Hence the claim.

In the next section authors proceed onto work the polynomial rings $R[x]$. R is a ring $\mathbb{Z}$ or $\mathbb{Z}_n$ or $\mathbb{Q}$ or even the reals $\mathbb{R}$.

NEUTROALGEBRA OF IDEALS IN POLYNOMIAL RINGS

In this section, polynomial rings $\mathbb{Z}[x]$ or $\mathbb{Z}_n[x]$ or $\mathbb{Q}[x]$ or $\mathbb{R}[x]$ and the collection of ideals in them are considered and NeutroAlgebra of ideals are defined.

An example of NeutroAlgebra of ideals in polynomial rings is provided.

Example 12: *Let* $T = \{Z[x],+,\times\}$ *be the ring of polynomials in the variable* x. *Now let* $B = \{$ *collection of all non-trivial ideals of* T *that is* B *does not contain the zero ideal (0) and the whole ring* $\mathbb{Z}[x]\}$. $\{B,+\}$ *is a NeutroAlgebra of ideals under + is established.*

For consider the ideal generated by $\langle x^2 \rangle = I_1$ and the ideal generated by $\langle x^3 + x^4 + 1 \rangle = I_2$. Clearly I_1 and $I_2 \in B$, I_1 is generated by x^2; so x^3 and $x^4 \in I_1$, so $x^3 + x^4 \in I_1$ and $x^3 + x^4 + 1 \in I_2$. Consider in

$$I_1 + I_2; \left(x^3 + x^4 + 1\right) - \left(x^4 + x^3\right) = 1$$

which generate the whole of the ring Z[x] and $\mathbb{Z}\left[x\right] \notin B$. So $\left\{B, +\right\}$ is a NeutroAlgebra of ideals under + as Z[x] is undefined element of B.

Now consider the product of the ideals I_1 with I_2; clearly $I_1 \times I_2$ is generated by $x^5 + x^6 + x^2$ and no other lesser degree. $I_1 \times I_2 \in B$. So undefined under product, for if $p\left(x\right)$ generates I_1 of degree n and $q\left(x\right)$ is of degree m which generates I_2 then $I_1 \times I_2$ has a $s\left(x\right)$ which is of degree $m + n$ and no lesser degree so $I_1 \times I_2$ is well defined and is in B. Hence $\left\{B, \times\right\}$ is not a NeutroAlgebra of ideals.

Example 13: *Let $\mathbb{Z}_{12}\left[x\right]$ be the ring of polynomials $S = \left\{Z_{12}, +, \times\right\}$ under usual product × and sum +. Let B be the collection of all ideals of S excluding 0 and $\mathbb{Z}_{12}$. Consider the two ideals I_1 and I_2 generated by $I_1 = \langle 4x^3 \rangle$ and $I_2 = \langle 3x^4 \rangle$.*

Now $I_1 \times I_2 = \left\{0\right\}$ and

$$I_1 + I_2 = \left\{\langle 4x^3, 3x^4, x^4 \ldots \rangle\right\} \neq Z_{12}\left[x\right].$$

Consider $J_1 = \langle 2x^2 + 1 \rangle$ and $J_2 = \langle x + 6 \rangle \in B$,

$$J_1 + J_2 = \left\{\langle 2x^2 + 1 + 10x\left(x + 6\right) \rangle\right\} = \langle 2x^2 + 1 + 12x^2 + 60 = 1 \rangle$$

so ideal generated by $J_1 + J_2 = Z_{12}\left[x\right]$.

So $J_1 + J_2 \notin B$. Thus composite number is used then $S = \left\{Z_n\left[x\right], +, \times\right\}$ can have both NeutroAlgebra of ideals both under + and ×.

Definition 14: *Let $S = \left\{R\left[x\right], +, \times\right\}$ be a polynomial ring over a commutative ring R finite or infinite order with identity. Let $B = \{$collection of all proper ideals of S excluding $\left\{0\right\}$ and $R\left[x\right]\}$. $\left\{B, \times\right\}$ is a NeutroAlgebra of polynomial ideals* *under usual product if there exists at least two ideals $I_1, I_2 \in B$ such that $I_1 \times I_2$ is not an ideal in B that is outer defined in B.*

Likewise define $\{B,+\}$ is a NeutroAlgebra of polynomial ideals under usual addition if there exists at least two ideals I_1 and I_2 in B such that $I_1 + I_2$ is not in B that is outerdefined in B.

Some examples to this effect are provided.

Theorem 15 *Let $S = \{R[x],+,\times\}$ be a polynomial ring ($\mathbb{R}$ = reals or rationals $\mathbb{Q}$ or integers $\mathbb{Z}$ or $\mathbb{Z}_p$; p a prime. B = {Collection of all proper ideals of S excluding $R[x]$, the whole ideal and (0), the zero ideal}.*

1. $\{B,+\}$ *is a NeutroAlgebra of polynomial ideals of S.*
2. $\{B,\times\}$ *is not a NeutroAlgebra of polynomial of ideals.*

Proof: Let $B = \{$ Collection of proper polynomial ideals of S excluding $R[x]$ the whole ring and the zero ideal (0)$\}$. Now to show $\{B,+\}$ is NeutroAlgebra of ideals it is sufficient to show there are at least two ideals I_1 and I_2 in *B* such that $I_1 + I_2$ is outerdefined in B. To this end take $I_1 = \langle x^3 \rangle$ and $I_2 = \langle x^4 + 1 \rangle$ be the polynomial ideals of B. Now $x^4 \in I_1$ and $x^4 + 1 \in I_2$ and $I_1 + I_2$ has $x^4 + 1 - x^4$ so that $1 \in I_1 + I_2$ so the ideal generated by $I_1 + I_2$ is $R[x]$ so is outerdefined in *B*. Hence $\{B,+\}$ is a NeutroAlgebra of polynomial ideals of S under the usual operation +.

To prove $\{B,\times\}$ is not a NeutroAlgebra of polynomial ideal under ×. Consider any two ideals of B say I_1 and I_2. If I_1 is generated by $\langle p(x) \rangle$ of degree n and that of I_2 is generated by $\langle q(x) \rangle$ of degree m then $I_1 \times I_2$ can only be an ideal generated by a polynomial of degree $m + n$. So $I_1 \times I_2$ is well defined at all times. Hence $\{B, \times\}$ is not a NeutroAlgebra of ideals under × in case R is Q or Reals $\mathbb{R}$ or integers $\mathbb{Z}$or $\mathbb{Z}_p$; p a prime..

In case of $\mathbb{Z}_n[x]$ the result is different as n is a composite number and as Z_n has zero divisors so does the polynomial ring $Z_n[x]$.

Example 16: *Let $S = \{Z_{15}[x],+,\times\}$ be the polynomial rings. Let B be the collection of non-trivial ideals of S, $I_1 = \langle 3x^3 + 1 \rangle$ and $I_2 = \langle x^2 \rangle$ be ideals of B. $I_1 + I_2 = \langle 1 \rangle$ as*

$\left(\left(3x^3+1\right)+12x^3=15x^3+1=1\right).$

since $x^3 \in I_2$ *and* $12x^3 \in I_2$ *as* I_2 *is generated by* x^2. *Thus* $I_1 + I_2$ *is outerdefined in* B. *Consider* $I_1 = \langle 6x+3\rangle$ *and* $I_2 = \left(5x^2+10\right) \in B$; $I_1 \times I_2 = \{0\}$ *so is again outerdefined.*

In view of this the following theorem is proved.

Theorem 17: *Let* $S = \left\{Z_n\left[x\right],+,\times\right\}$, *(n a composite number) be a polynomial ring of modulo integers in the variable x.* B *= collection of all proper ideals of S excluding S and (0).*

1. $\left(B,\times\right)$ is a NeutroAlgebra of ideals under ×.
2. $\left\{B,+\right\}$ is a NeutroAlgebra of ideals under +.

Proof:

Proof of (1): Given $B=\{$collection of all proper ideals of the polynomial ring S excluding $\mathbb{Z}_n\left[x\right]$ and the zero ideal$\}$. To prove $\left(B,\times\right)$ is a NeutroAlgebra of ideals. Clearly $\mathbb{Z}_n\left[x\right]$ is an infinite ring. Let I_1, $I_2 \in B$; $I_1 = \langle rx^3\rangle$ and $I_2 = \langle sx^2\rangle$ where $s.r \equiv 0\left(mod\, n\right)$ (r and s are in $\mathbb{Z}_n$ such that the product is zero divisor in Z_n). Hence the claim.

Proof of (2): Given n is a composite number. Let n = a + b, Take the two ideals I_2 generated by ; $\langle ax^2\rangle$ and I_2 generated by ; $\langle bx^2+1\rangle$. Clearly

$$I_1 + I_2 = ax^2 + bx^2 + 1$$

$$= ax^2 + bx^2 + 1$$

$$= ax^2 + bx^2 + 1 = \left(a+b\right)x^2 + 1 = \langle 0+1\rangle = 1 = Z_n\left[x\right].$$

Hence, {B, +} is a NeutroAlgebra of ideals under +.

CONCLUSION

In this chapter the authors have defined NeutroAlgebra of ideals of a ring R and the neutro operations sum of ideals of the ring R and product of ideals of a ring R. (In reals or rationals which are fields has non-trivial ideals cannot exist in rings.)

In case of *Z*, it is proved. The set of ideals *J* of *Z* is NeutroAlgebra under + (sum) of ideals and in case of ×(product) of ideals is not a NeutroAlgebra.

In case $\mathbb{Z}_n$ (n a composite number for if n is prime, no ideals exists in $\mathbb{Z}_n$); the set of ideals of $\mathbb{Z}_n$ under + and × are NeutroAlgebras.

In case of polynomial rings $\mathbb{Z}[x], Q[x]$ and R[x] and $\mathbb{Z}_n[x]$ are NeutroAlgebras under + (sum of ideals). However in case of $Z[x], Q[x]$ and $R[x]$, the product of ideals is not a NeutroAlgebra. But in $\mathbb{Z}_n[x]$ the ideals under × is a NeutroAlgebra of ideals. The following is proposed as an open problem.

Problem 1: Can there exist a NeutroAlgebra of ideals S which has every proper subset to be also a NeutrosubAlgebra?

For future study, special type of rings in which these properties are not true can be found.

REFERENCES

Agboola, A., Ibrahim, M., & Adeleke, E. (2020). Elementary examination of neutroalgebras and antialgebras viz-a-viz the classical number systems. *International Journal of Neutrosophic Science*, *4*, 16–19.

Kandasamy, V., Kandasamy, I., & Smarandache, F. (2020). NeutroAlgebra of Neutrosophic Triplets using $\{Z_n, \times\}$. *Neutrosophic Sets and Systems*, *38*, 510–523.

Smarandache, F. (1999). *A unifying field in logics. neutrosophy: Neutrosophic probability, set and logic*. Academic Press.

Smarandache, F. (2019). Introduction to NeutroAlgebraic Structures and AntiAlgebraic Structures (revisited). *Neutrosophic Sets and Systems*, *31*, 1–16.

Smarandache, F. (2020). NeutroAlgebra is a Generalization of Partial Algebra. *International Journal of Neutrosophic Science*, *2*, 8–17.

Smarandache, F., & Hamidi, M. (2020). Neutro-BCK-algebra. *International Journal of Neutrosophic Science*, *8*, 110–117.

Smarandache, F., & Rezaei, A. (2020). On neutro-BE-algebras and anti-BE-algebras. *International Journal of Neutrosophic Science*, *4*, 8–15.

Chapter 16
Quadripartitioned Single-Valued Neutrosophic Hypergroup

Carlos Granados
Universidad de Antioquia, Medellín, Colombia

ABSTRACT

The study of neutroalgebra has become one of the topics of great interest for researchers from different areas, where neutroGroup, neutro BCK-algebra, neutro Lie algebras, single valued neutro algebra, and others have been studied. This is why, in this work, the authors define the concept of quadripartitioned single-valued neutrosophic hypergroup as the initial concept of quadripartitioned single-valued neutrosophic hyper-algebra. They prove some properties on quadripartitioned single-valued neutrosophic hypergroup. The notion of NeutroQuadripartitionedSingleValuedHypergroup was defined, and some future research directions were proposed.

INTRODUCTION

To manage vulnerability based genuine and logical issues, Zadeh (1965) presented the fuzzy set as a useful apparatus. Later on Atanassov (1986) broadened the idea of fuzzy set hypothesis to the intuitionistic fuzzy set (Uncertainties), in which each component has both a membership degree and a non-membership degree. It is very evident that Uncertainties are more useful than fuzzy set hypothesis to bargain the different sorts of vulnerability model. Neutrosophy is a new branch of philosophy which was introduced by (Smarandache, 2002) which has been of great interested of researchers who study different topics (applied or pure science) and it studies the

DOI: 10.4018/978-1-6684-3495-6.ch016

origin, nature and scope of neutralities, as well as their interactions with different ideational spectra: (B) is an idea, proposition, theory, event, concept or entity; anti (B) is the opposite of (B); and (neut-B) means neither (B) nor anti (B), i.e. neutrality between the two extremes (Bal eta l., 2018).

As a generalization of classical algebraic structure, Algebraic hyper structure was introduced by F. Marty (1934). Corsini and Leoreanu-Fotea (2003) developed the applications of hyper structure. Algebraic hyperstructures has many applications in fuzzy sets, lattices, artificial intelligence, automation, combinatorics. Corsini (1993)introduced hypergroup theory. After while the hyperstructure theory has seen broader applications in many fields. Many papers have been written for many researchers related to neutroalgebra. A neutroalgebra is an algebra which has at least one neutrooperation or one neutroaxiom (axiom that is true for some elements, indeterminate for other elements, and false for the other elements) (Smarandache, 2020).

Neutrosophic set is a piece of neutrosophy which concentrates on the beginning, nature, and extent of neutralities, as well as their collaborations with various ideational spectra. Neutrosophic set is a strong general formal structure that has been proposed by Smarandache (2020). In any case,n eutrosophic set should be determined from a specializedperspective. With this impact, Wand et al. (2012) characterized the set-theoretic administrators on an occasion of neutrosophic set, they called it single-valued neutrosophic set (SVNS). This, it FULFILL different properties of SVNS, which are associated with the tasks and relations over SVNS. With a view to device a practical tool for inference, Belnap (1977) introduced the notion of a four valued logic. In his work, corresponding to a certain information he considered four possibilities namely T: True, F: False, none: neither true nor false and both: both true and false. He symbolized these four truth values as {T, F, both, none}, for more notions derived from this paper, we refer the reader to (Das, et al., 2021; Mohanasundari and Mohana, 2020).The possibility of quadripartitioned single-valued neutrosophic set is really extended from Smarandache, four mathematical neutrosophic valued rationale (Smarandache 2002) and Belnap (1977), four esteemed rationale, where the indeterminacy is isolated into two sections, in particular, “obscure” i.e., neither genuine nor bogus. Moreover, “logical inconsistency” i.e., both valid and bogus. With regards to neutrosophic concentrate be that as it may, the quadripartitioned single-valued neutrosophic set was presented by Chatterjee et al. (2016), in which they investigated a genuine model for a better comprehension of a quadripartitioned single-valued neutrosophic set climate and showed that such circumstances happen normally. They have likewise tackled a dynamic issue relating to design acknowledgment showing the application capacity of quadripartitioned single-valued neutrosophic set. On the other hand, the study of NeutroAlgebra by using quadripartitioned single-valued neutrosophic sets has been for a great interested of mathematicians and researcher who study this topic, for example Mohan and Krishnaswamy (2020) said that quadripartitioned single-valued neutrosophic set esteemed neutrosophic is a strong construction where we have four parts Truth T, False F, CONTRADICTION C and IGNORANCE I. Furthermore, they used the idea of fluffy, INTUITIONISTIC and single-valued neutrosophic set. And they proposed the idea of K-algebras on quadriprtitioned single-valued neutrosophic set, level subset of quadriprtitioned single-valued neutrosophic set and concentrated on a portion of the outcomes. Notwithstanding this they additionally explored the attributes of quadriprtitioned single-valued neutrosophic set K-subalgebras under homomorphism. Radha and Mary (2021) in their this article, they talked about above quadriprtitioned single-valued neutrosophic set subalgebra and quadriprtitioned single-valued neutrosophic set Falsehood beliefs of a quadriprtitioned single-valued neutrosophic set Falsehood Variable based math. Additionally, they examined a portion of its properties of quadriprtitioned single-valued neutrosophic set Pythagorean Untruthgoals.

On the other hand, several studies related to quadripartitioned single-valued neutrosophic set have been carried out. For example, the thought of straightforward bipolar quadripartittioned was introduced by Sinha and Majumdar (2020) as an important neutrosophic set. A few fundamental set hypothetical wordings, activities and properties of bipolar quadripartitioned single-valued neutrosophic set were given here. Additionally, various sorts of distances, closeness measures and entropy measure were talked about. At last a dynamic issue utilizing the closeness measure procedure of bipolar quadripartitioned single-valued neutrosophic sets were issued. Numerous numerical activities like normal, total, total, count, max, MINIMUM are performed with the assistance of total operations. Multicriteria Trait navigation isa methodology which is utilized to choose a best one when a few choices are incorporated getting looked at of many credits. So many analyst have focused on tackle the Multicriteria Property dynamic issues utilizing the idea of different connection coefficients of the various sets like fluffy set, Uncertainties, single valued neutrosophic sets, quadripartitioned single-valued neutrosophic sets. Furthermore, numerous scientists involved collection administrators as one of the apparatus to take care of a Multicriteria dynamic issue and furthermore concentrated on its properties. Majumdar et al. (2020) bipolar quadripartitioned single-valued neutrosophic set (BQSVNS) was presented. A few essential set hypothetical wordings like steady bipolar quadripartitioned single-valued neutrosophic set, sub-sethood of two bipolar quadripartitioned single-valued neutrosophic sets were shown. Mathematical activities like association, crossing point and supplement likewise were characterized. Various sorts of measure like comparability measure, semi likeness measure and distance measures between two bipolar quadripartitioned single-valued neutrosophic sets were examined with their properties. Besides, different proportions of similitude to be specific distance based closeness measure, cosine similarly measure, and participation work based comparability measure were presented in their paper. Following the idea of bipolar neutrosophic set, Roy et al. (2020) in their paper, a meaning of bipolar quadripartitioned single-valued neutrosophic set (QSVBNS) was presented as a speculation of both quadripartitioned single-valued neutrosophic set (QSVNS) and bipolar neutrosophic sets (BNS). There was an innate evenness in the meaning of bipolar quadripartitioned single-valued neutrosophic set. A few procedure on them were characterized and a set hypothetical review was refined. Different closeness measures and distance measures were characterized on bipolar quadripartitioned single-valued neutrosophic set. A calculation connecting with multi-models dynamic issue was introduced in light of quadripartitioned bipolar weighted closeness measure. A model was displayed to confirm the adaptability of the given strategy and the upside of thinking about bipolar quadripartitioned single-valued neutrosophic set instead of fluffy sets and bipolar fluffy sets.

In this work, we carry out the notion of quadripartitioned single-valued neutrosophic hypergroup which is an extension of the paper presented by (Rajareega et al., 2020). Besides, we show some properties on quadripartitioned single-valued neutrosophic hypergroup.

PRELIMINARIES

In this section, we show some well-known notions which will be useful for the development of this work.

Definition 2.1: Let Y be a space of points (objects), with a generic element in Y denoted by y. A quadripartitioned neutrosophic set B in Y is characterized by a truth-membership function T_B, a contradiction membership function C_B, an ignorance membership function I_B and a falsity membership function F_B. $T_B(y)$, $C_B(y)$, $I_B(y)$ and $F_B(y)$ are real standard or non-standard subsets of $]0-, 1+[$.There is no restriction on the sum of $T_B(y)$, $C_B(y)$, $I_B(y)$ and $F_B(y)$, thus

$$0^- \leq sup\, T_B(y) + sup\, C_B(y) + sup\, I_B(y) + sup\, F_B(y) \leq 4^+$$

(Chatterjee et al., 2016).

Definition **2.2: Let Y be** a space of points (objects),with a generic element of Y denoted by y. A quadripartitioned single-valued neutrosophic set (QSVNS) B in Y is characterized by TB, CB, IB and FB. For each point y in Y, and TB, CB, IB, FB $\in [0,1]$ (Chatterjee et al., 2016).

Definition **2.3: The complement** of a QSVNS B is denoted by C(B) and is defined by

1. TC(B)(y)=$_{FB(y}$),
2. CC(B)(y)$_{=1-C}$B(y),
3. IC(B)(y$_{)=1-}$IB(y),
4. FC(B)($_{y)=T}$B(y).

For all y in Y (Chatterjee et al., 2016).

Definition 2.4: A QSVNS B is contained in the other QSVNS D, B⊆D, if

1. TB($_y$)≤TD($_y$)
2. CB$_($y) ≤CD$_($y)
3. $I_B(y) \geq I_D(y)$
4. $F_B(y) \geq F_D(y)$

For all y in Y (Chatterjee et al., 2016).

Definition 2.5: The union of two QSVNS's B and D is a QSVNS E, written as E=B∪D, whose truth, contradiction, ignorance and falsity-membership functions are given by,

1. $T_E(y)=\max(T_B(y),T_D(y))$,
2. $C_E(y)=\max(C_B(y),C_D(y))$,
3. $I_E(y)=\min(I_B(y),I_D(y))$,
4. $F_E(y)=\min(F_B(y),F_D(y))$.

For all y in Y (Chatterjee et al., 2016).

Definition 2.6: The intersection of two QSVNS's B and D is a QSVNS E, written as E=B∩D, whose truth, contradiction, ignorance and falsity-membership functions are given by,

1. $T_E(y)=\min(T_B(y),T_D(y))$,
2. $C_E(y)=\min(C_B(y),C_D(y))$,
3. $I_E(y)=\max(I_B(y),I_D(y))$,
4. $F_E(y)=\max(F_B(y),F_D(y))$.

For all Y in y (Chatterjee et al., 2016).

Definition 2.7: The falsity-favorite of a QSVNS D, written as D∇B, whose truth, contradiction, ignorance and falsity-membership functions are given by,

1. $T_D(y)=T_B(y)$
2. $C_D(y)=0$
3. $I_D(y)=0$

4. $F_D(y)=\min\{F_B(y)+I_B(y)+C_B(y),1\}$.

For all y in Y (Chatterjee et al., 2016).

Definition 2.8: A hypergroup $\langle M,*\rangle$ is a set M equipped with an associative hyper-operation $(*):M\times M\to P(M)$ which satisfies $x*M=M*x=M$ for all $x\in M$ (Reproduction axiom) (Selvachandran and Salleh, 2015)

Definition 2.9: A hyperstructure $\langle M,*\rangle$ is called an M_v-group if the following axioms are satisfied:

1. $x*(y*z)\cap(x*y)*z\neq\varnothing$ for all $x,y,z\in M$,
2. $x*M=M*x=M$ for all $x\in M$.

If $\langle M,*\rangle$ only satisfies (I), then $\langle M,*\rangle$ is said to be a M_v- semigroup (Selvachandran and Salleh, 2015).

Definition 2.10: A subset L of M is said to be a subhypergroup if $\langle L,*\rangle$ is a hypergroup. Selvachandran and Salleh, 2015).

QUADRIPARTITIONED SINGLE-VALUED NEUTROSOPHIC HYPERGROUP

For the development of this section, M denotes the hypergroup $<M,*>$

Definition 3.1: Let B be a quadripartitioned single-valued neutrosophic set over M. Then, B is said to be a quadripartitioned single-valued neutrosophic hypergroup over M, if the following statements hold:

1. For all $q,w \in M$,
    a. $\min\{T_B(q),T_B(w)\} \leq \inf\{T_B(j): j \in q*w\}$,
    b. $\min\{C_B(q),C_B(w)\} \leq \inf\{C_B(j): j \in q*w\}$,
    c. $\max\{I_B(q),I_B(w)\} \geq \sup\{I_B(j): j \in q*w\}$,
    d. $\max\{F_B(q),F_B(w)\} \geq \sup\{F_B(j): j \in q*w\}$.
2. For all $q,w \in M$, there exists $r\in M$ such that $w\in q*r$ and
    a. $\min\{T_B(q),T_B(w)\} \leq T_B(r)$,
    b. $\min\{C_B(q),C_B(w)\} \leq C_B(r)$,
    c. $\max\{I_B(q),I_B(w)\} \geq I_B(r)$,
    d. $\max\{F_B(q),F_B(w)\} \geq F_B(r)$.
3. For all $q,w \in M$, there exists $u\in M$ such that $w\in u*q$ and
    a. $\min\{T_B(q),T_B(w)\} \leq T_B(u)$,
    b. $\min\{C_B(q),C_B(w)\} \leq C_B(u)$,
    c. $\max\{I_B(q),I_B(w)\} \geq I_B(u)$,

d. $\max\{F_B(q),F_B(w)\} \geq F_B(u)$.

Remark 3.2: If B satisfies condition (I), then B is a quadripartitioned single-valued neutrosophic semihypergroup over M. If conditions (II) of Definition 3.1 represents the left reproduction axioms and (III) of Definition 3.1 represents the right reproduction axioms. Then, B is a quadripartitioned single-valued neutrosophic subhypergroup of M.

Example 3.3: If the family of t-level sets of QSVNS B over M

$$B_t=\{q\in M: T_B(q)\geq t,\ C_B(q)\geq t,\ I_B(q)\leq t \text{ and } F_B(q)\leq t\}$$

is a subhypergroup of M. Therefore, B is a quadripartitioned single-valued neutrosophic hypergroup over M.

Theorem 3.4: Let B be a QSVNS over M. Then, B is a quadripartitioned single-valued neutrosophic hypergroup over M, if and only if, B is a quadripartitioned single-valued neutrosophic semihypergroup over M. Besides, B satisfies the left and right reproduction axioms.

Proof: It follows from Definition: 3.1.

Theorem 3.5: Let B be a QSVNS over M. If B is a quadripartitioned single-valued neutrosophic hypergroup over M. Then, for all $t\in[0,1]$ $B_t\neq\emptyset$ is a subhypergroup of M.

Proof: Let B be a quadripartitioned single-valued neutrosophic hypergroup over M and let $q,w \in B_t$. Then, $T_B(q)\geq t$, $T_B(w)\geq t$, $C_B(q)\geq t$, $C_B(w)\geq t$, $I_B(q)\leq t$, $I_B(w)\leq t$, $F_B(w)\leq t$ and $F_B(q)\leq t$. Therefore, we have

$$\inf\{T_B(u): u\in q*w\} \geq \min\{T_B(q),T_B(w)\} \geq \min\{t,t\}=t,$$

$$\inf\{C_B(u): u\in q*w\} \geq \min\{C_B(q),C_B(w)\} \geq \min\{t,t\}=t,$$

$$\sup\{I_B(u): u\in q*w\} \leq \max\{T_B(q),T_B(w)\} \leq \max\{t,t\}=t,$$

$$\sup\{F_B(u): u\in q*w\} \leq \max\{F_B(q),F_B(w)\} \leq \max\{t,t\}=t.$$

This implies, $u\in B_t$. Therefore, for all $u\in q*w$, $q*w \subseteq B_t$. Thus, for all $u\in B_t$, we have $u*B_t\subseteq B_t$.

Let $a,q\in B_t$, there exists $w\in M$ such that $q \in a*w$ and

$$\{T_B(w)\} \geq \min\{T_B(a),T_B(q)\} \geq \min\{t,t\}=t,$$

$$\{C_B(w)\} \geq \min\{C_B(a),C_B(q)\} \geq \min\{t,t\}=t,$$

$\{I_B(w)\} \leq \max\{I_B(a),I_B(q)\} \leq \max\{t,t\}=t,$

$\{F_B(w)\} \leq \max\{F_B(a),F_B(q)\} \leq \max\{t,t\}=t.$

This shows that $B_t \subseteq u*B_t$. Therefore, we have that $B_t = u*B_t$ and this proves that B_t is a subhypergroup of M.

Theorem 3.6: Let B be a QSVNS over M. Then, the following statements are equivalent:

1. B is a quadripartitioned single-valued neutrosophic hypergroup over M.
2. For all $t \in [0,1]$, $B_t \neq \emptyset$ is a subhypergroup of M.

Proof:

1. → (II) It follows from Theorem 3.5.
2. → (I) Let us consider B_t is a subhypergroup of M. We begin showing that conditions (I) of Definition 3.1 is satisfied.

Let $q,w \in B_{t1}$ and let

$\min\{T_B(q),T_B(w)\}= \min\{C_B(q),C_B(w)\}=\max\{I_B(q),I_B(w)\}=\max\{F_B(q),F_B(w)\}=t_1$

.

Since $q*w \subseteq B_{t1}$, then for every

$u \in q*w,\ T_B(u) \geq t_{1,}\ C_B(u) \geq t_1,\ I_B(u) \leq t_1,\ F_B(u) \leq t_1$

$\min\{T_B(q),T_B(w)\} \leq \inf\{T_B(u): u \in q*w\},$

$\min\{C_B(q),C_B(w)\} \leq \inf\{C_B(u): u \in q*w\},$

$\max\{I_B(q),I_B(w)\} \geq \sup\{I_B(u): u \in q*w\},$

$\max\{F_B(q),F_B(w)\} \geq \sup\{F_B(u): u \in q*w\}.$

Condition (I) is verified.

Next, we will show that condition (II) of Definition 3.1 is satisfied.

Let $j,q \in B_{t2}$, for every $t_2 \in [0,1]$ and let

$\min\{T_B(j),T_B(j)\}= \min\{C_B(j),C_B(j)\}=\max\{I_B(j),I_B(j)\}=\max\{F_B(j),F_B(j)\}= t_2$.

Then, there exists w $\in B_{t2}$ such that q$\in$ j*w $\subseteq B_{t2}$. Since w$\in B_{t2,}$

$T_B(w) \geq t_2 = \min\{T_B(j), T_B(q)\}$

$C_B(w) \geq t_2 = \min\{C_B(j), C_B(q)\}$

$I_B(w) \leq t_2 = \max\{I_B(j), I_B(q)\}$

$F_B(w) \leq t_2 = \max\{F_B(j), F_B(q)\}$

Condition (II) is verified. Similarly, (III).

Proposition 3.7: Let B be a QSVNS over M. Then,B is a quadripartitioned single-valued neutrosophic hypergroup over M, if and only if, for all

$\alpha,\beta,\gamma,\varepsilon \in [0,1]$, $B_{(\alpha,\beta,\gamma,\varepsilon)}$

is a subhypergroup of M.

Proof: It follows from Theorem 3.6.

Theorem 3.8: Let B be a quadripartitioned single valued neutrosophic hypergroup over M and for all $t_0,t_1 \in [0,1]$ B_{t0} and B_{t1} be the t-level sets of B with $t_0 \geq t_1$. Then,B_{t0} is a subhypergroup of B_{t1}.

Proof: For all $t_0,t_1 \in [0,1]$ B_{t0} and B_{t1} be the t-level sets of B with $t_0 \geq t_1$. This implies that $B_{t0} \subseteq B_{t1}$. Therefore, By Theorem 3.5,B_{t0} is a subhypergroup of B_{t1}.

Theorem 3.9: Let B and D be quadripartitioned single-valued neutrosophic hypergroups over M. Then, B∩D is a quadripartitioned single-valued neutrosophic hypergroup over M if it is non-null.

Proof: By Definition: 2.6.

$B \cap D = \{ <q, T_B \cap_D(q), C_B \cap_D(q), I_B \cap_D(q), F_B \cap_D(q)>: q \in M\}$

where

$T_B \cap_D(q) = T_B(q) \wedge T_D(w)$,

$C_B \cap_D(q) = C_B(q) \wedge C_D(w)$,

$I_D \cap_B(q) = I_B(q) \vee I_D(w)$

and

$F_D\cap_B(q)=F_B(q)\vee F_D(w)$.

Now, For all q,w∈M, we have

$\min\{T_B\cap_D(q),T_B\cap_D(w)\}=\min\{T_B(q)\wedge T_D(q),T_B(w)\wedge T_D(w)\}$

$=\text{in}\{T_B(q),T_B(w)\}\wedge\min\{T_D(q),T_D(w)\}$

$\leq\inf\{T_B(u): u\in q*w\}\wedge\inf\{T_D(u): u\in q*w\}$

$\leq\inf\{T_B(u)\wedge T_D(u): u\in q*w\}$

$=\inf\{T_B\cap_D(u): u\in q*w\}$.

In similar way, the other statements are satisfied.
Now, for all j,q∈M, there exists w∈M such that q∈j*w, thus

$\min\{T_D\cap_B(j),T_D\cap_B(q)\}=\min\{T_B(j)\wedge T_D(j)\},\{T_B(q)\wedge T_D(q)\}$

$=\min\{T_B(j),T_B(q)\}\wedge\min\{T_D(j),T_D(q)\}$

$\leq T_B(w)\wedge T_D(w)=T_D\cap_B(w)$.

Therefore, B∩D is a quadripartitioned single-valued neutrosophic hypergroup over M

Theorem 3.10: Let B and D be quadripartitioned single-valued neutrosophic hypergroups over M. Then, B∪D is a quadripartitionedsingle-valued neutrosophic hypergroup over M.

Proof: By Definition: 2.5.

$B\cup D=\{<q,T_{B\cup D}(q), C_{B\cup D}(q), I_{B\cup D}(q), F_{B\cup D}(q)>: q\in M\}$

where

$T_{B\cup D}(q)=T_B(q)\vee T_D(w)$,

$C_{B\cup D}(q)=C_B(q)\vee C_D(w)$,

$I_{D\cup B}(q)=I_B(q)\wedge I_D(w)$

and

$F_{DUB}(q)=F_B(q)\wedge F_D(w)$.

Now, For all q,w∈M, we have

$\min\{T_{BUD}(q),T_{BUD}(w)\}=\min\{T_B(q)\vee T_D(q),T_B(w)\vee T_D(w)\}$

$\leq\min\{T_B(q),T_B(w)\}\vee \min\{T_D(q),T_D(w)\}$

$\leq\inf\{T_B(u): u\in q*w\}\vee\inf\{T_D(u): u\in q*w\}$

$\leq\inf\{T_B(u)\vee T_D(u): u\in q*w\}$

$=\inf\{T_{BUD}(u): u\in q*w\}$.

In similar way, the other statements are satisfied. Therefore, B∪D is a quadripartitioned single-valued neutrosophic hypergroup over M.

Theorem 3.11: Let B be a quadripartitioned single-valued neutrosophic hypergroup over M. Then the falsity-favorite of B (i.e.,∇B) is also a quadripartitioned single-valued neutrosophic hypergroup over M.

Proof: By Definition: 2.7, D=∇B, where the membership values are

$T_D(y)=T_B(y)$, $C_D(y)=0$, $I_D(y)=0$

and

$F_D(y)=\min\{F_B(y)+I_B(y)+C_B(y),1\}$.

Then, we have to show for F_B, for all q,w∈M,

$\max\{F_D(q),F_D(w)\} = \max\{F_B(q)+C_B(q)+I_B(q)\wedge 1, F_B(w)+C_B(w)+I_B(w)\wedge 1\}$

$=\max\{ F_B(q)+C_B(q)+I_B(q), F_B(w)+C_B(w)+I_B(w)\}\wedge 1$

$=(\max\{F_B(q),F_B(w)\}+ \max\{C_B(q),C_B(w)\}+ \max\{I_B(q),I_B(w)\})\wedge 1$

$\geq(\sup\{F_B(u) :ur\in q*w\}+ \sup\{C_B(u) :ur\in q*w\}+ \sup\{I_B(u) :ur\in q*w\})\wedge 1$

$=\sup\{F_B(u)+ C_B(u)+ I_B(u)\wedge 1: u\in q*w\}$

=sup{F_D(u): u∈q*w}.

In similar way, the other statements are satisfied.

Definition 3.13: Let $(B,*)$ be a quadripartitioned single-valued neutrosophic hypergroup. Then $(B,*)$ is a NeutroQuadripartitionedSingleValuedNeutrosophicHypergroup (NQSVNH) if some (or all) of the following conditions hold in a way that

$$(\mathrm{T},\mathrm{C},\mathrm{I},\mathrm{F}) \notin \{(1,0,0,0),(0,0,0,1)\}.$$

1. There exist $j \in q * w \in M$ with $q \neq w$ such that min{q,w}≤ j, there exists $r \in M$ with w∈ $q * r$ such that min{q,w}≤r and there exists u ∈M with w∈u*q such that min{q,w}≤u (This condition is called degree of truth, "T").
2. There exist $j \in q * w \in M$ with $q \neq w$ such that max{q,w}≥ j, there exists $r \in M$ with w∈ $q * r$ such that max{q,w}≥r and there exists u ∈M with w∈u*q such that max{q,w}≥u (This condition is called degree of falsity, "F").
3. There exist $x, y \in M$ with $z * x$ or $z * y$ or $x * z$ or $y * z$ are contradiction in B, or the relation between $z * x$ and $z * y$, or the relation between $x * z$ and $y * z$ are contradiction. (This condition is called degree of contradiction, "C").
4. There exist $x, y \in M$ with $z * x$ or $z * y$ or $x * z$ or $y * z$ are ignorance in B, or the relation between $z * x$ and $z * y$, or the relation between $x * z$ and $y * z$ are ignorance. (This condition is called degree of ignorance, "I").

Example 3.14: Let W be the set of real numbers and define "$\# : W \times W \to P^*(W)$" on W as follows.

$$q \# w = \begin{cases} [q,w] & \text{if } \min\{q,w\} < y, \\ [w,q] & \text{if } \max\{q,w\} > y, \\ 0 & \text{if } q = w = 0 = y, \\ \dfrac{1}{y} & \text{otherwise.} \end{cases}$$

Then $(W,\#)$ is a NQSVNH.

Theorem 3.15: Let $(B,*_1)$ and $(D,*_2)$ be two NQSVNH's. Then $\left(B \times D, *\right)$ is an NQSVNH. Here for all $\left(x,y\right),\left(z,w\right) \in B \times D$, $(x,y)*(z,w)=\{(a,m):a\in x*_1z$ and $m\in y*_2w\}$ and $(x,y)*(z,w)$ is equivalent to $x*_1$ zand $y*_2w$.

Proof: It follows form the fact that since $(B,*_1)$ and $(D,*_2)$ are two NQSVNH's, it follows from (Adbul and Davvaz, 2013) that $\left(B \times D, *\right)$ is a neutrosophic hypergroup and Definition 3.13.

CONCLUSION

In this work, we have introduced the notion of hypergroup for the quadripartitioned single-valued neutrosophic set by defining several hyperalgebraic structures and some properties were proved. For future work associated to this paperinvolve the development of hyperalgebraic for the pentapartitioned or heptapartitioned single-valued neutrosophic sets and pentapartitioned or heptapartitioned interval-valued neutrosophic sets. On the other hand, the notion of NeutroQuadripartitionedSingleValuedNeutrosophicHypergroup and future works directions were proposed.

FUTURE RESEARCH DIRECTIONS

The results in this chapter may be considered as a base for any possible study in the field of NeutroQuadripartitionedSingleValuedNeutrosophicHypergroup. The author suggests some research ideas in this direction:

1. Find all NeutroQuadripartitionedSingleValuedOrderhypergroup of small order.
2. Define other NeutroQuadripartitionedSingleValuedOrderHypergroup such as NeutroQuadripartitionedSingleValuedOrderPolygroup, etc.
3. Find applications of NeutroQuadripartitionedSingleValuedOrderHypergroup in different fields.
4. Define AntiAlgebra.

FUNDING

This research received no specific grant from any funding agency in the public, commercial, or not-for-profit sectors.

CONFLICTS OF INTEREST

The author declares that there is no conflict of interest.

ACKNOWLEDGMENT

The author is grateful to the referees for their valuable comments and suggestions.

REFERENCES

Abdul, A., & Davvaz, B. (2013). Introduction to neutrosophic hypergroups. *Romai J.*, *9*(2), 1–10.

Atanassov, K. (1986). Intuitionistic fuzzy sets. *Fuzzy Sets and Systems*, *20*(1), 87–96.

Bal, M., Shalla, M., & Olgun, N. (2018). Neutrosophic Triplet Cosets and Quotient Groups. *Symmetry*, *10*(4), 126.

Belnap, N.D. (1977). A useful four valued logic. *Modern Uses of Multiple Valued Logic*, 9–37.

Chatterjee, R., Majumdar, P., & Samanta, S. K. (2016). On some similarity measures and entropy on quadripartitioned single valued neutrosophic sets. *Journal of Intelligent & Fuzzy Systems*, *30*, 2475–2485.

Corsini, P. (1993). Prolegomena of Hypergroup Theory. Aviani.

Corsini, P., & Leoreanu-Fotea, V. (2003). *Application of Hyperstructures Theory: Advances in Mathematics*. Kluwer Academic.

Das, S., Das, R., & Granados, C. (2021). Topology on quadripartitioned neutrosophic sets. *Neutrosophic Sets and Systems*, *45*, 54–61.

Majumdar, P., Pinaki, C., Sinha, K., & Kalyan, U. (2020). Bipolar quadripartitioned single-valued neutrosophic rough set. *Neutrosophic Sets and Systems*, *38*, 244–257.

Marty, F. (1934). Sur une Generalisation de la Notion de Groupe. *Proceedings of the 8th Congress Mathematiciens Scandinaves*, 45-49.

Mohan, M., & Krishnaswamy, M. (2020). K-algebras on quadripartitioned single valued neutrosophic sets. *Journal of Fuzzy Extension and Applications*, *1*(4), 325–339.

Mohanasundari, M., & Mohana, K. (2020). Quadripartitioned Single valued Neutrosophic Dombi Weighted Aggregation Operators for Multiple Attribute Decision Making. *Neutrosophic Sets and Systems*, *32*, 107–122.

Radha, R., & Arul Mary, A. S. (2021). Quadripartitioned neutrosophic pythagorean lie subalgebra. *Journal of Fuzzy Extension and Application,* 2(3), 283-296.

Rajareega, S., Preethi, D., Vimala, J., Selvachandran, G., & Smarandache, F. (2020). Some Results on Single Valued Neutrosophic Hypergroup. *Neutrosophic Sets and Systems*, *31*, 80–85.

Roy, S., Lee, J. G., Pal, A., & Samanta, S. K. (2020). Similarity Measures of Quadripartitioned Single Valued Bipolar Neutrosophic Sets and Its Application in Multi-Criteria Decision Making Problems. *Symmetry*, *12*(6), 1012.

Sinha, K., & Majumdar, P. (2020). Bipolar quadripartitioned single-valued neutrosophic sets. *Proyecciones (Antofagasta)*, *39*(6), 1597–1614.

Smarandache, F. (2002). *Neutrosophy, a new Branch of Philosophy*. Infinite Study.

Smarandache, F. (2020). Neutroalgebra is a generalization of partial algebra. *International Journal of Neutrosophic Science*, 2(1), 8–17.

Zadeh, L. (1965). Fuzzy sets. *Information and Control*, *8*(3), 338–353.

Chapter 17
On NeutroSemigroups

Abdul Akeem Adesina Agboola
Federal University of Agriculture, Abeokuta, Nigeria

Muritala Abiodun Ibrahim
Federal University of Agriculture, Abeokuta, Nigeria

Zulaihat Hassan-Ibrahim
Auburn University, USA

Emmanuel O. Adeleke
Federal University of Agriculture, Abeokuta, Nigeria

ABSTRACT

The focus of this chapter is on the study of NeutroSemigroups. NeutroSemigroups are alternatives to the classical semigroups where the closure law is allowed to be either partially well defined (with the degree of truth T) or partially indeterminate (with the degree of indeterminacy I) or partially outer defined (with the degree of falsehood F), and associative axiom is equally allowed to be either partially true (with the degree of truth T) or partially indeterminate (with the degree of indeterminacy I) or partially outer defined (with the degree of falsehood F). Elementary properties of NeutroSemigroups with interesting examples are presented.

INTRODUCTION

The concept of semigroup is natural, and the theory of semigroups cannot be reduced to either the theory of groups or the theory of rings, notwithstanding a close connection between these theories. In various fields of science and technology one encounters the systems S = {f1, f2, f3 · · ·, fn} of the sets of functions closed

DOI: 10.4018/978-1-6684-3495-6.ch017

under the composition o, of fi where fi may be of the general form: not necessarily univalent, or everywhere defined, or even single-valued. Every system of functions closed under composition is a semigroup and every semigroup is isomorphic to such a system of functions. Neither the concept of group, nor the concept of ring is sufficient to study general systems of functions, hence the need for the concept of semigroup and the theory of semigroups. The theory of semigroups has applications in many areas such as automata, electromechanical, machines, computers, money changing devises, telephone switch boards, neural networks, formal languages, biology, biochemistry, sociology, psychology, anthropology and so on.

PRELIMINARIES

A classical operation is an operation well defined for all the set's elements, a NeutroOperation is an operation partially well defined, partially indeterminate, and partially outer defined on the given set while an AntiOperation is an operation that is outer defined for all set's elements.

A classical law/axiom defined on a nonempty set is a law/axiom that is totally true (i.e. true for all set's elements), a NeutroLaw/NeutroAxiom (or Neutrosophic Law/ Neutrosophic Axiom) defined on a nonempty set is a law/axiom that is true for some set's elements [degree of truth (T)], indeterminate for other set's elements [degree of indeterminacy (I)], or false for the other set's elements [degree of falsehood (F)], where $T, I, F \in [0, 1]$, with $(T, I, F) \neq (1, 0, 0)$ that represents the classical axiom, and $(T, I, F) \neq (0, 0, 1)$ that represents the AntiAxiom. An AntiLaw/AntiAxiom defined on a nonempty set is a law/axiom that is false for all set's elements, a NeutroAlgebra is an algebra that has at least one NeutroOperation or one NeutroAxiom (axiom that is true for some elements, indeterminate for other elements, and false for other elements), and no Anti-Operation or AntiAxiom while an AntiAlgebra is an algebra endowed with at least one AntiOperation or at least one AntiAxiom.

Theorem 2.1: (Agboola, 2020b) Let U be a nonempty finite or infinite universe of discourse and let S be a finite or infinite subset of U. If n classical operations (laws and axioms) are defined on S where $n \geq 1$, then there will be $(2^n - 1)$ NeutroAlgebras and $(3^n - 2^n)$ AntiAlgebras.

Definition 2.2: [Classical group] (Gilbert & Gilbert, 2015) Let G be a nonempty set and let $*: G \times G \to G$ be a binary operation on G. The couple $(G, *)$ is called a classical group if the following conditions hold:

(G1) $x * y \in G \; \forall x, y \in G$ [closure law].
(G2) $x * (y * z) = (x * y) * z \; \forall x, y, z \in G$ [axiom of associativity].

(G3) There exists $e \in G$ such that $x * e = e * x = x\ \forall x \in G$ [axiom of existence of neutral element].

(G4) There exists $y \in G$ such that $x * y = y * x = e\ \forall x \in G$ [axiom of existence of inverse element] where e is the neutral element of G. If in addition $\forall x, y \in G$, we have

(G5) $x * y = y * x$, then $(G, *)$ is called an abelian group.

Definition 2.3: [NeutroSophication of the law and axioms of the classical group] (Agboola, 2020a)

(NG1) There exist some duplets $(x, y),(u, v),(p, q), \in G$ such that $x* y \in G$ (inner-defined with degree of truth T) and [$u * v$ = indeterminate (with degree of indeterminacy I) or $p * q \notin G$ (outer-defined/falsehood with degree of falsehood F)] [NeutroClosureLaw].

(NG2) There exist some triplets

$(x, y, z),(p, q, r),(u, v, w) \in G$

such that

$x * (y * z) = (x * y) * z$

(inner-defined with degree of truth T) and

$[[p * (q * r)]$or $[(p * q) * r]$ = indeterminate

(with degree of indeterminacy I) or

$u * (v * w) \neq (u * v) * w$

(outer-defined/falsehood with degree of falsehood F)] [NeutroAxiom of associativity (NeutroAssociativity)].

(NG3) There exists an element $e \in G$ such that $x * e = e * x = x$ (inner-defined with degree of truth T) and

$[[x * e]$or$[e * x]$ = indeterminate

(with degree of indeterminacy I) or $x * e \neq x \neq e * x$ (outer-defined/falsehood with degree of falsehood F)] for at least one $x \in G$ [NeutroAxiom of existence of neutral element (NeutroNeutralElement)].

(NG4) There exists an element $u \in G$ such that $x * u = u * x = e$ (inner-defined with degree of truth T) and

$$[[x * u]\text{or}[u * x)] = \text{indeterminate}$$

(with degree of indeterminacy I) or

$$x * u \neq e \neq u * x$$

(outer-defined/falsehood with degre of falsehood F)] for at least one $x \in G$ [NeutroAxiom of existence of inverse element (NeutroInverseElement)] where e is a NeutroNeutralElement in G.

(NG5) There exist some duplets

$$(x, y),(u, v),(p, q) \in G$$

such that $x*y = y*x$ (inner-defined with degree of truth T) and [[u * v]or[v * u] = indeterminate (with degree of indeterminacy I) or $p * q \neq q * p$ (outer-defined/falsehood with degree of falsehood F)] [NeutroAxiom of commutativity (NeutroCommutativity)].

Definition 2.4: [NeutroGroup] (Agboola, 2020a)

A NeutroGroup NG is an alternative to the classical group G that has at least one NeutroLaw or at least one of {NG1, NG2, NG3, NG4} with no AntiLaw or AntiAxiom.

Definition 2: [NeutroAbelianGroup] (Agboola, 2020a)

A NeutroAbelianGroup NG is an alternative to the classical abelian group G that has at least one NeutroLaw or at least one of {NG1, NG2, NG3, NG4} and NG5 with no AntiLaw or AntiAxiom.

Proposition 2.6: (Agboola, 2020a) Let (G, *) be a finite or infinite classical non abelian group. Then:

1. there are 15 types of NeutroNonAbelianGroups,
2. there are 31 types of NeutroAbelianGroups.

Example 2.7: (Agboola, 2020a) Let

NG = N = {1, 2, 3, 4 · · ·, }.

Then (NG, .) is a finite NeutroGroup where "." is the binary operation of ordinary multiplication.

Example 2.8: (Agboola, 2020a) Let

U = {a, b, c, d, e, f}

be a universe of discourse and let

NG = {e, a, b, c}

be a subset of U. Let $*$ be a binary operation defined on NG as shown in the Cayley table (Table 1).

Table 1.

*	e	a	b	c
e	e	a	b	c
a	a	b	a	b
b	b	c	f	c
c	c	d	c	e

Then (NG, $*$) is a finite NeutroAbelianGroup.

NEUTROSEMIGROUPS

Definition 3.1: [Classical Semigroup]Let S be a nonempty set and let

$*: S \times S \to S$

be a binary operation on S. The couple (S, $*$) is called a classical semigroup if the following conditions hold:

(S1) $x * y \in S \ \forall x, y \in S$ [closure law].

(S2) $x * (y * z) = (x * y) * z \ \forall x, y, z \in S$ [associative axiom].

Definition 3.2: [NeutroSophication of the law and axiom of the classical semigroup]

(NS1) There exist some duplets

$$(x, y),(u, v),(p, q), \in S$$

such that $x * y \in S$ (inner-defined with degree of truth T) and [$u * v$ = indeterminate (with degree of indeterminacy I) or $p * q \notin S$ (outer-defined/falsehood with degree of falsehood F)] [NeutroClosureLaw].

(NS2) There exist some triplets

$$(x, y, z),(p, q, r),(u, v, w) \in S$$

such that

$$x * (y * z) = (x * y) * z$$

(inner-defined with degree of truth T) and [[$p * (q * r)$]or [$(p * q) * r$] = indeterminate (with degree of indeterminacy I) or $u * (v * w) \neq (u * v) * w$ (outer-defined/falsehood with degree of falsehood F)] [NeutroAxiom of associativity (NeutroAssociativity)].

Definition 3.3: [NeutroSemigroup]A NeutroSemigroup NS is an alternative to the classical semigroup S that has at least one of {NS1, NS2} with no AntiLaw or AntiAxiom.

Proposition 3.4: Let $(S, *)$ be a classical semigroup. Then there are 3 types of NeutroSemigroups.

Proof: Since there is only 1 law and 1 axiom in S it follows that $n = 1 + 1 = 2$ and from Theorem 2.1, we have number of types of NeutroSemigroups = 22 − 1 = 4 − 1 = 3.

Example 3.5: Let $NS = N \cup \{0\}$. Then $(NS, \div)$ is a NeutroSemigroup. To see this, let $0, 1, 2 \in NS$. Then

$1 \div 0$ = indeterminate,

$1 \div 2 = 1/2$ (outer-defined),

Table 2.

*	x	y	z
x	x	u	y
y	u or v	?	x
z	y	?	w

2 ÷ 1 = 2 (inner-defined),

2 ÷ 2 = 1 (inner-defined),

2 ÷ (1 ÷ 0) = indeterminate, (2 ÷ 1) ÷ 0 = indeterminate,

2 ÷ (2 ÷ 2) = 2 but (2 ÷ 2) ÷ 2 = 1/2 (outer-defined),

1 ÷ (1 ÷ 1) = (1 ÷ 1) ÷ 1 = 1.

Example 3.6: Let U = {u, v, w, x, y, z} be a universe of discourse and let NS = {x, y, z} be a subset of U. Let ∗ be a binary operation defined on NS as shown in the Cayley table (Table 2).

Then (NS, ∗) is a NeutroSemigroup. To see this, consider the following:

y ∗ x = indeterminate,

z ∗ y = indeterminate,

x ∗ y = u (outer-defined),

x ∗ z = y (inner-defined),

z ∗ x = y (inner-defined),

x ∗ (z ∗ y) = indeterminate,(x ∗ z) ∗ y = indeterminate,

x ∗ (y ∗ z) = x but (x ∗ y) ∗ z = indeterminate,

x ∗ (x ∗ x) = (x ∗ x) ∗ x = x.

Proposition 3.7: Let (NS1, ∗) and (NS2, ∘) be two NeutroSemigroups and let NS1 × NS2 be the cartesian productuct of NS1 and NS2. For all (x, u),(y, v) ∈ NS1 × NS2,

let * be a binary operation defined on NS1 × NS2 by

(x, u) * (y, v) = (x ∗ y, u ∘ v).

Then (NS1 × NS2, ⊗) is a NeutroSemigroup.

Proof. Follows from Definition 3.3.

Example 3.8: Let (NS, ∗) be the NeutroSemigroup of Example 3.6. Then (NS × NS, *) is a Neu-troSemigroup.

Definition 3.9:

1. A NeutroSemigroup (NS, ∗) is a NeutroSemigroup with NeutroIdetity if there exists an element e ∈ NS such that x ∗ e = e ∗ x = x (inner-defined with degree of truth T) and [[u ∗ e]or[e ∗ u] = indeterminate (with degree of indeterminacy I) or v ∗ e ≠ v ≠ e ∗ v (outer-defined/falsehood with degree of falsehood F)].
2. A NeutroSemigroup (NS, ∗) is a NeutroSemigroup with adjoined NeutroIdetity if there exists an element e ∈ NS such that x ∗ e = e ∗ x = x (inner-defined with degree of truth T) and [[u ∗ e]or[e ∗ u] = indeterminate (with degree of indeterminacy I) or v ∗ e ≠ v ≠ e ∗ v (outer-defined/falsehood with degree of falsehood F)]. In this case we write NSe = NS ∪ {e}.

Definition 3.10: Let (NS, ∗) be a NeutroSemigroup.

1. An element z ∈ NS is called a NeutroZero element if x ∗ z = z ∗ x = z (inner-defined with degree of truth T) and [[u ∗ z]or[z ∗ u] = indeterminate (with degree of indeterminacy I) or v ∗ z ≠ z ≠ z ∗ v (outer-defined/falsehood with degree of falsehood F)] for at least one x ∈ NS.If a NeutroZero element is adjoined to NS, then NS ∪ {z} is called a NeutroSemigroup with adjoined NeutroZero.
2. An element x ∈ NS is called an idempotent element if x ∗ x = x. NS is called an idempotent NeutroSemigroup if all the elements of NS are idempotent.
3. Two elements x, y ∈ NS are said to be commutative if x∗y = y ∗x. NS is called a commutative NeutroSemigroup if all the elements of NS are commutative.

Example 3.11: Let U = {a, b, c, d, e, f} be a universe of discourse and let NS = {e, a, b, c} be a subset of U. Let ∗ be a binary operation defined on NS as shown in the Cayley table (Table 3).

It can be shown that (NS, ∗) is a NeutroSemigroup with NeutroIdentity element e. The elements e and b are idempotent elements.

Table 3.

*	e	a	b	c
e	e	a	b	a or c
a	a	?	a	b
b	b	c	b	c
c	d	e	c	e

Table 4.

*	a	b	c
a	a	b	c
b	b	?	?
c	c	c	c

Example 3.12: Let

U = {a, b, c, d, e, f}

be a universe of discourse and let NS = {a, b, c} be a subset of U. Let $*$ be a binary operation defined on NS as shown in the Cayley table (Table 4).

It can be shown that (NS, $*$) is a commutative NeutroSemigroup with NeutroIdentity. The elements a and c are idempotent elements.

Example 3.13: Let U = {a, b, c, d} be a universe of discourse and let

NS = {a, b, c} ∪ {e}.

Let $*$ be a binary operation defined on NS as shown in the Cayley table (Table 5).

Table 5.

*	e	b	c
e	e	b	c
b	b	?	a
c	c	c	c

It can be shown that (NS, $*$) is a commutative NeutroSemigroup with NeutroIdentity. The elements e and c are idempotent elements.

Definition 3.14: Let (NS, $*$) be a NeutroSemigroup. A relation τ on NS is called a NeutroEquiva-lenceRelation if the following conditions hold:

1. $x\tau x$ for at least a duplet $(x, x) \in NS$.
2. $x\tau y$ implies $y\tau x$ for at least a duplet $(x, y) \in NS$.
3. $x\tau y$ and $y\tau z$ implies $x\tau z$ for at least a triplet $(x, y, z) \in NS$.

Example 3.15: Let (NS, $*$) be the NeutroSemigroup of Example 3.6.

1. 1. Let τ be a relation on NS such that $x\tau y$ if and only if $x * y$ = inner-defined $\forall x, y \in NS$. Then

$\tau = \{(x, x),(x, z),(y, z),(z, x)\}$.

It follows from Definition 3.14 that τ is a NeutroEquivalenceRelation on NS.

2. 2. Let ρ be a relation on NS such that $x\rho y$ if and only if $x * y$ = outer-defined $\forall x, y \in NS$. Then

$\rho = \{(x, y),(z, z)\}$.

It follows from Definition 3.14 that τ is not a NeutroEquivalenceRelation on NS.

3. 3. Let μ be a relation on NS such that $x\mu y$ if and only if $x * y$ = indeterminate $\forall x, y \in NS$. Then

$\mu = \{(y, x),(y, y),(z, y)\}$.

It follows from Definition 3.14 that μ is not a NeutroEquivalenceRelation on NS.

Definition 3.16: Let τ be a NeutroEquivalenceRelation on the NeutroSemigroup (NS, $*$). τ is called a left(right) NeutroCongruence on NS if for at least a triplet

$(x, y, z) \in NS$, $y\tau z$

implies $(x * y)\tau (x * z)$ (respectively $(y * x)\tau (z * x)$). τ is called a NeutroCongruence on NS if it is both a left and a right NeutroCongruence on NS.

Example 3.17: Let τ be the NeutroEquivalenceRelation of Example 3.15. Since for the triplet $(x, x, x) \in$ NS we have

$$x\tau x \Rightarrow (x * x)\tau (x * x),$$

it follows that τ is a NeutroCongruence on NS.

Proposition 3.18: Let τ be a NeutroEquivalenceRelation on the NeutroSemigroup (NS, $*$). Then τ is a NeutroCongruence on NS if and only if for at least a quadruplet $(x, y, u, v) \in$ NS, $x\tau y$ and $u\tau v$ imply that $(x * u)\tau (y * v)$.

Proof: Suppose that τ is a NeutroCongruence on NS. Then there exits at least a quadruplet $(x, x, x, x) \in$ NS such that $x\tau x$ and $x\tau x$ will imply that $(x * x)\tau (x * x)$.

The converse is obvious.

Definition 3.19: Let τ be a NeutroCongruence on the NeutroSemigroup (NS, $*$) and let $[x]$ be the NeutroClass containing the element $x \in$ NS that is

$$[x] = \{y \in NS: y\tau x\}.$$

If NS/τ is the set of all NeutroClasses induced by τ, we define

$$NS/\tau = \{x\tau: x \in NS\}.$$

For $x\tau$, $y\tau \in$ NS/τ, let $*$ be a binary operation defined on NS/τ as

$$(x\tau) * (y\tau) = (x * y)\tau.$$

We will study the nature of the algebraic structure (NS/τ, $*$) in Example 3.20.

Example 3.20: Let τ be the NeutroCongruence of Example 3.17. We have the following:

$$[x] = \{x, z\}, [y] = \{\}, [z] = \{x, y\},$$

$$\therefore NS/\tau = \{[x], [z]\}.$$

It is noted that the NeutroClasses induced by τ do not partition NS since

$$[x]\cup[y]\cup[z] = \{x, y, z\} = NS$$

but $[x] \cap [z] = \{x\} \neq \emptyset$. Easy computations show that $(NS/\tau, *)$ is a NeutroSemigroup which we call NeutroQuotientSemigroup induced by τ .

Definition 3.21: Let (NS, $*$) be a NeutroSemigroup. A nonempty subset NI of NS is called a left(right) NeutroIdeal of NS if for all $x \in NS$ there is at least a $k \in NI$ such that

$$x*k \in NI(k*x \in NI).$$

NI is called a two-sided NeutroIdeal or simply a NeutroIdeal of NS if it is both a left and right NeutroIdeal of NS.

Example 3.22: Let NS = {x, y, z} be the NeutroSemigroup of Example 3.6. It can be shown that NI1 = {x, y}, NI2 = {x, z} and NI3 = {y, z} are NeutroIdeals of NS. It can also be shown that NI1 $\cap$ NI2 = {x} is a NeutroIdeal of NS but NI1 $\cap$ NI3 = {y} and NI2 $\cap$ NI3 = {z} are neither left(right) NeutroIdeals nor NeutroIdeals of NS. However,

$$NI1 \cup NI2 = NI1 \cup NI3 = NI2 \cup NI3 = \{x, y, z\}$$

are NeutroIdeals of NS.

Definition 3.23: Let NI be a NeutroIdeal of the NeutroSemigroup (NS, $*$). ρ is called a NeutroReesCongruence on NS relative to NI if ρ is a NeutroCongruence defined by

$$x\rho y \text{ if } x, y \in NI \text{ or } x = y \text{ for at least a duplet } (x, y) \in NS.$$

NS/ρ is called a NeutroReesQuotientSemigroup and it is denoted by NS/NI.

Example 3.24: Let NI1, NI2, NI3 be NeutroIdeals of Example 3.22. It can be shown that NS/NI1, NS/NI2, NS/NI3 are NeutroReesQuotientSemigroups.

Definition 3.25: Let (NS1, $*$) and (NS2, $\circ$) be two NeutroSemigroups. The mapping φ: NS1 $\rightarrow$ NS2 is called a NeutroSemigroupHomomorphism if for at least a duplet $(x, y) \in NS1$, we have

$$\varphi(x * y) = \varphi(x) \circ \varphi(y).$$

Example 3.26: (i) Let (NS $\times$ NS, $\otimes$) be the NeutroSemigroup of Example 3.8 and let φ: NS $\times$ NS $\rightarrow$ NS be a projection defined by

$$\varphi(x, y) = x \;\forall\; x, y \in NS.$$

Then φ is a NeutroSemigroupHomomorphism. To see this, consider

$\varphi((x, x) \otimes (x, z)) = \varphi(x, y) = x = \varphi(x, x) * \varphi(x, z)$.

But then,

$\varphi((x, y) \otimes (z, x)) = \varphi(y, u \text{ or } v) =?$ but $\varphi(x, y) * \varphi(z, x) = y$.

(ii) Let $(NS/\tau, \otimes)$ be the NeutroQuotientSemigroup of Example 3.20 and let ψ: $NS \rightarrow NS/\tau$ be a mapping defined by

$\psi(x) = x\tau \ \forall \ x \in NS$.

Then ψ is a NeutroSemigroupHomomorphism. To establish this, consider

$\psi(x * z) = y\tau = \psi(x) \otimes \psi(z)$.

But then,

$\psi(y * x) = \psi(u \text{ or } v) =?$ and $\psi(y) \otimes \psi(x) = (u \text{ or } v)\tau =?$.

CONCLUSION

In this paper, we have focused on the study of NeutroSemigroups. Several properties of NeutroSemi-groups and interesting examples were provided. More properties of NeutroSemigroups and applications will be presented in our future papers. Also, AntiSemigroups will be studied in our future papers.

REFERENCES

Agboola, A. A. A. (2020a). On Finite NeutroGroups of Type-NG [1,2,4]. *International Journal of Neutrosophic Science*, *10*(2), 84–95. doi:10.5281/zenodo.4006602

Agboola, A. A. A. (2020b). Introduction to NeutroRings. *International Journal of Neutrosophic Science*, *7*(2), 62–73. doi:10.5281/zenodo.3877121

Gilbert, L., & Gilbert, J. (2015). *Elements of Modern Algebra* (8th ed.). Cengage Learning.

Compilation of References

Agboola, A. A. A. (2020e). Introduction to AntiRings. *Neutrosophic Science and System*, *36*, 293–307.

Ibrahim, M. A., & Agboola, A. A. A. (2020b). Introduction to NeutroHyperGroups. *Neutrosophic Sets and Systems*, *38*, 15–32.

Smarandache, F., Rezaei, A., & Kim, H. S. (2020). A New Trend to Extensions of CI-algebras. *International Journal of Neutrosophic Science*, *5*(1), 8–15. doi:10.54216/IJNS.050101

Al-Tahan, M., Davvaz, B., Smarandache, F., & Anis, O. (2021b). Properties of Productional NeutroOrderedSemigroups. *Neutrosophic Sets and Systems*, *42*, 178–190.

Jun, Y. B., Song, S. Z., & Kim, S. J. (2019). Neutrosophic Quadruple BCI-Positive Implicative Ideals. *Mathematics*, *7*(5), 385. doi:10.3390/math7050385

Smarandache, F. (2020d). Extension of HyperGraph to n-SuperHyperGraph and to Plithogenic n-SuperHyperGraph, and Extension of HyperAlgebra to n-ary (Classical-/Neutro-/Anti-) HyperAlgebra. *Neutrosophic Sets and Systems*, *33*, 290–296. doi:10.5281/zenodo.3783103

Agboola, A. A. A. (2020c). On Finite NeutroGroups of Type-NG. *International Journal of Neutrosophic Science*, *10*(2), 84–95. doi:10.5281/zenodo.4277243

Davvaz, B., & Leoreanu-Fotea, V. (2007). *Hyperring Theory and Applications*. International Academic Press.

Jun, Y., Song, S. Z., Smarandache, F., & Bordbar, H. (2018). Neutrosophic quadruple BCK/BCI-algebras. *Axioms*, *7*(2), 41. doi:10.3390/axioms7020041

Agboola, A. A. A. (2020d). On Finite and Infinite NeutroRings of Type-NR. *International Journal of Neutrosophic Science*, *11*(2), 87–99. doi:10.5281/zenodo.4276366

Atanassov, K. (1989). More on intuitionistic fuzzy sets. *Fuzzy Sets and Systems*, *33*(1), 37–46. doi:10.1016/0165-0114(89)90215-7

Davvaz, B. (2012). Chemical Hyperalgebra: Dismutation Reactions. *MATCH Commun. Math. Comput.Chem.*, *67*, 55–63.

Li, Q., Ma, Y., Zhang, X., & Zhang, J. (2019). Neutrosophic Extended Triplet Group Based on Neutrosophic Quadruple Numbers. *Symmetry, 11*(5), 696. doi:10.3390ym11050696

Shortliffe, E. H. (1976). *Computer-Based Medical Consultations: MYCIN.* Elsevier.

Agboola, A. A. A. (2020e). Introduction to AntiGroups. *International Journal of Neutrosophic Science, 12*(2), 71-80. http://fs.unm.edu/IJNS/IntroductionAntiGroups.pdf

Al-Tahan, & Davvaz. (2020a). *On Some Properties of Neutrosophic Quadruple Hv-rings* (Vol. 36). Neutrosophic Sets and Systems.

Duda, R. O., Hart, P. E., & Nilsson, N. J. (1981). Subjective Bayesian methods for rule-based inference systems. In M. Kaufmann (Ed.), *Readings in artificial intelligence* (pp. 192–199). Elsevier. doi:10.1016/B978-0-934613-03-3.50017-9

Gonzalez, R. C., & Woods, R. E. (2002). *Digital Image Processing, seconded.* Prentice-Hall.

Al-Tahan, & Davvaz. (2020b). On neutrosophic quadruple Hv-groups and their properties. *Southeast Asian Bulletin of Mathematics, 44*, 613–625.

IbrahimM. A.AgboolaA. A. A. (2020). Introduction to NeutroHyperGroups. *Neutrosophic Sets and Systems, 38*, 15-32. http://fs.unm.edu/NSS/IntroductionToNeutroHyperGroups2.pdf doi:10.5281/zenodo.4300363

Ibrahim, M. A., & Agboola, A. A. A. (2020b). Introduction to Neutrosophic Hypernear-rings. *International Journal of Neutro-sophic Science, 10*(1), 9–22.

Zhang, M. (2010). *Novel approaches to image segmentation based on neutrosophic logic.* Academic Press.

Cheng, H. D., & Guo, Y. (2008). A new neutrosophic approach to image thresholding. *New Mathematics and Natural Computation, 4*(3), 291–308. doi:10.1142/S1793005708001082

MohammadzadehRezaei. (2020). On NeutroNilpotentGroups. *Neutrosophic Sets and Systems, 38*, 33-40. http://fs.unm.edu/NSS/OnNeutroNilpotentGroups3.pdf doi:10.5281/zenodo.4300370

Muhiuddin, G. (2019). Neutrosophic quadruple ideals in neutrosophic quadruple BCI- algebras. *Neutrosophic Sets and Systems, 25*, 161–173.

Sahin, M., & Kargın, A. (2019). Neutrosophic triplet groups based on set valued neutrosophic quadruple numbers. *Neutrosophic Set and Systems, 30*, 122-131.

Smarandache, F. (2020e). Structure, NeutroStructure, and AntiStructure in Science. *International Journal of Neutrosophic Science, 13*(1), 28-33. http://fs.unm.edu/IJNS/NeutroStructure.pdf

Zhang, M., Zhang, L., & Cheng, H. D. (2010, May). A Neutrosophic approach to image segmentation based on Watershed method. *Signal Processing, 90*(5), 1510–1517. doi:10.1016/j.sigpro.2009.10.021

Jiménez, D. S., Juan, A. V. M., Mara, E. R. U., & Hernández, N. B. (2021). NeutroAlgebra for the evaluation of barriers to migrants' access in Primary Health Care in Chile based on PROSPECTOR function. *Neutrosophic Sets and Systems, 39*, 1–9. doi:10.5281/zenodo.4444189

Sahin, M., & Kargın, A. (2019). Single valued neutrosophic quadruple graphs. *Asian Journal of Mathematics and Computer Research, 26*(4), 243–250.

Salama, A., & Samarandache, F. (2014). *Neutrosophic Crisp Set theory*. Educational Publisher Columbus.

Madeleine Al-Tahan, F. (2021). NeutroOrderedAlgebra: Applications to Semigroups. *Neutrosophic Sets and Systems, 39*, 133–147. doi:10.5281/zenodo.4444331

Rezaei, A., & Smarandache, F. (2020). On Neutro-BE-algebras and Anti-BE-algebras (revisited). *International Journal of Neutrosophic Science, 4*(1), 8–15.

Salama, A. A., Smarandache, F., & Eisa, M. (2014). Introduction to Image Processing via Neutrosophic Techniques. *Neutrosophic Sets and Systems, 5*, 59–64.

Smarandache, F. (2019). Introduction to NeutroAlgebraic Structures. In *Advances of Standard and Nonstandard Neutro-sophic Theories* (pp. 240–265). Pons Publishing House.

Agboola, A.A.A. (2020b). On Finite and Infinite NeutroRings of Type-NR [8,9]. *International Journal of Neutrosophic Science, 11*(2), 87-99.

Smarandache, F. (2019). Introduction to NeutroAlgebraic Structures and AntiAlgebraic Structures. In *Advances of Standard and Nonstandard Neutrosophic Theories*. Pons Publishing House. http://fs.unm.edu/AdvancesOfStandardAndNonstandard.pdf

Long, F., Zhang, H., & Feng, D. D. (2003). Fundamentals of content-based image retrieval. In *Multimedia Information Retrieval and Management Technological Fundamentals and Applications*. Springer-Verlag. doi:10.1007/978-3-662-05300-3_1

Smarandache, F. (2020b). NeutroAlgebra is a Generalization of Partial Algebra. *International Journal of Neutrosophic Science, 2*(1), 8-17.

Smarandache, F. (2021a). Universal NeutroAlgebra and Universal AntiAlgebra. In NeutroAlgebra Theory (Vol. 1). Educational Publ.

Al-Tahan, M. (2021). NeutroOrderedAlgebra: Theory and Examples. In *3rd International Workshop on Advanced Topics in Dynamical Systems*. University of Kufa.

Appel, O., Chiclana, F., Carter, J., & Fujita, H. (2017). Cross-ratio uninorms as an effective aggregation mechanism in Sentiment Analysis. *Knowledge-Based Systems, 124*, 16–22. doi:10.1016/j.knosys.2017.02.028

Kekre, H.B., & Sudeep, D. (2008a). Creating the Color Panoramic View using Medley of Grayscale and Color Partial Images. *WASET International Journal of Electrical Computer and System Engineering,* 2-3.

Smarandache, F. (2003). *A Unifying Field in Logics: Neutrosophic Logic, Neutrosophy, Neutrosophic Set, Neutrosophic Probability*. American Research Press.

Kekre, H. B., & Sudeep, D. (2008b). *Scaling Invariant Fusion of Image Pieces in Panorama Making and Novel Image Blending Technique. International Journal on Imaging, 1(A08).*

Pís, P., & Mesiar, R. (1989). Fuzzy model of inexact reasoning in medicine. *Computer Methods and Programs in Biomedicine*, *30*(1), 1–8. doi:10.1016/0169-2607(89)90117-X PMID:2582743

Smarandache, F. (2014). *Introduction to Neutrosophic Statistics*. Sitech and Education Publishing.

Smarandache, Rezaei, Agboola, Jun, Borzooei, Davvaz, Broumand Saeid, Akram, Hamidi, & Mirvakilii. (2021). On NeutroQuadrupleGroups. *51st Annual Mathematics Conference*.

Al-Tahan, M., Davvaz, B., Smarandache, F., & Anis, O. (2021). On Some NeutroHyperstructures. *Symmetry*, *13*(535), 1–12. doi:10.3390ym13040535

Kekre, H. B., & Sudeep, D. (2010). *Performance Comparison for Face Recognition using PCA, DCT &Walsh Transform of Row Mean and Column Mean. International Journal on Graphics, Vision and Image Processing.*

Smarandache, F. (1998). *Neutrosophic Probability, Set and Logic*. Am Res Press.

Sussner, P., & Schuster, T. (2018). Interval-valued fuzzy morphological associative memories: Some theoretical aspects and applications. *Information Sciences*, *438*, 127–144. doi:10.1016/j.ins.2018.01.042

Jarrah, K., Kyan, M., Krishnan, S., & Guan, L. (2006). Computational Intelligence Techniques and Their Applications in Content-Based Image Retrieval. *IEEE Int. Conf. on Multimedia & Expo*, 33-36.

Rezaei, A., Smarandache, F., & Mirvakili, S. (2021). Applications of (Neutro/Anti)sophications to Semihypergroups. *Journal of Mathematics*, 1–7. doi:10.1155/2021/6649349

Smarandache, F. (2015). Neutrosophic quadruple numbers, refined neutrosophic quadruple numbers, absorbance law, and the multiplication of neutrosophic quadruple numbers. *Neutrosophic Sets and Systems*, *10*, 96–98.

Kekre, H. B., & Sudeep, D. (2009). Improving the Performance of Image Retrieval using Partial Coefficients of Transformed Image. *International Journal of Information Retrieval, Serials Publications*, *2*(1).

Tsadiras, A. K., & Margaritis, K. G. (1998). The MYCIN certainty factor handling function as uninorm operator and its use as a threshold function in artificial neurons. *Fuzzy Sets and Systems*, *93*(3), 263–274. doi:10.1016/S0165-0114(96)00185-6

Vasantha Kandasamy & Smarandache. (2003). *Fuzzy Cognitive Maps and Neutrosophic Cognitive Maps*. Phoenix. http://fs.unm.edu/NCMs.pdf

Rehman, M., Iqbal, M., Sharif, M., & Raza, M. (2012). Content Based Image Retrieval: Survey. *World Applied Sciences Journal*, *19*(3), 404–412.

Smarandache, F. (2019). Introduction to NeutroAlgebraic Structures. In *Advances of Standard and Nonstandard Neutro- sophic Theories* (pp. 240–265). Pons Publishing House Brussels.

SmarandacheF. (2021b). NeutroGeometry & AntiGeometry are alternatives and generalizations of the Non-Euclidean Geometries. *Neutrosophic Sets and Systems, 46*, 456-477. http://fs.unm.edu/NSS/NeutroGeometryAntiGeometry31.pdf doi:10.5281/zenodo.5553552

Kr, S. B., & Mazumdar, B. (2010). Content Retrieval from X-RAY Images Using Color &Texture Features. *International Journal of Electronics Engineering*, *2*(1), 25–28.

Smarandache, F. (2020b). NeutroAlgebra is a Generalization of Partial Algebra. *International Journal of Neutrosophic Science, 2*(1), 8-17.

Laaksonen, J., Oja, E., Koskela, M., & Brandt, S. (2000). Analyzing low-level visual features using content-based image retrieval. *Proc Int'l Conf. Neural Information Processing*.

Vasantha Kandasamy, W. B. (2019). Neutrosophic Quadruple Vector Spaces and Their Properties. *Mathematics*, *7*, 758–765. doi:10.3390/math7080758

Adiga, & Deepak. (2008). A Universal Model for Content-Based Image Retrieval. *World Academy of Science, Engineering and Technology*, 46.

Agboola, A. A. A. (2020b). On Finite and In nite NeutroRings of Type-NR. *International Journal of Neutrosophic Science*, *11*(2), 87–99.

Agboola, A. A. A. (2020c). Introduction to NeutroGroups. *International Journal of Neutrosophic Science*, *6*(1), 41–47.

Smarandache, F. (2020a). NeutroAlgebra is a Generalization of Partial Algebra. *International Journal of Neutrosophic Science, 2*(1), 8-17.

Smarandache, F. (2019). Introduction to NeutroAlgebraic Structures and AntiAlgebraic Structures. In *Advances of Standard and Nonstandard Neutrosophic Theories* (pp. 240–265). Pons Publishing House Brussels.

Smarandache, F. (2020a). NeutroAlgebra is a Generalization of Partial Algebra. *International Journal of Neutrosophic Science*, *2*, 8–17. doi:10.54216/IJNS.020103

Mohammadzadeh, E., & Rezaei, A. (2020). On NeutroNilpotent Groups. *Neutrosophic Sets and Systems*, *38*, 33–40.

Wang, J. Z., Li, J., & Wiederhold, G. (2001, September). SIMPLIcity: Semantics-Sensitive Integrated Matching for Picture Libraries. *IEEE Transactions on Pattern Analysis and Machine Intelligence*, *23*(9), 947–963. doi:10.1109/34.955109

Birgale, L., Kokare, M., & Doye, D. (2006). Colour and Texture Features for Content Based Image Retrieval. *Proceedings of the International Conference on Computer Graphics, Imaging and Visualisation*. 10.1109/CGIV.2006.30

Kurian, J., & Karunakaran, V. (2012, October). A Survey on Image Classification Methods. *International Journal of Advanced Research in Electronics and Communication Engineering, 1*(4).

Ingle, D., & Bhatia, S. (2012). Content Based Image Retrieval using Combined Features. *International Journal of Computer Applications, 44*(17).

Danish, M., Rawat, R., & Sharma, R. (2013). A Survey: Content –Based Image Retrieval Based on Color, Texture, Shape & Neuro Fuzzy. *Int. Journal of Engineering Research and Applications, 3*(5), 839-844.

Sifuzzaman, M., Islam, M. R., & Ali, M. Z. (2009). Application of Wavelet Transform and Its Advantage Compared to Fourier Transform. *The Journal of Physiological Sciences; JPS, 13*, 121–134.

Saxena, K., Richaria, V., & Trivedi, V. (2012, September). A Survey on Content Based Image Retrieval using BDIP, BVLC and DCD. *Journal of Global Research in Computer Science, 3*(9), 38–41.

Banerjee, M., & Kundu, M. K. (2003). Content Based Image Retrieval With Fuzzy Geometrical Features. *The 12th IEEE International Conference on Fuzzy Systems*, 932-937. 10.1109/FUZZ.2003.1206556

Tseng, V. S., Su, J. H., & Huang, W. J. (2007). FCBIR: A Fuzzy Matching Technique for Content-Based Image Retrieval. In Theoretical Advances and Applications of Fuzzy Logic and Soft Computing. Springer Berlin Heidelberg.

Manjunath, B. S., Salembier, P., & Sikora, T. (2003). *Introduction to MPEG-7 – Multimedia Content Description Interface*. Wiley.

Agboola, A. A. A. (2020d). Introduction to NeutroRings. *International Journal of Neutrosophic Science, 7*(2), 62–73.

Agboola, A. A. A., & Akinleye, S. A. (2015). Neutrosophic Hypervector Spaces. *ROMAI Journal, 11*, 1–16.

Smarandache, F. (2020b). Introduction to NeutroAlgebraic Structures and AntiAlgebraic Structures (revisited). *Neutrosophic Sets and Systems, 31*, 1–16. doi:10.5281/zenodo.3638232

Smarandache, F. (2020b). Introduction to NeutroAlgebraic Structures and AntiAlgebraic Structures. *Neutrosophic Sets and Systems, 31*, 1–16.

Ji, R., Yao, H., & Liang, D. (2008). DRM: Dynamic Region Matching for Image Retrieval Using Probabilistic Fuzzy Matching and Boosting Feature Selection. *Springer Journal of Signal, Image and Video Processing, 2*(1), 59–71. doi:10.100711760-007-0037-0

Younes, A. A., Truck, I., & Akdag, H. (2007). Image Retrieval Using Fuzzy Representation of Colors. *Springer Journal of Soft Computing – A Fusion of Foundations, Methodologies and Applications, 11*(3), 287–298.

Wu, K., & Yap, K. H. (2007). Content-Based Image Retrieval Using Fuzzy Perceptual Feed-back. *Springer Journal of Multimedia Tools and Applications, 32*(3), 235–251. doi:10.100711042-006-0050-2

Li, J., Liang, H., Wang, L., & Zhang, J. (2007). The Medical Image Retrieval Based on the Fuzzy Feature. *International Conference on Mechatronics and Automation (ICMA)*, 1245–1250. 10.1109/ICMA.2007.4303727

Ajorloo, H., & Lakdashti, A. (2012). IRFUM: Image retrieval via fuzzy modeling. *Computer Information, 30*(5), 913–941.

Kang, S. J., Woo, C. H., & Hwang, H. S. (2000). Evolutionary Design of Fuzzy Rule Base for Nonlinear System Modeling and Control. *IEEE Transactions on Fuzzy Systems, 8*(1), 37–45. doi:10.1109/91.824766

Lakdashti, A., Moin, M. S., & Badie, K. (2008a). Reducing Feature Space Dimensionality for Image Retrieval. *The 3rd International Symposium on Communications, Control and Signal Processing (ISCCSP),* 305–308. 10.1109/ISCCSP.2008.4537239

Lakdashti, A., Moin, M. S., & Badie, K. (2008b). A Novel Semantic-Based Image Retrieval Method. *The 10th International Conference on Advanced Communication Technology (ICACT),* 17–20. 10.1109/ICACT.2008.4493928

Lakdashti, A., & Moin, M. S. (2007). A New Content-Based Image Retrieval Approach Based on Pattern Orientation Histogram. In *LNCS 4418* (pp. 587–595). Springer-Verlag. doi:10.1007/978-3-540-71457-6_54

Zhang, G., Ma, Z. M., & Cai, Z. P. (2007). Directed Filter for Dominant Direction Fuzzy Set in Content-based Image Retrieval. In *Proceedings of the 22th ACM Symposium on Applied Computing (SAC 2007)*. ACM Press. 10.1145/1244002.1244019

Agboola, A.A.A., Davvaz, B., & Smarandache, F. (2017). Neutrosophic quadruple algebraic hyperstructures. Ann. Fuzzy Math. Inform., 14, 29–42.

Agboola, A. A. A. (2014). On Neutrosophic Canonical Hypergroups and Neutrosophic Hyperrings. *Neutro-sophic Sets and Systems., 2*, 34–41.

Smarandache, F. (2020c). *Neutroalgebra is a generalization of partial algebra*. Infinite Study.

Smarandache, F. (1991). *A Unifying field in logic: Neutrosophic Logic, Neutrosophy, Neutrosophic Set, Neutrosophic Probability*. American Research press.

Smarandache, F. (2020c). Generalizations and Alternatives of Classical Algebraic Structures to NeutroAlgebraic Structures and AntiAlgebraic Structures, Journal of Fuzzy Extension and Applications (JFEA), J. Fuzzy. *Ext. Appl., 1*(2), 85–87. doi:10.22105/jfea.2020.248816.1008

Xiaoling, W., & Kanglin, X. (2003, April). Application of the Fuzzy Logic in Content-based Image Retrieval. *JCS&T*, *5*(1), 19–24.

Lin, H. C., Chiu, C. Y., & Yang, S. N. (2003). Finding textures by textual descriptions, visual examples and relevance feedback. *Pattern Recognition Letters*, *24*(14), 2255–2267. doi:10.1016/S0167-8655(03)00052-7

Tripathy, S. S., Shekhar, R., & Kumar, R. S. (2011). Texture retrieval system using intuitionistic fuzzy set theory. In *Devices and Communications (ICDeCom), International Conference*. IEEE. 10.1109/ICDECOM.2011.5738490

Albergsten, F. (2008). *Statistical texture measures computed from gray level co-occurrence matrices.* Image Processing Laboratory, Department of Informatics, University of Oslo.

Hu, K. M. (1962). Visual pattern recognition by moment invariants. *I.R.E. Transactions on Information Theory*, *8*(2), 179–187. doi:10.1109/TIT.1962.1057692

Gupta, D., & Choubey, S. (2013, March). Discrete wavelet transform for image processing. *International Journal of Emerging Technology and Advanced Engineering*, *4*(3), 598–602.

Sridevi, Y., & Babu, R. M. (2011). Content Based Image Retrieval Using Color and Texture Features. *International Journal of Computer Science and Information Technologies*, *2*(5), 2384–2389.

Kannan, R., & Andres, F. (2012). Content Based Image Retrieval Using Various Distance Metrics. *ICDEM*, 154–161.

Kokare, M., Chatterji, B. N., & Biswas, P. K. (2003). Comparison of Similarity Metrics for Texture Image Retrieval. *TENCON 2003. Conference on Convergent Technologies for the Asia-Pacific Region*, *2*, 571-575. 10.1109/TENCON.2003.1273228

Ramamurthy, B. (2012). CBMIR: Content Based Medical Image Retrieval System Using Texture and Intensity for Dental Images. In *Eco-friendly Computing and Communication Systems.* Springer Berlin Heidelberg.

Agboola, A. A A. (2013). *Introduction to Neutrosophic Hypergroups. ROMAI J.*, *9*(2), 1–10.

Agboola, A. A. A., Ibrahim, M. A., & Adeleke, E. O. (2020). Elementary Examination of NeutroAlgebras and AntiAlgebras viz-a-viz the Classical Number Systems. *International Journal of Neutrosophic Science*, *4*, 16–19. doi:10.54216/IJNS.040102

Kchang, S., & Hsu, A. (1992). Image Information System: Where Do We Go From Here? *IEEE Trans. on Knowledge and Data Engineering.* doi:10.1109/69.166986

Puviarasan, N., Bhavani, R., & Vasanthi, A. (2014, March). Image Retrieval Using Combination of Texture and Shape Features. *International Journal of Advanced Research in Computer and Communication Engineering*, *3*(3), 5873–5877.

Sudhakar, P., Prasad, P., & Edara, S. R. (2013). Content Based Image Retrieval Using Fuzzy Texton and Shearlet Transform. In *Industrial Engineering, Management Science and Applications* (pp. 791–80). Springer.

Kothyari, P., Dwivedi, S., & Mandoria, H. L. (2016, June). Content-Based Image Retrieval Using Statistical Feature and Shape Extraction. *IUP Journal of Information Technology*, *12*(3), 62–69.

Amin, Salama, El-Henawy, Mahfouz, & Gafar. (2021). Intelligent Neutrosophic Diagnostic System for Cardiotocography Data. Computational Intelligence and Neuroscience.

Yasser, El-Khalek, Twakol, Abo-Elsoud, Salama, & Khalifa. (n.d.). A hybrid Automated Intelligent COVID-19 Classification System Based on Neutrosophic Logic and Machine Learning Techniques Using Chest X-ray Images, Studies in Systems, Decision and Control. *Advances in Data Science and Intelligent Data Communication Technologies for COVID-19, 378*.

Alhasan, Salama, & Smarandache. (2021). Introduction to Neutrosophic Reliability Theory. *International Journal of Neutrosophic Science*, *15*(1), 52–61.

Yasser, I., & Twakol, A. (2020). COVID-X: Novel Health-Fog Framework Based on Neutrosophic Classifier for Confrontation Covid-19. *Neutrosophic Sets and Systems*, *35*, 1–21.

Salama, A. A. (2020). Study on the Suez Canal Port, Egypt. *Neutrosophic Sets and Systems*, *35*, 22–44.

Salama, A. A., Eisa, M., ElGhawalby, H., & Fawzy, A. E. (2019). A New Approach in Content-Based Image Retrieval Neutrosophic Domain. In C. Kahraman & İ. Otay (Eds.), *Fuzzy Multi-criteria Decision-Making Using Neutrosophic Sets. Studies in Fuzziness and Soft Computing* (Vol. 369, pp. 361–369). Springer. doi:10.1007/978-3-030-00045-5_14

Amin, B., & Gamall, M. (2019). Classifying Cardiotocography Data based on Rough Neural Network. *International Journal of Advanced Computer Science and Applications*, *10*(8), 352–356. doi:10.14569/IJACSA.2019.0100846

Agboola, A.A.A., Ibrahim, M.A., & Adeleke, E.O. (2020). Elementary Examination of NeutroAlgebras and AntiAlgebras viz-a-viz the Classical Number Systems. International Journal of Neutrosophic Science, 4(1), 16–19.

Agboola, A. A. A. (2020a). Introduction to AntiRings. *Neutrosophic Sets and Systems*, *36*, 293–307.

Agboola, A. A. A. (2020a). Introduction to NeutroGroups. *International Journal of Neutrosophic Science*, *6*, 41–47.

Chan, Y. K., & Chen, C. Y. (2004). Image Retrieval System Based on Color Complexity and Color Spatial Features. *Journal of Systems and Software*, *71*(1), 65–70. doi:10.1016/S0164-1212(02)00140-1

ElWahsh, H., Gamal, M., Salama, A., & El-Henawy, I. (2018a). A novel approach for classifying MANETs attacks with a neutrosophic intelligent system based on genetic algorithm. *Security and Communication Networks*, *2018*, 1–7. doi:10.1155/2018/5828517

ElWahsh, H., Gamal, M., Salama, A., & El-Henawy, I. (2018b). Intrusion detection system and neutrosophic theory for MANETs: A comparative study. *Neutrosophic Sets and Systems*, *23*, 16–22.

Salama, A., Smarandache, F., & ElGhawalby, H. (2018). Neutrosophic approach to grayscale images domain. *Neutrosophic Sets and Systems*, *21*, 13–19.

Salama, ElGhawalby, & Nasr. (2018). Retract neutrosophic crisp system for gray scale image. *Asian Journal of Mathematics and Computer Research*, *3*, 104–117.

Salama, Eisa, ElGhawalby, & Fawzy. (2018). Neutrosophic Image Retrieval with Hesitancy Degree. In *New Trends in Neutrosophic Theories and Applications*. Pons Editions.

El-Nakeeb, ElGhawalby, Salama, & El-Hafeez. (2017). Neutrosophic Crisp Mathematical Morphology. *Neutrosophic Sets and Systems, 16*, 57-69.

Salama, A. A. (2017, July). A Neutrosophic Image Retrieval Classifier. *International Journal of Computers and Applications*, *170*(9), 1–6. doi:10.5120/ijca2017914798

ElWahsh, H., Gamal, M., Salama, A. A., & El-Henawy, I. M. (2017). Modeling Neutrosophic Data by Self-Organizing Feature Map: MANETs Data Case Study. *Procida Computer*, *121*, 152–157. doi:10.1016/j.procs.2017.11.021

Agboola, A. A. A. (2020b). Introduction to NeutroRings. *International Journal of Neutrosophic Science*, *7*, 62–73.

Akinleye, S. A., Smarandache, F., & Agboola, A. A. A. (2016). On neutrosophic quadruple algebraic structures. *Neutrosophic Sets System.*, *12*, 122–126.

Hwang, K. H., Lee, H., & Choi, D. (2012). Medical Image Retrieval: Past and Present, Healthc. *Information Research*, *18*(1), 3–9. PMID:22509468

Gilbert, L., & Gilbert, J. (2015). *Elements of Modern Algebra* (8th ed.). Cengage Learning.

Ibrahim, M. A., & Agboola, A. A. A. (2020a). Introduction to NeutroVector Space I. *Neutrosophic Science and System*, *36*, 329–350.

Rezaei, A., & Smarandache, F. (2020). On Neutro-BE-algebras and Anti-BE-algebras. *International Journal of Neutrosophic Science*, *4*, 8–15.

Smarandache, F. (2003). *A Unifying Field in Logics, Neutrosophic Logic, Neutrosophy, Neutrosophic Set, Neutrosophic Probability and Statistics*. American Research Press.

Hamidi, M., & Smarandache, F. (2020). Neutro-BCK-Algebra. *International Journal of Neutrosophic Science*, *8*, 110–117. doi:10.54216/IJNS.080204

Ibrahim, M. A., Agboola, A. A. A., Adeleke, E. O., & Akinleye, S. A. (2020). On Neutrosophic Quadruple Hypervector Spaces. *International Journal of Neutrosophic Science*, *4*(1), 20–35. doi:10.54216/IJNS.040103

Abdul, A., & Davvaz, B. (2013). Introduction to neutrosophic hypergroups. *Romai J.*, *9*(2), 1–10.

Abobala, M. (2021). Semi Homomorphisms and Algebraic Relations between Strong Refined Neutrosophic Modules and Strong Neutrosophic Modules. *Neutrosophic Sets and Systems, 39*, 107–119.

Abobala, M. (2021). Semi Homomorphisms and Algebraic Relations Between Strong Refined Neutrosophic Modules and Strong Neutrosophic Modules. *Neutrosophic Sets and Systems, 39*, 107–119.

Abobala, M., & Hatip, A. (2020, July). AH-Substructures in Strong Refined Neutrosophic Modules. *International Journal of Neutrosophic Science, 18*, 110–116.

Abobala, M., Hatip, A., & Alhamido, R. K. (2019, October). A Contribution to Neutrosophic Groups. *International Journal of Neutrosophic Science, 14*, 67–76. doi:10.54216/IJNS.000202

Agboola, A. A. A. (2020a). Introduction to neutrogroups. *International Journal of Neutrosophic Science, 6*, 41–47.

Agboola, A. A. A. (2020b). Introduction to neutrorings. *International Journal of Neutrosophic Science, 7*, 62–73.

Agboola, A. A. A. (2020b). On finite neutrogroups of type-NG [1,2,4]. *International Journal of Neutrosophic Science, 10*(2), 84–95.

Agboola, A. A. A. (2020b). On Finite NeutroGroups of Type-NG [1,2,4]. *International Journal of Neutrosophic Science, 10*(2), 84–95.

Agboola, A. A. A. (2020c). Introduction to anti groups. *International Journal of Neutrosophic Science, 12*(2), 71–80.

Agboola, A. A. A. (2021). Introduction to antigroups. *International Journal of Neutrosophic Science, 12*, 71–80.

Agboola, A. A. A., & Akinleye, S. A. (2015). Neutrosophic hypervector spaces. *ROMAI Journal, 11*, 1–16.

Agboola, A. A. A., & Davvaz, B. (2013). Introduction to neutrosophic hypergroups. *ROMAI Journal, 9*(2), 1–10.

Agboola, A. A. A., & Davvaz, B. (2014). On neutrosophic canonical hypergroups and neutrosophic hyperrings. *Neutrosophic Sets and Systems, 2*, 34–41.

Agboola, A. A. A., & Ibrahim, M. A. (2020). Introduction to antirings. *Neutrosophic Sets and Systems, 36*, 293–307.

Agboola, A., Ibrahim, M., & Adeleke, E. (2020). Elementary examination of neutroalgebras and antialgebras viz-a-viz the classical number systems. *International Journal of Neutrosophic Science, 4*, 16–19.

Al Tahan, M., Hoskova-Mayerova, S., & Davvaz, B. (2019). Some results on (generalized) fuzzy multi-Hv-ideals of Hv-rings. *Symmetry, 11*(11), 1376. doi:10.3390ym11111376

Al Tahan, M., Hoskova-Mayerova, S., & Davvaz, B. (2020). Fuzzy multi-polygroups. *Journal of Intelligent & Fuzzy Systems, 38*(2), 2337–2345. doi:10.3233/JIFS-191262

Al-Tahan, M. (2021). NeutroOrderedAlgebra: Theory and Examples. In *International Workshop on Advanced Topics in Dynamical Systems*. University of Kufa.

Al-Tahan, M., Davvaz, B., Smarandache, F., & Anis, O. (2021). On Some NeutroHyperstructures. *Symmetry, 13*(535), 1–12.

Al-Tahan, M., Smarandache, F., & Davvaz, B. (2021). NeutroOrderedAlgebra: Applications to semigroups. *Neutrosophic Sets and System, 39*, 133–147.

Al-Tahan, M., Smarandache, F., & Davvaz, B. (2021). NeutroOrderedAlgebra: Applications to Semigroups. *Neutrosophic Sets and Systems, 39*, 133–147.

Anuradha. (2020). Neutrosophic Fuzzy Hierarchical Clustering for Dengue Analysis in Sri Lanka. *Neutrosophic Sets and Systems, 31*, 179-199.

Atanassov, K. (1986). Intuitionistic fuzzy sets. *Fuzzy Sets and Systems, 20*(1), 87–96.

Atanassov, K. T. (1986). Intuitionistic fuzzy sets. *Fuzzy Sets and Systems, 20*(1), 87–96. doi:10.1016/S0165-0114(86)80034-3

Bakbak, D., Uluçay, V., & Sahin, M. (2019). Neutrosophic Multigroups and Applications. *Mathematics, 7*(1), 95. doi:10.3390/math7010095

Bal, M., Shalla, M., & Olgun, N. (2018). Neutrosophic Triplet Cosets and Quotient Groups. *Symmetry, 10*(4), 126.

Barnett, J. H. (2011). *Applications of Boolean algebra: Claude Shannon and circuit design*. Available from the webpage http://www. cs. nmsu. edu/historical-projects

Basumatary, B. (2016). Towards forming the field of fuzzy closure with reference to fuzzy boundary. *Journal of Process Management. Nuclear Technology, 4*(1), 30–40.

Basumatary, B., & Broumi, S. (2020). Interval-Valued Triangular Neutrosophic Linear Programming Problem. *International Journal of Neutrosophic Science, 10*(2), 105–115.

Basumatary, B., Wary, N., Mwchahary, D. D., Brahma, A. K., Moshahary, J., Basumatary, U. R., & Basumatary, J. (2021). A Study on Some Properties of Neutrosophic Multi Topological Group. *Symmetry, 13*(9), 1689. doi:10.3390ym13091689

Belnap, N.D. (1977). A useful four valued logic. *Modern Uses of Multiple Valued Logic*, 9–37.

Bhattacharya, S. (2004). *A model to a Smarandache Geometry*. http://fs.unm.edu/ModelToSmarandacheGeometry.pdf

Birkhoff, G. D. (1932). A Set of Postulates for Plane Geometry (Based on Scale and Protractors). *Annals of Mathematics*, 33.

Blizard, W. (1989). Multiset theory. *Notre Dame Journal of Formal Logic, 30*, 36–66.

Bordignon, F., & Gomide, F. (2014). Uninorm based evolving neural networks and approximation capabilities. *Neurocomputing*, *127*, 13–20. doi:10.1016/j.neucom.2013.04.047

Broumi, S., Bakali, A., Talea, M., Smarandache, F., Uluçay, V., Sahin, S., & Pramanik, S. (2018). Neutrosophic sets: An overview. In F. Smarandache & S. Pramanik (Eds.), *New trends in neutrosophic theory and applications* (Vol. 2, pp. 403–434). Pons Editions.

Burmeister, P. (1982). Partial algebras-survey of a unifying approach towards a two-valued model theory for partial algebras. *Algebra Universalis*, *15*(1), 306–358. doi:10.1007/BF02483730

Burmeister, P. (1993). Partial algebras-An introductory survey. In I. G. Rosenberg & G. Sabidussi (Eds.), *Algebras and orders*. Springer. doi:10.1007/978-94-017-0697-1_1

Calvo, S. C. (n.d.). *The Shape of Space: The Beginning of Non-Education Geometry*. Retrieved from https://www.encyclopedia.com/science/encyclopedias-almanacs-transcripts-and-maps/shape-space-beginning-non-euclidean-geometry

Cetica, M. (2021). Fractals - A Non-Euclidean Geometry. *Marco Cetica*. Retrieved from https://marcocetica.com/posts/non_euclidean_geometry/

Chakraborty, A., Banik, B., Mondal, S. P., & Alam, S. (2020). Arithmetic and Geometric Operators of Pentagonal Neutrosophic Number and its Application in Mobile Communication Service Based MCGDM Problem. *Neutrosophic Sets and Systems*, *32*, 61–79.

Chang, C. L. (1968). Fuzzy Topological Space. *Journal of Mathematical Analysis and Applications*, *24*(1), 182–190. doi:10.1016/0022-247X(68)90057-7

Chatterjee, R., Majumdar, P., & Samanta, S. K. (2016). On some similarity measures and entropy on quadripartitioned single valued neutrosophic sets. *Journal of Intelligent & Fuzzy Systems*, *30*, 2475–2485.

Coker, D. (1997). An introduction to intuitionistic fuzzy topological spaces. *Fuzzy Sets and Systems*, *88*(1), 81–89. doi:10.1016/S0165-0114(96)00076-0

Corsini P. (2016). HX-groups and Hypergroups. *Analele Univer sitatii Ovidius Constanta*, *24*(3), 101-121.

Corsini, P. (1993). Prolegomena of Hypergroup Theory. Aviani.

Corsini, P. (1993). Prolegomena of Hypergroup Theory. Udine Aviani Editore.

Corsini, P. (2016). HX-Hypergroups associated with the direct product of some Z/nZ. *Journal of Algebraic Structures and Their Applications*, (3), 1–15.

Corsini, P. (2016). On Chinese hyperstructures Z/nZ. *Journal of Discrete Mathematical Science Cryptography*, (6), 133–137.

Corsini, P. (2017). Hypergroups associated with HX-groups. *Analele Universitaţii Ovidius Constanţa*, *25*(2), 49–64. doi:10.1515/auom-2017-0020

Corsini, P., & Leoreanu-Fotea, V. (2003). *Application of Hyperstructures Theory: Advances in Mathematics*. Kluwer Academic.

Coxeter, H.S.M. (1942). *Non-Euclidean Geometry*. University of Toronto Press.

Cristea, I., Novák, M., & Onasanya, B. (2021). Link between HX-Groups and Hypergroups. *Algebra Colloquium*, *28*(3), 441–452. doi:10.1142/S1005386721000341

Das, S., Das, R., & Granados, C. (2021). Topology on quadripartitioned neutrosophic sets. *Neutrosophic Sets and Systems*, *45*, 54–61.

De-Baets, B., & Fodor, J. (1999). Van Melle's combining function in MYCIN is a representable uninorm: An alternative proof. *Fuzzy Sets and Systems*, *104*(1), 133–136. doi:10.1016/S0165-0114(98)00265-6

Deli, I. (2016). Refined Neutrosophic Sets and Refined Neutrosophic Soft Sets: Theory and Applications. Handbook of Research on Generalized and Hybrid Set Structures and Applications for Soft Computing, 321-343.

Deli, I., Broumi, S., & Ali, M. (2014). Neutrosophic Soft Multi-Set Theory and Its Decision Making. *Neutrosophic Sets and Systems*, *5*, 65–76.

Deschrijver, G., & Kerre, E. E. (2004). Uninorms in L*-fuzzy set theory. *Fuzzy Sets and Systems*, *148*, 243–262. doi:10.1016/j.fss.2003.12.006

Devi, R., Dhavaseelan, R., & Jafari, S. (2017). On Separation Axioms in an Ordered Neutrosophic Bitopological Space. *Neutrosophic Sets and Systems*, *18*, 27–36.

Dey, A., & Pal, M. (2015). Multi-fuzzy complex numbers and multi-fuzzy complex sets. *International Journal of Fuzzy System Applications*, *4*(2), 15–27. doi:10.4018/IJFSA.2015040102

Ejaita, O. A., & Asagba, P. (2017). An Improved Framework for Diagnosing Confusable Diseases Using Neutrosophic Based Neural Network. *Neutrosophic Sets and Systems*, *16*, 28–34.

El-Hefenawy, N., Metwally, M. A., Ahmed, Z. M., & El-Henawy, I. M. (2016). A review on the applications of neutrosophic sets. *Journal of Computational and Theoretical Nanoscience*, *13*(1), 936–944. doi:10.1166/jctn.2016.4896

Fan, E., Hu, K., & Li, X. (2019, March). Review of neutrosophic-set-theory-based multiple-target tracking methods in uncertain situations. In *2019 IEEE International Conference on Artificial Intelligence and Computer Applications* (pp. 19-27). IEEE. 10.1109/ICAICA.2019.8873497

Fathi, S., ElGhawalby, H., & Salama, A. A. (2016). A Neutrosophic Graph Similarity Measures. *New Trends in Neutrosophic Theories and Applications, 1*.

Fodor, R., Yager, R., & Rybalov, A. (1997). Structure of uninorms. *International Journal of Uncertainty, Fuzziness and Knowledge-based Systems*, *5*(04), 411–427. doi:10.1142/S0218488597000312

Fuchs, L. (1963). *Partially Ordered Algebraic Systems*. Pergamon Press.

Gayen, S., Smarandache, F., Jha, S., Singh, M. K., Broumi, S., & Kumar, R. (2020). Soft Subring Theory Under Interval-valued Neutrosophic Environment. *Neutrosophic Sets and Systems, 36*, 193–219.

González-Caballero, E., Leyva-Vázquez, M., & Smarandache, F. (2021). On Neutrosophic Uninorms. *Neutrosophic Sets and Systems, 45*, 340–348.

González-Caballero, E., Smarandache, F., & Leyva-Vázquez, M. (2019). On Neutrosophic Offuninorms. *Symmetry, 11*(9), 1136. doi:10.3390ym11091136

González-Hidalgo, M., Massanet, S., Mir, A., & Ruiz-Aguilera, D. (2014). *A new edge detector based on uninorms*. Paper presented at the 15th International Conference on Information Processing and Management of Uncertainty in Knowledge-Based Systems. 10.1007/978-3-319-08855-6_19

Hailperin, T. (1981). Boole's algebra isn't Boolean algebra. *Mathematics Magazine, 54*, 173–184.

Hajék, P. (1985). Combining functions for certainty degrees in consulting systems. *International Journal of Man-Machine Studies, 22*(1), 59–76. doi:10.1016/S0020-7373(85)80077-8

Hanf, W. (1975). The Boolean algebra of logic. *Bulletin of the American Mathematical Society, 81*(3), 587–589. doi:10.1090/S0002-9904-1975-13747-5

Hart, P. E., Duda, L. O., & Einaudi, M. T. (1978). PROSPECTOR-A Computer-Based Consultation System for Mineral Exploration. *Mathematical Geology, 10*(5), 589–610. doi:10.1007/BF02461988

Hatip, A. (2020). *The Special Neutrosophic Functions. International Journal of Neutrosophic Science.*

Heidari, D., & Davvaz, B. (2011). On ordered hyperstructures. *Politehn. Univ. Bucharest Sci. Bull. Ser. A Appl. Mathematics and Physics, 73*(2), 85–96.

Hongxing L., Qinzi D., & Peizhuang W. (1985), *Hypergroup (I), Busefal*, (23). Academic Press.

Hongxing L., Qinzi D., & Peizhuang W. (1986), *Hypergroup (I), Busefal*, (25). Academic Press.

Hu, K., Ye, J., Fan, E., Shen, S., Huang, L., & Pi, J. (2017). A novel object tracking algorithm by fusing color and depth information based on single valued neutrosophic cross-entropy. *Journal of Intelligent & Fuzzy Systems, 32*(3), 1775–1786. doi:10.3233/JIFS-152381

Huntington, E. V. (1904). Sets of independent postulates for the algebra of logic. *Transactions of the American Mathematical Society, 5*(3), 288–309. doi:10.1090/S0002-9947-1904-1500675-4

Ibrahim, M. A., & Agboola, A. A. A. (2020). Introduction to NeutroHyperGroups. *Neutrosophic Sets and System, 38*, 15–32.

Ibrahim, M. A., & Agboola, A. A. A. (2020). Introduction to neutrohyperGroups. *Neutrosophic Sets and Systems, 38*, 15–32.

Ibrahim, M. A., & Agboola, A. A. A. (2020). NeutroVector Spaces I. *Neutrosophic Sets and Systems, 36*, 328–351.

Ibrahim, M. A., & Agboola, A. A. A. (2020a). Introduction to neutro hypergroups. *Neutrosophic Sets and Systems*, *38*, 15–32.

Ibrahim, M. A., & Agboola, A. A. A. (2020b). Introduction to Neutrosophic Hypernear-rings. *International Journal of Neutrosophic Science*, *10*(1), 9–22.

Ibrahim, M. A., & Agboola, A. A. A. (2020c). Introduction to neutrovector space I. *Neutrosophic Sets and Systems*, *36*, 329–350.

Isaacs, I. M. (2008). *Finite Group Theory: Graduate Studies in Mathematics* (Vol. 92). American Mathematical Society.

James, A. W. (2005). *Hyperbolic Geometry* (2nd ed.). Springer.

Jiménez, D. S., Mayorga, J. A. V., Ubilla, M. E. R., & Hernández, N. B. (2021). Neutroalgebra for the evaluation of barriers to migrants' access in Primary Health Care in Chile based on PROSPECTOR function. *Neutrosophic Sets and Systems*, *39*, 1–9.

Kandasamy, V., Kandasamy, I., & Smarandache, F. (2020). NeutroAlgebra of Neutrosophic Triplets using $\left\{Z_n, \times\right\}$. *Neutrosophic Sets and Systems*, *38*, 510–523.

Kandasamy, V., Kandasamy, I., & Smarandache, F. (2020). NeutroAlgebra of Neutrosophic Triplets using {Zn, ×} NeutroAlgebra of Neutrosophic Triplets using {Zn, ×}. *Neutrosophic Sets and Systems*, *38*, 509–523.

Kandil, A., Nouth, A. A., & El-Sheikh, S. A. (1995). On fuzzy bitopological spaces. *Fuzzy Sets and Systems*, *74*(3), 353–363. doi:10.1016/0165-0114(94)00333-3

Kapoor, P., & Prem Kumar Singh, C. (2020). IT act Crime Pattern Analysis Using Regression and Correlation Matrix. *Proceedings of 8th International Conference on Reliability, Infocom Technologies and Optimization*, 1102-1106. doi: 10.1109/ICRITO48877.2020.9197835

Karaaslan, F. (2016). Correlation Coefficient between Possibility Neutrosophic Soft Sets. *Math. Sci. Lett.*, *5*(1), 71–74. doi:10.18576/msl/050109

Karaaslan, F. (2016). Correlation coefficients of single-valued neutrosophic refined soft sets and their applications in clustering analysis. *Neural Computing & Applications*, 1–13.

Kelly, J. C. (1963). Bitopological spaces. *Proceedings of the London Mathematical Society*, *3*(1), 71–89. doi:10.1112/plms3-13.1.71

Khan, M., Son, L. H., Ali, M., Chau, H. T. M., Na, N. T. N., & Smarandache, F. (2018). Systematic review of decision making algorithms in extended neutrosophic sets. *Symmetry*, *10*(8), 314. doi:10.3390ym10080314

Koundal, D., Gupta, S., & Singh, S. (2016). Applications of neutrosophic sets in medical image denoising and segmentation. In F. Smarandache & S. Pramanik (Eds.), *New trends in neutrosophic theory and application* (pp. 257–275). Pons Editions.

Kuciuk, L., & Antholy, M. (2005). An introduction to the Smarandache geometries. *JP Journal of Geometry & Topology*, *5*(1), 77–81. http://fs.unm.edu/IntrodSmGeom.pdf

Linfan, M. (2006). *Pseudo-Manifold geometries with applications*. Cornell University. https://arxiv.org/pdf/math/0610307

Lobachevsky, N. (2010). Pangeometry (A. Papadopoulos, Ed. & Trans.). European Mathematical Society.

Lotfy, M. M. (2015). Review of Recommender Systems Algorithms Utilized in Social Networks based e-Learning Systems & Neutrosophic System. *Neutrosophic Sets and Systems*, *8*, 32–41.

Majumdar, P., Pinaki, C., Sinha, K., & Kalyan, U. (2020). Bipolar quadripartitioned single-valued neutrosophic rough set. *Neutrosophic Sets and Systems*, *38*, 244–257.

Marík, V., & Kouba, Z. (1991). Some knowledge-acquisition methods for Prospector-like systems. *Knowledge-Based Systems*, *4*(4), 225–230. doi:10.1016/0950-7051(91)90052-4

Marty, F. (1934). Sur une Generalisation de la Notion de Groupe. *Proceedings of the 8th Congress Mathematiciens Scandinaves*, 45-49.

Marty, F. (1934). Sur une generalization de la notion de groupe. *Proceedings of the 8th Congress on Mathmatics, Scandinaves*, 45–49.

Miyamoto, S. (2001). Fuzzy Multisets and Their Generalizations. In *Multiset Processing* (pp. 225–235). Springer. doi:10.1007/3-540-45523-X_11

Mohammadzadeh, E., & Rezaei, A. (2020). On neutronilpotent groups. *Neutrosophic Sets and Systems*, *38*, 33–40.

Mohanasundari, M., & Mohana, K. (2020). Quadripartitioned Single valued Neutrosophic Dombi Weighted Aggregation Operators for Multiple Attribute Decision Making. *Neutrosophic Sets and Systems*, *32*, 107–122.

Mohan, M., & Krishnaswamy, M. (2020). K-algebras on quadripartitioned single valued neutrosophic sets. *Journal of Fuzzy Extension and Applications*, *1*(4), 325–339.

Muller, D. E. (1954). Application of Boolean algebra to switching circuit design and to error detection. *Transactions of the IRE Professional Group on Electronic Computers*, *3*, 6-12. 10.1109/IREPGELC.1954.6499441

Muzaffar, A., Nafis, N. T., & Sohail, S. S. (2020). Neutrosophy logic and its classification: An overview. *Neutrosophic Sets and Systems*, *35*, 239–251.

Mwchahary, D. D., & Basumatary, B. (2020). A note on Neutrosophic Bitopological Space. *Neutrosophic Sets and Systems*, *33*, 134–144.

Nguyen, G. N., Son, L. H., Ashour, A. S., & Dey, N. (2019). A survey of the state-of-the-arts on neutrosophic sets in biomedical diagnoses. *International Journal of Machine Learning and Cybernetics*, *10*(1), 1–13. doi:10.100713042-017-0691-7

Olgun. (2020). *On Refined Neutrosophic R-module. International Journal of Neutrosophic Science*.

Onasanya, B. O., & Hoskova-Mayerova, S. (2018). Some Topological and Algebraic Properties of alpha-level Subsets Topology of a Fuzzy Subset. *An. St. Univ. Ovidius Constanta, 26*, 213–227.

Onasanya, B. O., & Hoskova-Mayerova, S. (2019). Multi-fuzzy group induced by multisets. *International Journal of Pure and Applied Mathematics, 41*, 597–604.

Pandey, L. K., Ojha, K. K., Singh, P. K., Singh, C. S., Dwivedi, S., & Bergey, E. A. (2016). Diatoms image database of India (DIDI): a research tool. *Environmental Technology & Innovation, 5*, 148-160. doi:10.1016/j.eti.2017.02.005

Peng, X., & Dai, J. (2020). A bibliometric analysis of neutrosophic set: Two decades review from 1998 to 2017. *Artificial Intelligence Review, 53*(1), 199–255. doi:10.100710462-018-9652-0

Popov, M. R. (1996). The Smarandache Non-Geometry. *Abstracts of Papers Presented to the American Mathematical Society Meetings, 17*(3), 595.

Pramanik, S. (2020). Rough neutrosophic set: An overview. In F. Smarandache & S. Broumi (Eds.), *Neutrosophic theories in communication, management and information technology* (pp. 275–311). Nova Science Publishers.

Pramanik, S., Mallick, R., & Dasgupta, A. (2018). Contributions of selected Indian researchers to multi-attribute decision making in neutrosophic environment. *Neutrosophic Sets and Systems, 20*, 108–131.

Pramanik, S., & Roy, T. K. (2014). Neutrosophic game theoretic approach to Indo-Pak conflict over Jammu-Kashmir. *Neutrosophic Sets and Systems, 2*, 82–101.

Radha, R., & Arul Mary, A. S. (2021). Quadripartitioned neutrosophic pythagorean lie subalgebra. *Journal of Fuzzy Extension and Application, 2*(3), 283-296.

Rajarajeswari, P., & Uma, N. (2014). Correlation Measure for Intuitionistic Fuzzy Multi Sets. *International Journal of Research in Engineering and Technology, 3*(1), 611–617. doi:10.15623/ijret.2014.0301104

Rajarajeswari, P., & Uma, N. (2014). Zhang and Fu's Similarity Measure on Intuitionistic Fuzzy Multi Sets, International Journal of Innovative Research in Science. *Engineering and Technology, 3*(5), 12309–12317.

Rajareega, S., Preethi, D., Vimala, J., Selvachandran, G., & Smarandache, F. (2020). Some Results on Single Valued Neutrosophic Hypergroup. *Neutrosophic Sets and Systems, 31*, 80–85.

Reichel, H. (1984). *Structural induction on partial algebras*. Akademie-Verlag.

Rezaei, A., & Smarandache, F. (2020). On neutro-BE-algebras and anti-BE-algebras (revisited). *International Journal of Neutrosophic Science, 4*(1), 8–15.

Rezaei, A., Smarandache, F., & Mirvakili, S. (2021). Applications of (Neutro/Anti)sophications to Semihypergroups. *Journal of Mathematics, 2021*, 1–7. https://doi.org/10.1155/2021/6649349

Roy, S., Lee, J. G., Pal, A., & Samanta, S. K. (2020). Similarity Measures of Quadripartitioned Single Valued Bipolar Neutrosophic Sets and Its Application in Multi-Criteria Decision Making Problems. *Symmetry, 12*(6), 1012.

Russell, B. (1897). *Introduction: An essay on the foundations of geometry.* Cambridge University Press.

Sahin, M., Deli, I., & Ulucay, V. (2016). Jaccard vector similarity measure of bipolar neutrosophic set based on multi-criteria decision making. *International conference on natural science and engineering (ICNASE'16).*

Sahin, M., Deli, I., & Ulucay, V. (2016). Jaccard Vector Similarity Measure of Bipolar Neutrosophic Set Based on Multi-Criteria Decision Making. *International Conference on Natural Science and Engineering (ICNASE'16).*

Sahin, M., Olgun, N., Ulucay, V., Kargin, A., & Smarandache, F. (2017). *A new similarity measure based on falsity value between single valued neutrosophic sets based on the centroid points of transformed single valued neutrosophic numbers with applications to pattern recognition.* Infinite Study.

Sahin, M., Deli, I., & Ulucay, V. (2017). Extension principle based on neutrosophic multi-sets and algebraic operations. *Journal of Mathematical Extension, 12*(1), 69–90.

Şahin, M., Kargın, A., & Yücel, M. (2021). Neutro-Topological Space and Anti-Topological Space. *NeutroAlgebra Theory, I,* 16.

Sahin, M., Olgun, N., Uluçay, V., Kargın, A., & Smarandache, F. (2017). A New Similarity Measure Based on Falsity Value between Single Valued Neutrosophic Sets Based on the Centroid Points of Transformed Single Valued Neutrosophic Numbers with Applications to Pattern Recognition. *Neutrosophic Sets and Systems, 15,* 31–48.

Sahin, M., Ulucay, V., & Menekse, M. (2018). Some New Operations of (α, β, γ) Interval Cut Set of Interval Valued Neutrosophic Sets. *Journal of Mathematical and Fundamental Sciences, 50*(2), 103–120. doi:10.5614/j.math.fund.sci.2018.50.2.1

Sahin, R. (2014). Neutrosophic Hierarchical Clustering Algoritms. *Neutrosophic Sets and Systems, 2,* 19–24.

Sala, F., Chami, I., Wolf, A., Gu, A., & Ré, C. (2019). *Into the Wild: Machine Learning In Non-Euclidean Spaces.* Stanford University. Retrieved from https://dawn.cs.stanford.edu/2019/10/10/noneuclidean/

Salama, A. A., & Alblowi, S. A. (2012). Neutrosophic Set and Neutrosophic Topological Spaces. *IOSR Journal of Mathematics, 3*(4), 31-35.

Salama, A. A., Smarandache, F., & Kroumov, V. (2014). Closed sets and Neutrosophic Continuous Functions. *Neutrosophic Sets and Systems, 4,* 4–8.

Sankari, H., & Abobala, M. (2020). AH-Homomorphisms in Neutrosophic Rings and Refined Neutrosophic Rings. *Neutrosophic Sets and Systems*, *38*, 524–536.

Sankari, H., & Abobala, M. (2020). n-Refined Neutrosophic Modules. *Neutrosophic Sets and Systems*, *36*, 1–11.

Sebastian, S., & Ramakrishnan, T. V. (2010). Multi-Fuzzy Sets. *International Mathematical Forum*, *5*(50), 2471–2476.

Sebastian, S., & Ramakrishnan, T. V. (2011). Multi-fuzzy sets: An extension of fuzzy sets. *Fuzzy Information and Engineering*, *3*(1), 35–43. doi:10.100712543-011-0064-y

Shahzadi, G., Akram, M., & Saeid, A. B. (2017). An Application of Single-Valued Neutrosophic Sets in Medical Diagnosis. *Neutrosophic Sets and Systems*, *18*, 80–88.

Shannon, C. E. (1949). The synthesis of two-terminal switching circuits. *The Bell System Technical Journal*, *28*(1), 59–98. doi:10.1002/j.1538-7305.1949.tb03624.x

Shinoj & John. (2012). Intuitionistic fuzzy multisets and its application in medical diagnosis. *World Academy of Science, Engineering and Technology*, *6*, 1–28.

Silva-Jiménez, D., Valenzuela-Mayorga, J. A., Roja-Ubilla, M. E., & Batista-Hernández, N. (2021). NeutroAlgebra for the evaluation of barriers to migrants' access in Primary Health Care in Chile based on PROSPECTOR function. *Neutrosophic Sets and Systems*, *39*, 1–9.

Singh, P. K. (2020). Multi–granular based n–valued neutrosophic contexts analysis. *Granular Computing*, *5*(3), 287-301. doi:10.1007/s41066-019-00160-y

Singh, P. K. (2021a). Turiyam set a fourth dimension data representation. *Zeitschrift für Angewandte Mathematik und Physik*, *9*(7), 1821–1828. doi:10.4236/jamp.2021.97116

Singh, P. K. (2021b). Fourth dimension data representation and its analysis using Turiyam Context. *Journal of Computer and Communications*, *9*(6), 222–229. doi:10.4236/jcc.2021.96014

Singh, P. K. (2021c). Single-valued Plithogenic graph for handling multi-valued attribute data its context. *International Journal of Neutrosophic Sciences*, *15*(2), 98–112. doi:10.5281/zenodo.5482171

Singh, P. K. (2021d). Complex multi-–fuzzy context analysis at different granulation. *Granular Computing*, *6*(1), 191–206. doi:10.100741066-019-00180-8

Singh, P. K. (2021e). Bipolarity in multi–way fuzzy context and its analysis using m–way granulation. *Granular Computing*, *2021*. Advance online publication. doi:10.100741066-021-00277-z

Singh, P. K. (2021f). AntiGeometry and NeutroGeometry Characterization of Non-Euclidean Data Sets. *Journal of Neutrosophic and Fuzzy Systems*, *1*(1), 24–33.

Singh, P. K. (2022a). Data with Non-Euclidean Geometry and its Characterization. *Journal of Artificial Intelligence and Technology*, *2*(1), 3–8. doi:10.37965/jait.2021.12001

Singh, P. K. (2022b). Complex Plithogenic Set. *International Journal of Neutrosophic Sciences, 18*(1), 57–72.

Sinha, K., & Majumdar, P. (2020). Bipolar quadripartitioned single-valued neutrosophic sets. *Proyecciones (Antofagasta), 39*(6), 1597–1614.

Smarandache, F. (1998). Neutrosophy. Neutrosophic Probability, Set, and Logic, 105.

Smarandache, F. (1999). *A unifying field in logics. neutrosophy: Neutrosophic probability, set and logic*. Academic Press.

Smarandache, F. (2002). *Neutrosophy, a new Branch of Philosophy*. Infinite Study.

Smarandache, F. (2017b). *Plithogeny, plithogenic set, logic, probability, and statistics.* Brussels: Pons ASBL.

Smarandache, F. (2019). Introduction to NeutroAlgebraic Structures and AntiAlgebraic Structures. In *Advances of Standard and Nonstandard Neutrosophic Theories.* Pons Publishing House. http://fs.unm.edu/AdvancesOfStandardAndNonstandard.pdf

Smarandache, F. (2020). NeutroAlgebra is a Generalization of Partial Algebra. *International Journal of Neutrosophic Science, 2*(1), 8-17.

Smarandache, F. (2020b). NeutroAlgebra is a Generalization of Partial Algebra. *International Journal of Neutrosophic Science, 2*(1), 8-17.

Smarandache, F. (2021). NeutroGeometry&AntiGeometryare alternatives and generalizations of the Non-Euclidean Geometries. *Neutrosophic Sets and Systems, 46*, 456-476. http://fs.unm.edu/NSS/NeutroGeometryAntiGeometry31.pdf

Smarandache, F., Rezaei, A., & Kim, H. S. (2020). *A New Trend to Extensions of CI-algebras.* Academic Press.

Smarandache, F., Rezaei, A., Agboola, A. A. A., Jun, Y. B., Borzooei, R. A., Davvaz, Borumand Saeid, A., Akram, M., Hamidi, M., & Mirvakili, S. (2021). *On Neutro Quadruple Groups.* Infinite Study.

Smarandache, F. (1998). *A Unifying Field in Logics Neutrosophy: Neutrosophic Probability, Set and Logic*. American Research Press.

Smarandache, F. (1998). *A unifying field in logics. Neutrosophy: neutrosophic probability, set and logic*. American Research Press.

Smarandache, F. (1998). *Neutrosophy: neutrosophic probability, set, and logic: analytic synthesis & synthetic analysis*. American Research Press.

Smarandache, F. (2002). Neutrosophy, a new branch of philosophy, multiple valued logic. *International Journal, 8*, 297–384.

Smarandache, F. (2005). *Unifying Field in Logics: Neutrosophic Logic. Neutrosophy Set, Neutrosophic Probability and Statistics*. American Research Press.

Smarandache, F. (2005a). Neutrosophic set-a generalization of the intuitionistic fuzzy set. *International Journal of Pure and Applied Mathematics*, *24*, 287–297.

Smarandache, F. (2013). n-Valued refined neutrosophic logic and its applications in Physics. *Progress in Physics*, *4*, 143–146.

Smarandache, F. (2013). n-Valued Refined Neutrosophic Logic and Its Applications in Physics. *Progress in Physics*, *4*, 143–146.

Smarandache, F. (2015b). Neutrosophic quadruple numbers, refined neutrosophic quadruple numbers, absorbance law, and the multiplication of neutrosophic quadruple numbers. *Neutrosophic Sets and Systems*, *10*, 96–98.

Smarandache, F. (2017a). *Neutrosophic perspectives: triplets, duplets, multisets, hybrid operators, modal logic, hedge algebras, and applications* (2nd ed.). Pons Publishing House.

Smarandache, F. (2019). *Advances of standard and nonstandard neutrosophic theories*. Pons Publishing House.

Smarandache, F. (2019). *Introduction to NeutroAlgebraic Structures and AntiAlgebraic Structures. In Advances of Standard and Nonstandard Neutrosophic Theories*. Pons Publishing House Brussels.

Smarandache, F. (2019). Introduction to NeutroAlgebraic Structures. In *Advances of Standard and Nonstandard Neutrosophic Theories* (pp. 240–265). Pons Publishing House.

Smarandache, F. (2019a). Introduction to neutroalgebraic structures and antialgebraic structures (revisited). *Neutrosophic Sets and Systems*, *31*, 1–16.

Smarandache, F. (2019b). *Introduction to neutroalgebraic structures, in advances of standard and nonstandard neutrosophic theories*. Pons Publishing House.

Smarandache, F. (2020). Introduction to NeutroAlgebraic structures and AntiAlgebraic structures (revisited). *Neutrosophic Sets and System*, *31*, 1–16.

Smarandache, F. (2020). Neutroalgebra is a generalization of partial algebra. *International Journal of Neutrosophic Science*, *2*(1), 8–17.

Smarandache, F. (2020). NeutroAlgebra is a generalization of partial algebra. *International Journal of Neutrosophic Science*, *2*, 8–17.

Smarandache, F. (2020a). Generalizations and Alternatives of Classical Algebraic Structures to NeutroAlgebraic Structures and AntiAlgebraic Structures. *Journal of Fuzzy Extension and Applications*, *1*(2), 85–87.

Smarandache, F. (2020a). Introduction to neutro algebraic structures and anti algebraic structures (revisited). *Neutrosophic Sets and Systems*, *31*, 1–16.

Smarandache, F. (2020a). Introduction to NeutroAlgebraic Structures and AntiAlgebraic Structures (revisited). *Neutrosophic Sets and Systems*, *31*, 1–16.

Smarandache, F. (2020b). Neutro algebra is a generalization of partial algebra. *International Journal of Neutrosophic Science*, *2*(1), 8–17.

Smarandache, F. (2020d). Structure, NeutroStructure, and AntiStructure in Science. *International Journal of Neutrosophic Science*, *13*(1), 28–33.

Smarandache, F. (2021). Plithogenic probability & statistics are generalizations of multivariate probability & statistics. *Neutrosophic Sets and Systems*, *43*, 280–289.

Smarandache, F., & Abobala, M. (2020). n-Refined Neutrosophic Vector Spaces. *International Journal of Neutrosophic Science*, *7*, 47–54. doi:10.54216/IJNS.070104

Smarandache, F., & Hamidi, M. (2020). Neutro-BCK-algebra. *International Journal of Neutrosophic Science*, *8*, 110–117.

Smarandache, F., & Pramanik, S. (Eds.). (2016). *New trends in neutrosophic theory and applications*. Pons Editions.

Smarandache, F., & Rezaei, A. (2020). On neutro-BE-algebras and anti-BE-algebras. *International Journal of Neutrosophic Science*, *4*, 8–15.

Sonea A. (2019). HX-Groups associated to the dihedral group D_n. *Journal of Multiple Valued Logic and Soft Computing*.

Sumathi, I. R., & Arockiarani, I. (2015). Fuzzy Neutrosophic Groups. *Advanced in Fuzzy Mathematics*, *10*(2), 117–122.

Sumathi, I. R., & Arockiarani, I. (2016). Topological Group Structure of Neutrosophic set. *Journal of Advanced Studies in Topology*, *7*(1), 12–20. doi:10.20454/jast.2016.1003

Tong, F. (2019). What is Geometric Deep Learning? *Medium*. https://flawnsontong.medium.com/what-is-geometric-deep-learning-b2adb662d91d

Ulucay, V., Sahin, M., & Hassan, N. (2015). Neutrosophic soft expert sets. *Applied Mathematics*, *6*(01), 116–127. doi:10.4236/am.2015.61012

Vougiouklis, T. (1991). The fundamental relation in hyperrings. The general hyperfield. *Proc. of the 4th Int. in Congress on Algebraic Hyperstructures and Appl.,* 203-211.

Vougiouklis, T. (1994). *Hyperstructures and Their Representations*. Hadronic Press, Inc.

Wang, H., Smarandache, F.Y., Zhang, Q., & Sunderraman, R. (2010). Single valued neutrosophic sets. *Multi Space and Multi Structure, 4*, 410-413.

Wang, H., Smarandache, F., Zhang, Y., & Sunderraman, R. (2010). Single valued neutrosophic sets. *Review of the Air Force Academy*, *1*, 10–14.

Wikipedia. (2022a). Non-Euclidean Geometry. In *Wikipedia*. Retrieved from https://en.wikipedia.org/wiki/Non-Euclidean_geometry

Wikipedia. (2022b). Anisotrophy. In *Wikipedia*. https://en.wikipedia.org/wiki/Anisotropy

Wikipedia. (2022c). Mandelbrot Set. In *Wikipedia*. https://en.wikipedia.org/wiki/Mandelbrot_set

Wikipedia. (2022d). Citation Analysis. In *Wikipedia*. https://en.wikipedia.org/wiki/Citation_analysis

Yager, R. R. (1986). On the theory of bags. *International Journal of General Systems*, *13*(1), 23–37. doi:10.1080/03081078608934952

Yager, R. R. (2001). Uninorms in fuzzy system modeling. *Fuzzy Sets and Systems*, *122*(1), 167–175. doi:10.1016/S0165-0114(00)00027-0

Yager, R. R., & Kreinovich, V. (2003). Universal approximation theorem for uninorm-based fuzzy systems modeling. *Fuzzy Sets and Systems*, *140*(2), 331–339. doi:10.1016/S0165-0114(02)00521-3

Yager, R. R., & Rybalov, A. (1996). Uninorm aggregation operators. *Fuzzy Sets and Systems*, *80*(1), 111–120. doi:10.1016/0165-0114(95)00133-6

Yang, Y., Tan, X., & Meng, C. (2013). The multi-fuzzy soft set and its application in decision making. *Applied Mathematical Modelling*, *37*(7), 4915–4923. doi:10.1016/j.apm.2012.10.015

Ye, J. (2014). Single valued neutrosophic minimum spanning tree and its clustering method. *Journal of Intelligent Systems*, *23*(3), 311–324. doi:10.1515/jisys-2013-0075

Ye, J. (2017). Single-valued neutrosophic similarity measures based on cotangent function and their application in the fault diagnosis of steam turbine. *Soft Computing*, *21*(3), 817–825. doi:10.100700500-015-1818-y

Yuhua, F. (2013). Neutrosophic Examples in Physics. *Neutrosophic Sets and Systems*, *1*, 26–33.

Yuhua, F. (2015). Examples of Neutrosophic Probability in Physics. *Neutrosophic Sets and Systems*, *7*, 32–33.

Zadeh, L. A. (1965). Fuzzy sets. *Information and Control*, *8*(3), 338–353. doi:10.1016/S0019-9958(65)90241-X

Zhang, C., & Luo, X. (1999). Isomorphic Transformations of Uncertainties for Incorporating EMYCIN-Style and PROSPECTOR-Style Systems into a Distributed Expert System. *Journal of Computer Science and Technology*, *14*(4), 386–392. doi:10.1007/BF02948741

About the Contributors

Saba Al-Kaseasbeh, Vice Dean of Student Affairs at Tafila Technical University, is an assistant professor in the department of mathematics, Tafila technical university, 2016. She received her Ph.D. in Mathematics from North Dakota State University, ND, USA in 2014. She joined the department of mathematics at the Tafila Technical University (TTU), Jordan, in 2015. Her research is in the field of Commutative Algebra, Combinatorics, Algebraic graph theory, and NeutroAlgebra.

Sonea Andromeda Cristina graduated with a Ph.D. in Algebra in Summer 2021. In the same year, she started to teach at the University of Life Sciences, Lasi, Romania, as an associated assistant. Her research area is hypertsructure theory and its applications.

Osman Anis received his Master degree in Mathematics from Saint Joseph University. Currently, he is a senior lecturer at Lebanese International University (LIU) and Math program coordinator at LIU, Akkar campus. He is an active researcher in Neutrosophic science.

Noel Batista Hernández has a degree in law, from the University of Holguín, Cuba, Mater in Public Administration, from the University of Havana, Doctor in Pedagogical Sciences from the Central Institute of Pedagogical Sciences, Cuba. Researcher accredited by the Ministry of Higher Education, Sciences, Technologies and Innovation of Ecuador, Author of multiple scientific articles, director and researcher of several research projects related to public management, pedagogy, and law.

Rakhal Das is a research scholar in Mathematics. He is now pursuing research in the Department of Mathematics, Tripura University, Tripura. He has published several papers in neutrosophic topological space, neutrosophic multi criteria decision making, etc.

Suman Das is a research scholar in Mathematics. He is now pursuing research in the Department of Mathematics, Tripura University, Tripura. He has published several papers in neutrosophic topological space, neutrosophic bitoplogical space, neutrosophic soft topological space, neutrosophic multi criteria decision making, etc.

Somen Debnath has earned his Ph.D degree in Mathematics from Tripura University (A central university), Suryamaninagar, Agartala, Tripura 799022, India. At present, he is working as a Post Graduate Teacher(in Mathematics) at Umakanta Academy, Agartala-799001, Tripura, India. His research interests include fuzzy sets, rough sets, soft sets, hypersoft sets, neutrosophic sets, Pythagorean fuzzy sets, matrix representation of interval-valued intuitionistic fuzzy sets, game theory, multi-criteria decision making, etc. He has over 10 years of teaching and over 6 years of research experience. He has published 20+ research articles in national and international peer-reviewed journals so far.

Erick González Caballero received his PhD (2013) in technical science from the Technological University of Havana in Cuba, and BS degree (1998) in Mathematics from the University of Havana in Cuba. After graduating, he works as a scientist or teacher at various universities and research centers in Cuba. In 2011 he continued his teaching and scientific activities at the Technological University of Havana. He was a posdoctoral visitor (2016) at the University of Oviedo, Spain. His researchs include, decision-making under uncertainty, rough sets, fuzzy sets and their generalizations, image processing, expert systems, cooperative games, soft computing, among others. He has published in important journals like "Group Decision and Negotiation", "International Journal of Computational Intelligence Systems", "Journal of Intelligent & Fuzzy Systems", "International Journal of Fuzzy Systems", and "Neutrosophic Sets and Systems", also in Atlantis Press and Springer book chapters.

Carlos Granadis got his master's degree at Universidad del Atlántico in 2019, later in 2021 he started his doctorate in Mathematics at Universidad de Antioquia. He has published many papers associated to general topology, mathematical education, sequences spaces, decision making and neutrosophic theory.

Mariam Hariri is currently a Senior Instructor of Mathematics in the College of Natural and Health Sciences at Zayed University, United Arab Emirates. She holds a Master's degree in Mathematics from Beirut Arab University, Lebanon. She is a Fellow of the Higher Education Academy (Advance HE), Certified Apple Teacher, and a Certified Expert in Digital Teaching and Learning from Blackboard Academy. During her 14-year career in higher education, she contributed to the field in several capacities, such as math instructor, course coordinator, Math placement exam creator and facilitator, curriculum leader, Math program coordinator, mentor, and researcher. In the past, she served as the Chair of the Science Math and Technology Department and the Assistant Dean for Student Affairs in University College at Zayed University. Before coming to the UAE, she worked as a Teaching Assistant of Mathematics in the Faculty of Science and Faculty of Engineering at Beirut Arab University. Her current research interests include Abstract Algebra, Neutrosophy, innovative teaching strategies, and the use of technology in math education.

Ahmed Hatip received his B.Sc. from Aleppo University, Syria and M.Sc. from Faculty of science Gaziantep University, Turkey. Currently, Ahmed Ph. D student in Faculty of science Gaziantep University, Turkey. His current research interests are Neutrosophic, Decision Making, and Intelligence.

Ilanthenral Kandasamy is an Assistant Professor in the School of Computer Science and Engg, VIT, India.

Darjan Karabasevic is an Associate Professor of Management and Informatics and a Vice dean for Scientific Research at the Faculty of Applied Management, Economics and Finance, University Business Academy in Novi Sad. He obtained his degrees at all the levels of studies (B.Sc. appl. in Economics, B.Sc. in Economics, Academic Specialization in the Management of Business Information Systems and PhD.in Management and Business) at the Faculty of Management in Zajecar, John Naisbitt University Belgrade. He is a member of Neutrosophic Science International Association (NSIA), Head of NSIA branch of Serbia and Co-President. His current research is focused on the informatics, management, decision-making theory and neutrosophy.

Jeevan Khaklary is an Assistant Professor, Department of Mathematics, Central Institute of Technology, Kokrajhar.

Necati Olgun studied and received the Ph.D. degree from Hacettepe University in Ankara, Turkey in 2005. Olgun worked as a Research Assistant at University of Gaziantep from 1996-1999 and Hacettepe University from 1999-2005 with respectively. Since 2020, Olgun is working as an Professor at the University of Gaziantep. Olgun's main research interests include universal differential modules, differential operator rings, Kaehler modules, regular rings, symmetric derivations, exterior derivations, projective dimension, fuzzy groups, rings and modules. Olgun published more than 30 papers.

Surapati Pramanik has been working as a teacher educator since 2006 in Nandalal Ghosh B.T. College, India. He received a Ph. D. in Mathematics from Indian Institute of Engineering Science and Technology (IIEST), Shibpur and Ph. D. in Education, an M. Sc. and an M. Ed. from University of Kalyani. He acts as a Ph. D. supervisor in Mathematics for Jadavpur University and IIEST, Shibpur. His research focuses on fuzzy Multi Criteria Decision Making (MCDM), mathematics education, Soft computing and neutrosophic sets. His research earned outstanding paper awards several times in West Bengal State Science and Technology Congress in mathematics (2011) and Social Sciences (2010, 2013, 2019). He is an editorial board member of "Neutrosophic Sets and Systems", and "Current Chinese Science: Artificial Intelligence & Robotics". His publication includes 27 book chapters, 137 research papers, two editorial books. He acts as a reviewer for more than 50 international journals and did more than 200 reviews reputed international journals. He is a life member of thirteen scientific organizations including Calcutta Mathematical Society, and ISI, Kolkata. He works on topology, neutrosophic hybrid sets, fuzzy multi criteria decision making.

Rafael Rojas is a professor of logistics and supply chain management at the University of the Andes in Bogotá, Colombia. He got his MSc in Industrial Engineering from the Universidad Industrial de Santander and is actually finishing his PhD in Engineering at the University of the Andes. He is editor in Neutrosophic Sets and Systems and other reputable journals. His major researches include supply chain management and neutrosophic theory applications.

Ahmed A. Salama is recipient of many international awards. Previous Position: 1) Dean of the Higher Institute of Business and Computer Sciences, Arish, Egypt. 2) Head of Mathematics and Computer Science Department, Faculty of Science, Port Said University, Egypt. PhD and DSc. in Mathematics and Computer Sciences. Published many articles and Books in mathematics, computer science and Statistics. The first Arab to use the Neutrosophic concepts in these areas (computer Sci., Statistics and Topology). Research Interests: Mathematics, Computer Science, Statistics, IT, Neutrosophic Research The following is involved in the fields of Neutrosophic. Founder the Neutrosophic Crisp Set Theory. Founder the Neutrosophic Topological spaces and many Applications in Computer Sci. and Information Systems.

Prem Kumar Singh has more than 11 years of experience in teaching, research, and academics. Currently, he is working as an Associate Professor at Department of Computer Science and Engineering, Gandhi Institute of Technology and Management-Visakhapatnam, Andhra Pradesh-India. Prior to that, he worked as capacity of as an Assistant Professor at Amity School of Engineering and Technology,

Amity University-Noida Sector 125. He has completed a Post-Doc in Computer Science from Faculty of Computer Science and Information Technology, University of Malaya-Kuala Lumpur which was fully funded by HIR Lab Malaysian Govt of Higher Education. He holds a PhD degree in Computer Science from School of Information Technology and Engineering, VIT University-Vellore which was fully funded by NBHM Department of Atomic Energy, Govt of India. He has completed a Master Degree in Computer Applications from the CSJM University, Kanpur –India after completion of Bachelor Degree in Computer Science from University of Allahabad-Uttar Pradesh, respectively. His current research interests are soft computing, complex data analysis, graph theory, mathematical modelling and machine intelligence. Towards this direction, he has introduced interval-valued, bipolar, three-polar, m-polar, and complex fuzzy graphs as frontier researchers in this field for knowledge processing tasks. He has published more than 75 research papers in peer reviewed conferences and journals which were cited more than 1350 with 23 H-index and more than 37 i10 papers. He has filed a Malaysian patent for query optimization. He has a copyright of Python based code for research impact analysis of any University on Several Metrics which echo with NIRF ranking 2019. He served as Technical Program Committee, Reviewer, Guest Editor, Editorial board member as well as Speaker for more than 150 peer reviewed conferences and journals. Recently, he is listed as Top 2% Scientist by Stanford University in consecutive years 2020 and 2021.

Nijwm Wary is a Research Scholar, Department of Mathematical Sciences, Bodoland University.

Index

L

M

N

O

P

Q

S

T

U

V

www.ingramcontent.com/pod-product-compliance
Lightning Source LLC
LaVergne TN
LVHW081344300926
849161LV00008B/2060
* 9 7 8 1 6 6 8 4 3 4 9 5 6 *